The Institute of Mathematics
and its Applications
Conference Series

The Institute of Mathematics and its Applications Conference Series

Previous volumes in this series were published by
Academic Press to whom all enquiries should be addressed.
Forthcoming volumes will be published by
Oxford University Press throughout the world.

NEW SERIES

Mathematics in Remote Sensing

Based on the proceedings of a conference organized by the Institute of Mathematics and its Applications on Mathematics and its Applications in Remote Sensing held at Danbury, Essex, in May 1986

Edited by

S. R. BROOKS
Marconi Research Centre, Chelmsford

CLARENDON PRESS · OXFORD · 1989

Oxford University Press, Walton Street, Oxford OX2 6DP

Oxford New York Toronto
Delhi Bombay Calcutta Madras Karachi
Petaling Jaya Singapore Hong Kong Tokyo
Nairobi Dar es Salaam Cape Town
Melbourne Auckland
and associated companies in
Berlin Ibadan

Oxford is a trademark of Oxford University Press

Published in the United States
by Oxford University Press, New York

British Library Cataloguing in Publication Data
Mathematics in remote sensing.
1. Remote sensing, Mathematical techniques
I. Brooks, S. R.   II. Series
621'.36'78
ISBN 0-19-853618-6

Library of Congress Cataloging in Publication Data
Conference on Mathematics and Its Applications in Remote Sensing (1986
Danbury, England)
Mathematics in remote sensing: based on the proceedings of a
Conference on Mathematics and Its Application in Remote Sensing held
at Danbury, Essex in May 1986 / edited by S. R. Brooks.
p. cm. – (Institute of Mathematics and its applications
conference series; new ser., 21)
Bibliography: p. Includes index.
1. Remote sensing – Mathematics – Congresses.   I. Brooks, S. R.
II. Title.   III. Series.   G70.4.C644   1986   621.36'78 – dc20   89–8620
ISBN 0-19-853618-6

Printed in Great Britain by
St. Edmundsbury Press, Bury St. Edmunds, Suffolk

PREFACE

Remote sensing has always placed severe demands on the
ability of users to extract information from the data generated.
These demands derive from a variety of sensing systems,
utilising different bands of the electromagnetic spectrum, and
they relate to a great diversity of applications within numerous
scientific and engineering disciplines.

Satellite systems of the future, carrying microwave remote
sensing payloads, are placing new requirements on the handling
and interpretation of remote sensing data. This Conference
addresses some of the quantitative techniques that are needed
to satisfy these requirements, particularly in the area of
information extraction and data interpretation.

This book contains the papers presented at the Conference
"Mathematics and its Applications in Remote Sensing". The
intention of the Conference was to approach remote sensing in
a thematic manner which ran through this process of information
extraction, rather than adopting an approach based on sensing
instruments, or on applications.

The Conference was divided into seven major topics. After
reviewing some of the problem areas associated with remote
sensing, the second topic considers aspects of scattering
theory and models. The presentation of inverse methods follows.
The emphasis in both these topics is towards the analysis of
microwave data. Statistical methods are reviewed under the next
topic, while the analysis and assimilation of data is adopted
as the fifth theme. The final topics present aspects of spatial
data, including spatial information systems and scene analysis.

We hope that this volume will prove of use to remote sensing
scientists. These scientists will confront the exciting
challenges involved in the realisation of the potential of
present and planned remote sensing systems.

S.R. Brooks,
Marconi Research Centre.

ACKNOWLEDGEMENTS

The Institute thanks the authors of the papers, the editor, Dr. S.R. Brooks, FIMA (Marconi Research Centre) and also Miss Pamela Irving for typing the papers.

CONTENTS

CONTRIBUTORS

S.B.M. BELL; *NERC Unit for Thematic Information Systems, Department of Geography, University of Reading, Whiteknights, P.O. Box 227, Reading, Berks., RG2 2AB.*

J.M. BLACKLEDGE; *Department of Physics, Imaging Group, Wheatstone Laboratory, King's College, London University, The Strand, London, WC2R 2LS.*

D. BLACKNELL; *GEC-Marconi Research Centre, West Hanningfield Road, Great Baddow, Chelmsford, Essex, CM2 8HN.*

H. BONDI; *Churchill College, Cambridge, CB3 0DS.*

J.F. BOYCE; *Wheatstone Laboratory, King's College, London University, The Strand, London, WC2R 2LS.*

R.E. BURGE; *Department of Physics, Imaging Group, Wheatstone Laboratory, King's College, London University, The Strand, London, WC2R 2LS.*

P.G. CHALLENOR; *Institute of Oceanographic Sciences, Brook Road, Wormley, Godalming, Surrey, GU8 5UB.*

D.G. CORR; *Systems Designers Scientific, Pembroke House, Pembroke Broadway, Camberley, Surrey, GU15 3XD.*

M. CUNNINGHAM; *Department of Electrical Engineering, University of Manchester, Manchester, M13 9PL.*

B.M. DIAZ; *Department of Computer Science, Liverpool University, Chadwick Building, P.O. Box, 147, Liverpool, L69 3BX.*

J. FENG; *Wheatstone Laboratory, King's College, London University, The Strand, London, WC2R 2LS.*

P.F. FISHER; *Kent State University, Kent, Ohio 44118, U.S.A.*

A. HENDRY; *GEC-Marconi Research Centre, West Hanningfield Road, Great Baddow, Chelmsford, Essex, CM2 8HN.*

F. HOLROYD; *Faculty of Mathematics, The Open University, Walton Hall, Milton Keynes, MK7 6AA.*

x

J. T. HOUGHTON; *Meteorological Office, London Road, Bracknell, Berkshire, RG12 2SZ.*

P.L.C. JEYNES; *The Blackett Laboratory, Imperial College of Science and Technology, Prince Consort Road, London, SW7 2BZ.*

J. JOHNSON; *Centre for Configurational Studies, The Open University, Milton Keynes, MK7 6AA.*

J. KITTLER; *Department of Electronic and Electrical Engineering, University of Surrey, Guildford, Surrey, GU2 5XH.*

D.P. LIDIARD; *Department of Physics, Imaging Group, Wheatstone Laboratory, King's College, London University, The Strand, London, WC2R 2LS.*

M.S. LONGUET-HIGGINS; *Department of Applied Mathematics and Theoretical Physics, University of Cambridge, Silver Street, Cambridge, CB3 9EW.*

S. LUTTRELL; *RSRE, St. Andrews Road, Great Malvern, Worcs., WR14 3PS.*

P.M. MATHER; *Remote Sensing Unit, Department of Geography, The University, Nottingham, NG7 2RD.*

R.J. MILLER; *GEC-Marconi Research Centre, West Hanningfield Road, Great Baddow, Chelmsford, Essex, CM2 8HN.*

J. OAKLEY; *Department of Electrical Engineering, University of Manchester, Manchester, M13 9PL.*

C.D. OBRAY; *Department of Statistics and Operational Research, Coventry Polytechnic, Priory Street, Coventry, CV1 5FB. CV1 5FB.*

C.J. OLIVER; *RSRE, St. Andrews Road, Great Malvern, Worcs., WR14 3PS.*

D. PAIRMAN; *Atmospheric Physics Department, Oxford University, Oxford, OX1 3PU.*

G. PEACEGOOD; *Department of Computer Science, University College London, Gower Street, London, WC1E 6BT.*

S. QUEGAN; *Department of Applied and Computational Mathematics, University of Sheffield, Sheffield, S10 2TN.*

M.S. SCIVIER; *Systems Designers Scientific, Pembroke House, Pembroke Broadway, Camberley, Suurey, GU15 3XD.*

J.J. SETTLE; *NERC Unit for Thematic Information Systems, Department of Geography, University of Reading, Whiteknights, P.O. Box 227, Reading, Berks., RG2 2AB.*

J.C. SIMON; *Institut de Programmation, Université Pierre et Marie Curie, (Paris VI), 4 Place Jussieu, 57252 Paris, Cedex 05, France.*

J. SKINGLEY; *GEC-Marconi Research Centre, West Hanningfield Road, Great Baddow, Chelmsford, Essex, CM2 8HN.*

M.A. SROKOSZ; *Institute of Oceanographic Sciences, Brook Road, Wormley, Godalming, Surrey, GU8 5UB.*

D.J. WEBB; *Institute of Oceanographic Sciences, Brook Road, Wormley, Godalming, Surrey, GU8 5UB.*

G.G. WILKINSON; *Department of Computer Science, University College London, Gower Street, London, WC1E 6BT.*

L.R. WYATT; *Department of Electronic and Electrical Engineering, University of Birmingham, P.O. Box 363, Birmingham, B15 2TT.*

INTRODUCTION

Sir Hermann Bondi
(Churchill College, Cambridge)

Remote sensing is something that has been talked about
for twenty years and more. One of the problems we have had is
that in the early days it was over-sold. In the early days
when instumentation was not as good as it now is, let alone
what it promises to be, it could deliver splendid pictures to
put on the Managing Director's wall but not a lot else. We
now therefore have the slightly more difficult task to convince
people that remote sensing is actively useful.

In one area of course there is no need for us to talk
about it: that is the defence area. Reconnaissance satellites
have made a major contribution to improving the international
climate for over twenty years. They are, in one sense, in a
very easy situation because the provider of the satellite
equals the operator of the satellite equals the user of the
information. In so many other fields of remote sensing what
we suffer from is that the user hardly yet exists. We know
that there could be great utilisation of this and that by
certain interests, but they have not organised themselves to
make use of it. Obviously where there is a single area of
interest like meteorology this has been well established for
a long time. When you go to other areas the users are highly
fragmented (or to use the awful modern term disaggegated)
and to get across that it is to their advantage to change their
mode of operation so they can take full advantage of what
remote sensing can now, and will in future, provide is going
to be a more of an up-hill task than many imagine. One of the
curious things about remote sensing outside the defence and
meteorology fields where it is so well established is that
amongst the earliest users were the geologists. I have nothing
at all against geology, but when you ask, for any purpose, why
go to the cost of putting your system up in space the answer
about space based systems is that what is so marvellous is

that once they are up, they are up, and go on functioning and therefore you get repeated overflights which is terribly useful for changing scenes. In geology one probably wants an overflight in good weather once every five million years. So that this long lasting base in space which for much of remote sensing is going to be the dominant feature, has been used first of all by resource hunting geologists (mainly of course in parts of the world where the over burden and vegetation were slight) is a remarkable accident of history again showing that new information, however valuable, can often only be absorbed where there is a user organised in such a way that he is able to use it, as is the case in meteorology, as is the case in certain areas of geology, but not when you come to the oceanic side. It is a much more difficult task to see how the very clear utility to the ocean-using communities can be made effective, for organisational reasons. But it is important that we keep in mind the significance of looking at time-dependent phenomena.

   The second point I want to make, and to which quite a number of papers are devoted, is about the difficulty of interpreting information. To make nice wall-paper for the Managing Director's office is easy, the aesthetics come out almost without trying. To see and extract the useful information that is wanted is going to be the centre piece of the future of remote sensing. I like to give a comparison with geophysical exploration, particularly I may say to my political friends, to whom the glamour of space hardware out-weighs everything else. In geophysical exploration you let off a big bang. Frankly, any fool can let off a big bang. You then receive the reflected and refracted waves on a lot of hydrophones. You need moderately intelligent people to place those hydrophones, to keep them in order, to record their data. But then where the money lies is not in making the bang, nor in gathering the return sound, where the money lies is in interpreting what is often publicly available information. The geophysical interpretation companies, even when the oil price is low, make a very good living indeed, because they have the intellectual capacity that makes use of the information that anybody can have but few can read. What they do equally applies to many areas in remote sensing. What they do is not to divorce the information coming from remote sensing from their other sources of information, but embed it in knowledge that has been and is being derived by other means. There are psychological problems in so using that should not be underestimated: when people are very skilful, highly trained in certain types of gathering knowledge, in equipment and using the existing instrumentation, in ways of getting useful knowledge from such data, then it is not easy for them to accept and adapt to

a new refined, sophisticated field of information flow with
its own complexities of interpretation. So I think it is
interpretation above all where the intellectual challenges
lie and may I say boastfully that if there are intellectual
challenges in the field they are at least in part usually
mathematical challenges.

Let me finish by pointing out one or two of the hazards
and hopes that there are in the future. When I stand on a
hilltop and look at the landscape below me on a nice sunny
but perhaps hazy day most of what I see may not be terribly
easy to interpret although I know what landscapes look like.
But one thing is quite obvious and outstanding to me: if I
get a flash of the reflection of the sun from the windscreen
of a car, I know for certain there is the windscreen of a car
there. It blots out everything else I see. It is a very
small object in that vast landscape but it stands out, and
I think there is an important challenge for us here to use
data processing to generate a similar visibility for an
important feature in remote sensing. I would like to spend
a moment to draw your attention to one potential such use.
There are many pests in forestry, in crops and the like,
that start in a very small area where they could easily be
eradicated with relatively modest applications of not
over-hazardous counter-agents. Once they have spread, the
whole difficulty of fighting them becomes much more serious
and usually requires the release into the environment of a lot
of not very desirable material. An early warning system for
pests would be enormously important; but if you want to see in
a vast forest a group of five diseased trees, you must perfect
a method like the flash of reflection of sunlight from the
moving car windscreen, and we are still a long way from this.

Let me finish by talking about money. It is a useful
commodity as most of us know, but where in the remote sensing
area are there likely to be the biggest financial returns, and
what are the risks that stand in their way? If by looking
from a satellite I found that the coffee crop in some part of
the world was not doing very well, I could make many fortunes
on the commodity exchange if I knew this sooner than somebody
else. The potential gains there are probably much larger
than in most remote sensing fields, but it raises problems
of the utmost legal and political complexity. Who owns the
information that comes from upstairs of the area of the
country other than that to whom the satellite belongs? What
should you do with it? Should you publish the raw information
rapidly so that is is useful to those who command the brains
needed to derive the interpretation that matters? Is it
private to the government of the country of which it is taken?

I think these are enormously difficult questions of law and
of politics. It is a matter of grave concern to me that
the amount of work that is going on in these fields strikes me
as quite insufficient compared with the needs. It may well be
that instead of getting real benefits for the world as a
whole, we will get major rows that do not do anybody a great
deal of good. So I feel here it is a non-mathematical area
to which I wish more thought was applied. But these are
random thoughts, and I am only somebody who calls himself
on occasions an itinerant administrator. These were only my
preliminary thoughts serving as an introduction to the real
meat to which I and all of us look forward in the talks that
will be given.

PROBLEMS IN THE RETRIEVAL OF USEFUL ATMOSPHERIC INFORMATION
FROM SATELLITE REMOTE SENSING OBSERVATIONS

J.T. Houghton
(Meteorological Office, Bracknell)

1.   INTRODUCTION

The advent of observations from orbiting satellites has
brought about something of a revolution in operational
meteorology. The ability to observe over all parts of the globe
almost continuously in time has enabled much more detailed
study to be made of individual weather systems and has also
enabled the atmospheric circulation to be viewed as a whole.
Observations of many different kinds can be made from space.
Radiation which has been emitted or reflected can be measured
over the whole range of wavelengths from the ultra-violet,
through the visible and infra-red to the microwave and radio
regions. A great deal can be learnt for instance from studies
of cloud images in the visible or infra-red. In this paper,
however, I do not wish to review the broad field. I shall
address a particular problem and restrict my remarks to a
consideration of the problems concerned with the measurement
from satellite based instruments of atmospheric temperature
structure and how the information from such measurements can be
incorporated in operational forecasting models.

2.   THE FORECASTING MODEL

First I need to describe briefly the numerical models of
the atmospheric circulation which are the basic tools required
for operational forecasting. For the global model of the
Meteorological Office, appropriate variables (the three
components of velocity, the geopotential height, the potential
temperature and the humidity) are specified on a three -
dimensional grid with about 150 km spacing between grid points
in the horizontal and with 15 levels in the vertical, the
vertical coordinate being specified in terms of pressure.
About one third of a million grid points are therefore involved.

The relevant equations for the global model are (1) the equations of motion for flow in the horizontal in which the acceleration (including the Coriolis term which is particularly important on the rotating earth) is balanced by the pressure gradient and by friction, the latter including a parametrisation to take account of all effects of motion on smaller scales than the grid spacing, (2) the hydrostatic equation, (3) the equation of continuity and, (4) the thermodynamic equation. The forecasting process consists of integration forwards in time of these equations from a specified initial state. Parametrisations of physical processes, for instance convection, the transfer of radiation, moist processes, effects involving clouds and the transfer of heat momentum and water-vapour across the surface also need to be included. About 10^{11} operations are involved in a 24-hour forecast taking about 4 minutes on a Cyber 205 computer. A forecast for six days ahead, a period now covered by a number of major forecasting centres, can therefore be accomplished in half an hour.

In addition to the global model the Meteorological Office also runs models over limited areas with a smaller grid size, one covering Europe and the Eastern Atlantic with a 75 km horizontal grid and an experimental model with a 15 km horizontal grid covering the British Isles. Further information regarding the models can be found in Gadd (1985), Golding (1984) and Houghton (1986).

3.   SOURCES OF DATA AND THEIR ASSIMILATION

To specify the initial state of the model a wide range of data has to be brought together. To be of any use for forecasting the data needs to be available within an hour or so of its observation, the times of observation being standardised so far as is possible at midnight and noon GMT. A typical mix of observations at one particular observation time will be from over two thousand surface synoptic stations on land, over one thousand ship observations, several thousand wind observations in the tropics obtained from clouds tracked from sequences of satellite images, several hundred observations from commercial aircraft, several hundred radiosonde balloon ascents from locations all over the globe, several hundred observations from buoys and other automatic stations interrogated from space and several thousand soundings from satellites of atmospheric temperature profiles. All this data is required to be assimilated into the model with great care and with attention to the error characteristics particular to each kind of observation. In the Meteorological Office models assimilation of the data is carried out gradually beginning with the model situation derived from the previous forecast 6 hours before

the time of observation. Details of the process are given by
Atkins and Woodage (1985). What needs particularly to be
pointed out in the context of this paper is the importance that
is attached to accuracy and high quality control in the
observations. Because the model possesses significant skill
in forecasting the atmospheric state up to about five days
ahead, at any given time a good description of the atmospheric
state exists within the model arising from the integrating
effect of the assimilation process on the data which has been
available during the previous few days. To be really useful
therefore, new data presented to the model for assimilation must
be of high quality and high accuracy.

4. REMOTE TEMPERATURE SOUNDING

The temperature structure of the atmosphere below an
orbiting satellite can be inferred from observations of the
radiation emitted by a suitable atmospheric constituent whose
distribution in the atmosphere is accurately known. Carbon
dioxide possesses suitable emission bands in the infra-red at
4.3 μm and 15 μm wavelength and oxygen suitable emission bands
in the microwave part of the spectrum near 5 mm in wavelength.
The principle is illustrated in Fig. 1 which illustrates the
emission spectrum in the infra-red of the atmosphere and the
underlying surface. Superimposed on the examples of spectra
shown are black-body emission curves at various temperatures.
Looking first at the region near 11 μm wavelength (900 cm^{-1})
where the atmosphere is largely transparent, it will be seen
that the temperature of the surface can be observed - around
320°K over the Sahara, 280°K over the Mediterranean and only 180°K
over the Antarctic plateau. Moving now the the central and
most absorbing part of the carbon dioxide band near 15 μm
(667 cm^{-1}) where the radiation originates from the stratosphere,
it will be seen that the temperature is around 220°K over all
three locations. Between 800 and 667 cm^{-1} the absorption by
carbon dioxide increases and the radiation observed by the
satellite originates at different levels in the troposphere.
A description of the temperature structure of the troposphere
and lower stratosphere is therefore contained within the spectrum
of emitted radiation in the carbon-dioxide band.

On the Tiros-N series of satellites (Fig. 2) three instruments
for temperature sounding are mounted, namely the High
Resolution Infra-Red Sounder (HIRS), the Stratospheric Sounding
Unit (SSU) and the Microwave Sounding Unit (MSU). Measurements
from these instruments are combined to provide temperature
soundings for operational meteorological models. Further details
of the instruments and the methods can be found in Houghton,
Taylor and Rodgers (1984).

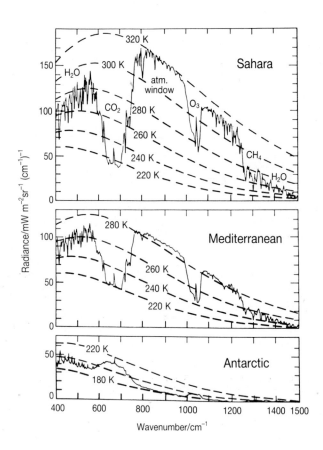

Fig. 1   Thermal emission from the earth plus atmosphere
         emitted vertically upwards and measured by the
         infra-red interferometer spectrometer on Nimbus 4,
         (a) over Sahara, (b) over Mediterranean, (c) over
         Atlantic.   The radiances of black bodies at various
         temperatures are superimposed.   (From Hanel et al
         (1971)).

TIROS-N Spacecraft

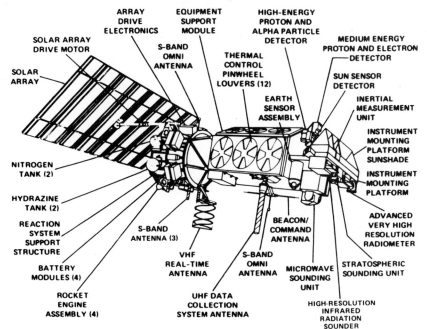

Fig. 2   The Tiros-N satellite. Note the three instruments
for remote sounding namely the high resolution infra-
red sounder (HIRS), the microwave sounding unit (MSU)
and the stratospheric sounding unit (SSU).

5.   THREE PROBLEMS OF REMOTE TEMPERATURE SOUNDING

Three problems associated with obtaining accurate measurements
of temperature structure from satellite observed radiances are
firstly the poor vertical resolution of the measurements,
secondly the interfering effect of clouds and thirdly the
difficulty of adequate quality control. I shall address these
problems briefly in turn.

The radiation which reaches the satellite at any given
wavelength originates from a region of the atmosphere
described by a "weighting function" whose vertical extent is
around 10 km. Figure 3 illustrates the family of weighting
functions appropriate to the various channels belonging to the
three instruments on the Tiros-N satellites. From such a set
of functions of broad vertical extent even though they overlap,
it is clearly difficult to recover information about the
structure within a temperature profile having a scale in the
vertical less than a few kilometres. This limitation is of

concern to the local forecaster who needs to know the detailed
vertical structure. For input to numerical models, however,
since what is required is the average temperature over layers a
few km thick, the information provided by satellite instruments
does not appear inappropriate. There is still, however, the
problem of retrieving from the satellite data the measurements
required by numerical models with sufficient accuracy.

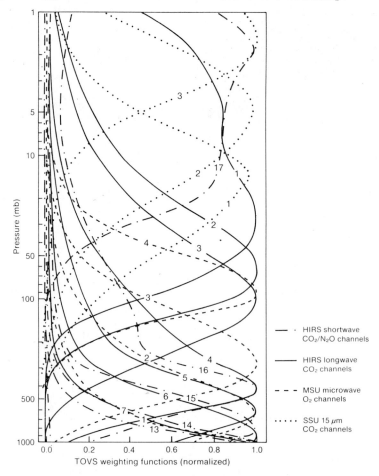

Fig. 3   Weighting functions (functions describing the region
         from which the radiation detected by different channels
         originates) for various channels of the HIRS, the MSU
         and the SSU (see Fig. 2).

In the infra-red, clouds of any significant thickness emit
radiation largely as though they were black-bodies and even
thin cirrus clouds interfere substantially with the radiation

field.   At microwave wavelengths clouds interfere much less with the
radiation streams - their effect may be ignored except for
clouds which are precipitating or which contain large droplets.
Using information available from a variety of infra-red
wavelengths and by comparing infra-red measurements with
microwave ones methods for allowing for the presence of clouds
have been developed.   The presence of cloud can also be
inferred from high resolution cloud images.   It is especially
important to identify regions where there is a complex or
broken cloud structure as in these cases it is especially
difficult to allow for cloud effects.   (For more details see
Houghton, Taylor and Rodgers (1984)).

I have already emphasised in §3 the importance of quality
control.   Observations which are seriously contaminated by
clouds or by varying surface conditions (for instance varying
albedo at microwave wavelengths) need to be eliminated.
Because of the difficulty of recognising such observations
early temperature soundings from satellites were difficult
to incorporate into operational forecasting models.   One way of
eliminating poor observations is to compare the satellite
measurements with predictions from the forecast model and to
discard those that deviate from the forecast field by more
than a given amount.   This process has the desired effect of
eliminating observations of poor quality.   Since however all
observations, whether or not they are correct, which disagree
significantly with the forecast field are removed many of the
observations of the most value are also eliminated.   The
residual observations will only possess a small information
content.   The overall effect is to throw out the baby with
the bath water.   This problem will be further addressed in the
next section.

6.   THE RETRIEVAL OF TEMPERATURE PROFILES

Given a set of radiance measurements from a number of
spectral channels of a satellite instrument, how can the best
temperature profile consistent with those observations be
produced?   There is, of course, other information available
in addition to the satellite measurements, namely, our knowledge
of typical atmospheric temperature profiles for different
seasons and locations (what is called climatological information)
or information which is available from the forecast model
appropriate to any given time and location.   Rodgers (1976)
(see also Houghton, Taylor and Rodgers (1984)) shows how
information from these two sources may be combined (it is
called an information retrieval process) in the expression

$$\hat{x} = x_o + (K.C)^T . (K.C.K + E)^{-1} (y_m - y_o (x_o))$$

where $\hat{x}$ is the retrieved temperature profile.

x_o is the a priori estimate (or first guess) of the profile obtained from climatological information or from the forecast model

C is the error covariance of x,

y_m is the vector of satellite measurements (expressed as brightness temperatures) from which the effect of clouds has been eliminated by some appropriate pre-processing,

E is the error covariance of y_m,

y_o (x_o) is the vector of brightness temperatures appropriate to the channels of the satellite instrument calculated from the first guess profile,

K is a matrix of the partial derivative of y with respect to x and contains the 'weighting function' information (cf Fig. 3). It is slightly dependent on temperature but for our purposes here we shall ignore that non-linearity.

Consider the errors of the retrieval process and how they depend on the first guess and on the errors in the measurements. The error covariance S of the retrieved profile $\hat{x}$ is given by (Eyre et al. 1986)

$$S = C - (K.C.)^T . (K.C.K^T + E)^{-1} . (K.C)$$

Eyre et al have evaluated S for a range of values of E corresponding to different noise levels in the satellite radiances and for values of C characteristic of climatological or forecast first guess information. The results are shown in Fig. 4 where a climatological first guess has been employed. The dotted line on the right in Fig. 4 is an estimate of the likely error in a climatological first guess; the similar line on the right in Fig. 5 is an estimate of the likely error in the forecast first guess.

The forecast first guess error statistics illustrated in Fig. 5 are appropriate to the area around the British Isles. The diagram emphasises rather clearly the importance of high accuracy in the satellite observations if significant impact on the forecast in the region of the UK is to result. For parts of the world, for instance the Southern Hemisphere, where information from other sources is much more sparse, the demand on the satellite observations is not quite so great. However, in all cases it is true to say that increased accuracy

in satellite measurements will lead to significant improvement
in their information content.

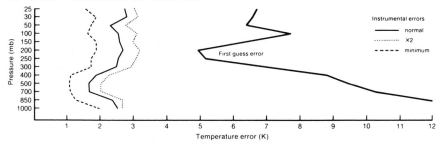

Retrieval and first guess errors (after Eyre) (Climatological first guess)

Fig. 4   Retrieval and first guess error for climatological
         first guess (after Eyre et al (1986)). Three different
         estimates of instrument noise have been used. What is
         described as 'normal' is for the NOAA 7 instruments
         allowing for errors induced by pre-processing and 'cloud-
         clearing' techniques. The dotted line is for twice
         'normal' noise and what is described as 'minimum' is
         for the NOAA 7 instrument assuming no pre-processing
         or 'cloud-clearing' errors.

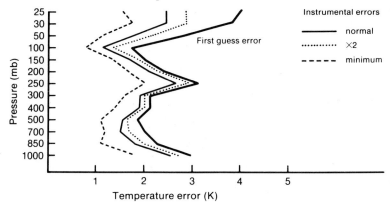

Retrieval and first guess errors (after Eyre) (Forecast first guess)

Fig. 5   Retrieval and first guess errors for forecast first
         guess (after Eyre et al (1986)). For key to
         instrumental errors see Fig. 4.

7.   THE UTILISATION OF TEMPERATURE PROFILES FROM SATELLITES

     It would be unfortunate if I finished this paper with what
might appear rather pessimistic results presented in Figs. 4
and 5. I would now like to describe briefly some of the ways
in which temperature information from satellite instruments is

currently utilised at the Meteorological Office at Bracknell
and to give an indication of plans for the future.

Temperature profiles over all parts of the globe derived
from satellite radiances as described in §6 (based on
climatological first guesses) are provided by the National
Oceanic and Atmospheric Administration (NOAA) of the U.S. and
incorporated routinely into the numerical forecasting models.
They have a particularly large influence on Southern Hemisphere
analyses. In addition the output of the radiometers on the
Tiros-N polar orbiting satellites is read out directly when
the satellites are within communication range of a ground
station at Lasham in the U.K. This local information is
retrieved in a similar manner. The density of observations
from the local source is greater; also they are available
considerably earlier than the global information from NOAA.
Some of the local retrievals are fed into the forecasting models.
Charts are also prepared from the locally received satellite
data; an example is in Fig. 6. These are made available to
the forecaster who is able to use them to give additional
information particularly in those areas where other sources of
information are sparse.

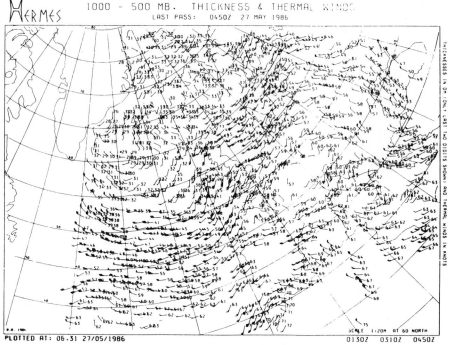

Fig. 6 Satellite radiance data received at Lasham, UK, during
        three passes of the NOAA satellite early on 27 May 1986.
        The radiances have been 'retrieved' in terms of 1000-500mb
        thickness (numbers are the thickness in decameters with
        the first digit (5) missing). From these thicknesses
        thermal winds have been plotted as an aid to the
        Meteorological Office forecaster.

Temperature soundings obtained from satellites are clearly proving their value in operational meteorology - in fact they have already become indispensable. But there is room for substantial improvement in their accuracy and in the ways in which they are utilised.

For the future, better satellite instruments are being developed. In particular, the Meteorological Office is contributing part of an instrument (the other part being provided by NOAA) known as the Advanced Microwave Sounding Unit (AMSU) which from the early 1990s will provide much improved microwave temperature and water-vapour soundings. Retrieval methods for operational use are also being improved, in particular by the proper use of forecast first guess profiles.

The problems of quality control, and of ensuring that the observations incorporated in the forecasting models have the maximum information content, remains a formidable one. It is here where human assessment and intervention can be most effective. What is required is that all sources of information be presented to an experienced meteorologist in an effective and timely manner so that he can make decisions regarding the quality and information content of data from various sources. Achieving the optimum man-machine mix is one of the major challenges for the future.

REFERENCES

1. Atkins, M.J. and Woodage, M.J., (1985)   Observations and Data Assimilation. *Meteor. Mag.* **114**, pp. 227-233.

2. Eyre, J.R., Pescod, R.W., Watts, P.D., Lloyd, P.E., Adams, W. and Allam, R.J., (1986)   TOVS Retrievals in the UK: Technical Proceedings of the 3rd International Study Conference 13-19 August 1986, Madison, Wisconsin, U.S.A.

3. Gadd, A.J., (1984)   The 15-level weather prediction model. *Meteor. Mag.* **114**, pp. 222-226.

4. Golding, B.W., (1984)   The Meteorological Office mesoscale model: its current status. *Meteor. Mag.* **113**, pp. 288-313.

5. Hannel, R.A., Schlachman, B., Rogers, D. and Vanous, D., (1971)   Nimbus - 4 Michleson Interferometer. *Appl. Opt.* **10**, 1376-1382.

6. Houghton, J.T., Taylor, F.W. and Rodgers, C.D., (1984) Remote Sounding of Atmospheres. Cambridge University Press.

7. Houghton, J.T., (1986)   The Physics of Atmospheres.
    Cambridge University Press.

8. Rodgers, C.D., (1976)   Retrieval of atmospheric temperature
    and composition from remote measurements of thermal radiation.
    Rev. Geophys. Space Phys. **14**, 609-624.

PROBLEMS IN PASSIVE REMOTE SENSING
OF THE EARTH'S LAND SURFACE

P.M. Mather
*(Remote Sensing Unit, Department of Geography,
The University, Nottingham)*

1. INTRODUCTION

Passive remote sensing of the Earth's land surface entails
the observation, recording and interpretation of measurements of
reflected Solar radiation and of thermal energy emitted by the
Earth. Such measurements are made at a distance, without
physical contact between the sensing device and the target.
Measurements from sensors carried onboard orbiting Earth
satellites will be considered here. These platforms carry a
range of instruments, both profiling and imaging. Attention
will be restricted here to the outputs from imaging sensors
which generate a two-dimensional set of observations of the
surface of the Earth beneath their flight path. In general,
simultaneous observations of ground radiance are made in several
regions of the electromagnetic spectrum so that a remotely-
sensed image of a given area can be considered as a three-
dimensional array of numerical observations with two
geographical dimensions and a third dimension representing
spectral wavelength. A common use to which these data are put
is the generation of thematic maps which purport to show the
geographical distribution of land-surface cover types over the
area of the image. In this paper some of the fundamental problems
that arise in such applications of remotely-sensed data will be
considered. One class of problem can be described as practical,
since it involves the use of mathematical and statistical
techniques. The difficulties experienced in the manipulation
and analysis of very large datasets are formidable. A second,
and perhaps more basic, problem area is that of understanding
and modelling the physical interactions and processes that
interpose themselves between the data (which are quantized
measures of spectral radiance) and the analyst's interpretation
in terms of real-world phenomena (such as forests, agricultural
crops, geological and geomorphological features). Such problems

need to be resolved if thematic maps derived from remotely-
sensed data are to be used routinely as inputs to geographical
databases for subsequent manipulation and analysis alongside and
in conjunction with other spatial data using spatial information
systems technology.

Time and space do not permit a detailed account of the wide-
ranging problem areas just outlined; I propose to identify
problems rather than define solutions to them. The first
section of this paper provides a brief summary of the
characteristics of remote-sensing platforms and the imaging
sensors they carry. This section will include brief details of
the nature of the data collected by such sensors. The second
section is devoted to a survey of the class of problem I have
described as practical, and is concerned with numerical methods
of processing the large datasets which comprise a digital image.
The final section contains a consideration of the physical
problems encountered whenever remotely-sensed data are
interpreted in terms of Earth-surface phenomena.

2. PLATFORMS AND SENSORS

Remote sensing from Earth satellites began on April 1st 1960
with the launch of the first TIROS (Television Infra-Red
Observation Satellite) and subsequent members of the TIROS
(later NOAA) family have contributed to the collection of data
concerning the Earth's atmosphere and surface up to the present
day. Other satellite series, principally Landsat and SPOT, have
followed, and more are expected during the remainder of the
1980s from the Japanese and Indian governments. The principal
characteristics of these satellites are (i) their near-polar
orbits, (ii) the multispectral nature of their sensors and
(iii) their spatial resolving power.

A near-polar orbit allows observation of the Earth's surface
between approximately 82°N and 82°S over an area extending to
either side of the ground track of the satellite (Fig. 1). The
width of the area viewed by the imaging sensors onboard the
satellite is called the swath width; this is a function of the
satellite altitude and the angular field of view of the sensor.
The orbital path moves successively westwards so that over a
period of time depending on the swath width of the instruments
the entire Earth is imaged. This cycle time ranges from one
day in the case of the TIROS/NOAA satellites to 26 days for the
recently-launched French SPOT satellite. Thus, the near-polar
orbit gives repetitive coverage of much of the Earth's surface.

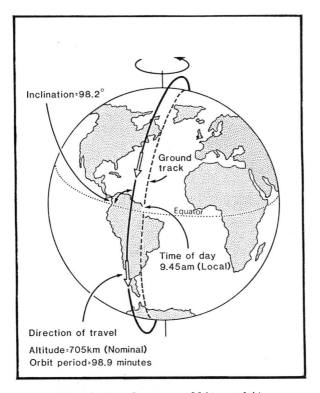

Fig. 1   Landsat satellite orbit

The passive imaging instruments carried by Earth Observation
Satellites measure reflected Solar energy and, in some cases,
emitted terrestrial energy at each of a large number of
contiguous cells on the Earth's surface. The dimensions of each
cell (or pixel, for picture element) depend upon the spatial
resolving power of the instrument, which is discussed below.
Reflection and emission are measured in a number of wavebands
across the electromagnetic spectrum. Passive remote sensing
devices, with which this paper is concerned, measure reflection
and emission in the 0.4 - 12.5μm region of the spectrum.
Reflection takes place in the visible (0.4 - 0.7μm), near-infrared
(0.7 - 1.1μm) and middle-infrared (1.1 - 3.0μm) wavebands, while
emittence of heat energy occurs in the thermal region (3.0 to
12.5μm). The choice of waveband is influenced by (i) the spectral
reflectance and emittence characteristics of the target of
interest, and (ii) the spectral characteristics of the interaction
between electromagnetic energy in the 0.4 - 12.5μm waveband and
the Earth's atmosphere. These interactions fall into two groups;
scattering, which predominantly affects the shorter wavelengths,
and adsorption, which is a function of the characteristics of
the molecular components of the atmosphere. All incoming and

outgoing electromagnetic energy in the spectral region under
consideration is affected to some degree by these processes,
though the intensity of the interaction is such that certain
wavebands are unusable for Earth observation from orbital
altitudes. The wavebands that are less affected by scattering
and absorption are termed atmospheric windows (Fig. 2). The
characteristics of the sensor will be determined by the position
of these windows and by the reflectance and emittence
characteristics of the targets of interest.

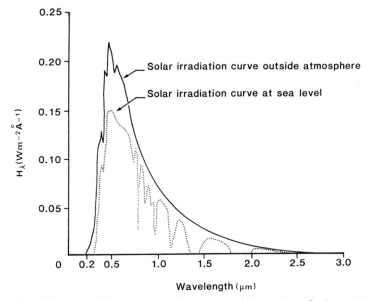

Fig. 2   Solar irradiance at the top and bottom of the atmosphere,
         showing regions of high and low attenuation.

Fig. 3 shows the wavebands utilised by two imaging systems
designed for quite different purposes. The first is the Coastal
Zone Colour Scanner (CZCS) carried by the US Nimbus-7 satellite;
this instrument was intended to gather information concerning
(i) the distribution and concentration of organic and inorganic
substances present in the surface waters of the coastal oceans
and (ii) the distribution of sea-surface temperatures. In order
to measure ocean colour a number of narrow wavebands in the
visible region of the spectrum, centred at 0.44, 0.52, 0.56
and 0.67μm, were chosen so as to characterise the shape of the
reflectance spectrum of chlorophyll. A fifth channel, centred
at 0.75μm in the near-infrared, was included to provide land/sea
discrimination, while the heat emission from the ocean surface
was measured in a sixth channel centred on 11.5μm. In contrast,
the Landsat series of satellites carry an instrument known as the
Multispectral Scanner (MSS) which was designed to collect

information concerning the distribution and nature of the
vegetation covering the Earth's land surfaces. Four relatively-
wide wavebands were chosen, covering the 0.4-0.5, 0.5-0.6, 0.6-0.7
and 0.7-1.1μm regions. The choice of these wavebands was based
on knowledge of the spectral reflectivity characteristics
of vegetation, which show a strong visible/near-infrared contrast,
and of the location of suitable atmospheric windows. Despite
the fact that the Landsat MSS was designed for use in crop
identification and vegetation studies, the data it has
generated has found widespread applications in disciplines as
varied as geology, coastal engineering and hydrology. Thus, the
characteristics of the target determine the actual choice of
waveband within the constraints of atmospheric effects and
of sensor technology.

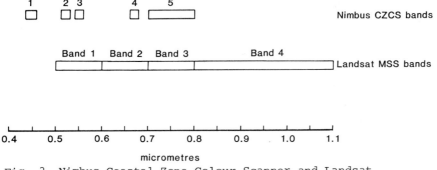

Fig. 3   Nimbus Coastal Zone Colour Scanner and Landsat
         Multispectral Scanner spectral bands.

Data from the Landsat MSS and Thematic Mapper (TM) instruments
are perhaps the most widely-used in passive remote sensing
of the Earth's land surface and in the remainder of this paper
MSS and TM data will be used for illustrative purposes. The
Landsat series of satellites dates from 1972, and the earlier
Landsats (numbered 1 to 3) are now defunct. Landsat-4 (1982)
developed a power supply problem early in its life and is not in
routine use. Landsat-5 (1983) is the current operational Landsat,
and it carries both an MSS and a TM sensor. The principles on
which the two sensors operate are similar; they differ in that
the TM has seven wavebands ranging from the blue-green region
of the visible spectrum to the thermal infrared while the MSS
has only four, as described earlier. In addition, the spatial
resolution of the MSS is 80m while that of the TM is 30m.
The spatial resolution of a sensor is a complex property
(Townshend, 1980) but it can be thought of simply as the ground
surface dimensions of the instantaneous angular field of view of
the sensor (Fig. 4).

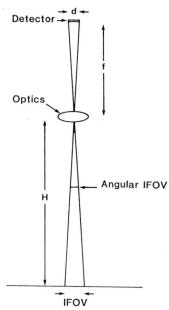

Fig. 4   Instantaneous field of view of the Landsat Multispectral
         Scanner. The height of the satellite is H and the
         focal length of the optics is f. The dimension of the
         ground area projected onto the detector d is 80m on the
         ground.

Landsat follows a circular, near-polar orbit at a nominal
altitude of 700km as shown in Fig. 1. This orbit takes the
satellite over each point on the Earth's surface between 82°N
and 82°S once every 16 days. As the satellite moves in a
southerly direction over the illuminated side of the Earth a
scanning mirror sweeps from side to side and directs upwelling
radiance through a series of filters onto a set of detectors.
The outputs from these detectors are sampled and then converted
from analogue to digital form onboard the satellite. The sample
values are quantised on a 0-63 or 0-255 scale and relayed back
to ground receiving stations on Earth. After some corrections
for systematic errors the data for a single image covering a
ground area of 185 x 185km are written to a computer-readable
magnetic tape. A TM image thus consists of seven values per
30m x 30m pixel while a MSS image has four values per pixel.
Although the spatial resolution of the MSS is 80m, the pixel
values are sampled to give a ground dimension of 80m across the
scan direction and 57m along the scan direction. A simple
calculation shows that there are approximately 38 million
pixels per band in a TM image (giving a storage requirement of
266Mb) and 7.5 million pixels per band in a MSS image, giving
a storage requirement of 30Mb. These very large dimensions

represent a serious constraint on the nature of the mathematical
and statistical operations that can be carried out on such data,
and on the computer hardware needed to apply such operations.

3. PRACTICAL PROBLEMS

The generation of thematic maps from multispectral image
data such as that provided by Landsat TM and MSS requires (i)
geometric fidelity, in that the elements of the image should be
capable of being represented cartographically and (ii) a pattern-
recognition capability which can recognise the nature of the land-
cover type at a particular pixel location. Unless the
multivariate observations of surface reflectance and emission
that make up a multispectral image can be correlated with
particular geographical phenomena and their location expressed
in terms of an acceptable coordinate system (or map projection)
then their use in the solution of practical problems will be
limited. In the future, as spatial information technology
develops and sources of digital geographical data increase, a
major use of remotely-sensed data will be to provide information
concerning observable characteristics of the Earth's surface,
in particular those subject to change, to be used in conjunction
with these other data types within a geographical information
system. This aim will not be realised, however, unless the two
problems mentioned above can be resolved.

Fig. 1 shows the orbital path of Landsat. The columns of the
image matrix are orientated parallel to this track, while the
rows of the matrix run in the direction of the scan-mirror
sweep. It is clear from this figure that the departure of the
row/column orientation from a north-south/east-west configuration
increases with latitude, giving a raw MSS or TM image its
characteristic skew. Other factors, such as the variation in the
altitude and attitude of the spacecraft, contribute other
errors. Panoramic distortion (Fig. 5), though negligible for
Landsat MSS and TM because of the narrow angular field of view
of these instruments ($\pm 5.8°$) must also be taken into account
if, for example, AVHRR data from the TIROS/NOAA satellites is
used; the AVHRR has an angular field of view of $\pm 56°$. Two
approaches to the definition of a transformation from image
to geographic coordinates are possible. The first utilises a
model of the satellite's orbital geometry and requires accurate
measurements of the attitude and altitude of the satellite
as well as details of the Earth's ellipsoid. Some progress is
being made in this area, particularly with lower-resolution
data such as that produced by the AVHRR (1.1km resolution at
nadir) (Brush, 1985). However, the more widely-used method
for higher-resolution data such as that produced by Landsat TM
and MSS involves the derivation of an empirical least-squares
transform to relate the two coordinate systems. Polynomial
least-squares functions of low order (usually no higher than

three) are used to define a statistical relationship between the
image column/row and the map easting/northing coordinate systems
for selected ground control points that are accurately
measured on image and map (Bernstein, 1976; Davison, 1984). The
disadvantage of this method is its inability to generate an
image map of an area unless a suitable conventional map is
available unless satellite position-fixing techniques (Ashkenazi,
1984) can be used to determine the coordinates of control points
that are detectable on an image. In the hands of the uninitiated
user the relationship between probable error of estimation and
the number and spatial distribution of the ground control points
may be overlooked, giving spuriously high goodness-of-fit
values. Numerical instability might also be involved, unknown
to the user, if the matrix of sum of squares and cross-products
is ill-conditioned; this is more likely to occur with polynomial
functions than in the case of a standard regression analysis.

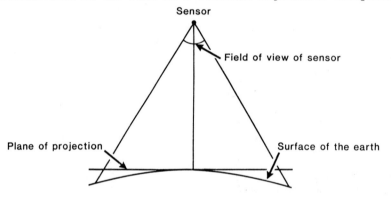

Fig. 5   Panoramic distortion resulting from the projection of
         a curved surface onto a horizontal plane.

   Given a suitable transformation function, the second part of
the geometric correction procedure can be implemented. At this
stage the positions of the image elements in map-coordinate
space are computed. These positions rarely correspond to the
locations of the pixels in image-coordinate space, for pixels in
the raw image are defined at integer row/column intersections.
An interpolation procedure is therefore needed. Most
interpolators are high-pass filters, hence the output image is
generally somewhat smoothed. This may have some effect on its
subsequent use, for smoothing implies blurring with consequent
loss of detail. As noted earlier, the size of the remotely-
sensed image is a constraint on the choice of interpolator.
If a seven-band TM image is to be transformed then around 266
million pixel values must be interpolated. The temptation to
go for a quick method such as the nearest-neighbour interpolator
may well be irresistible, for more sophisticated

techniques such as the use of piecewise bicubic polynomials make
correspondingly greater demands on computer time. The cost-
effectiveness of each technique must be carefully considered
against the fidelity of the output.

The problems faced at the geometrical transformation stage
can therefore be summarised as (i) the choice of an efficient
and accurate means of deriving a set of functions to map from
image to geographic coordinates (and vice-versa) and (ii) the
choice of a suitable interpolator to resample the image in terms
of geographic coordinates. Developments in mathematical
techniques capable of carrying out these operations of
transformation and interpolation are required. It may also be
possible to make use of developments in expert systems to build
in some statistical knowledge relating to the number and
location of ground-control points required, and the condition
of the matrix of sums of squares and cross-products, in the
least-squares methods. Artificial Intelligence and computer
vision techniques might also be exploited whenever an existing
set of ground-control points must be related to a newly-
acquired image. Clearly, there are opportunities for interesting
developments in this area.

A digital multispectral image consists of quantised
measurements of radiance upwelling into the field of view of
the sensor. From these indirect measurements we wish to
identify Earth surface cover types. The pattern recognition
problem can be defined as the determination of an obscure
(i.e. not directly-observable) property from a set of indirect
measurements (Varmuza, 1980). The set of indirect measurements
of the obscure property is called a pattern, and the problem
can be restated as the recognition of the class of category
to which the pattern belongs. Sometimes the individual
elements of the pattern are termed features; each feature can
be considered as one axis of a p-dimensional feature space.
Each pattern maps to a point in this feature space, and the
basic hypothesis is that similar patterns are close together
in feature space. A number of algorithms are in use, each
implementing a particular decision rule on the basis of which
the feature space is subdivided into non-overlapping regions.
Each region is then labelled in terms of a real-world object
such as "row crop", urban area" or "water". Exploratory
methods of pattern recognition attempt a bootstrap solution to
the problem by attempting to determine the number and nature
of the groups present from the distribution of points in feature
space, while confirmatory methods begin with an initial
hypothesis that k groups of patterns are present, with an
initial estimate of the statistical properties of those groups.
For illustrative purposes a commonly-used method, that of
maximum likelihood, will be summarised briefly.

   The maximum likelihood method requires that the user is
able to estimate the number k of groups present in the data, and
has obtained sample estimates of the mean vector and covariance
matrices (assumed positive-definite) for each group. It is
assumed that the frequency distribution of the measurements in
each group can be closely approximated by the multivariate
normal density. On the basis of the definitional formula of
the multivariate normal density the probability that an
observation vector belongs to each of the k categories in turn
is computed and the observation vector is allocated to that
class for which the probability is a maximum. In practice the
logarithm of the probability is used to eliminate the need to
compute exponentials, and other savings in computation time
are made by removing constants. Even so, the amount of computer
time needed to process a full Landsat MSS or TM image using this
method is considerable, and may not be cost-effective. Problems
of a statistical and mathematical nature are encountered in the
estimation of the properties of the classes from the samples
or "training classes". Spatial autocorrelation among the
sample values reduces the degrees of freedom and leads to
inefficient estimates of the mean and covariances (J.B. Campbell,
1981; Labovitz and Masuoka, 1984) while outliers (for example,
pixels included in a training sample which represent a mixture
of different land-surface cover types) will have an undue
influence on the estimates. Robust methods of estimating the
mean and covariance could be used (N.A. Campbell, 1980) in
order to reduce the effect of such outliers.

   The number of possible combinations of seven observations,
each of which can take on an integer value in the range 0-255
is very large - the magnitude can be appreciated if it is
realised that if the combinations could be enumerated at the
rate of 1000 per second it would take around two million years
to count the total. A method based upon precalculation of class
allocations for each possible combination is impossible.
However, experience shows that for a given digital image only
a relatively small subset of the possible combinations
actually occurs, so that a hash-table method becomes feasible.
The four (MSS) or seven (TM) values measured at a given pixel
location are concatenated to give a four- or seven-byte integer,
which can be operated upon to give the hash table address. The
operation is a single division. Once the hash table address
is known it is possible either to read back the class
identification associated with the given pixel vector or, if this
is the first occurrence, to calculate the class label and store
it in the hash table (Mather, 1985a,b).

   While such techniques can be used to speed up the computations
involved in pattern recognition using a general-purpose computer,
they cannot overcome a basic problem, that of feature selection.

Pattern recognition using traditional photointerpretation
methods uses two image properties termed tone and texture.
Tone is the shade of grey used to depict a particular object in
a specified spectral band whereas texture is the variation in
greyness around a particular point; it is a property of the
context of the pixel rather than of the pixel itself. Pattern
recognition methods applied to digital remotely-sensed images
have, in the main, made use only of tonal or spectral
information, that is, the pixel values have been considered
individually without regard to the properties of the
neighbourhood of the pixel. The use of textural information
can be expected to lead to an improvement in classification
accuracy. Texture, or the pattern of greylevel variation in a
neighbourhood, could be measured directly in the image domain
or alternatively in the frequency domain. Weska et al. (1976)
showed that image-domain or statistical methods such as those
proposed by Haralick et al. (1973) were superior to Fourier-
based methods. Unfortunately, Haralick's measures, based
upon counts of pixel greylevel values in a neighbourhood, are
computationally expensive and consequently have not been
widely used. Kittler and Foglein (1984) provide an excellent
recent overview of the subject, and note that textural
information can be taken into account either during the
classification phase (with measures of texture being treated
as additional features) or following a conventional pixel-by-
pixel classification for example by the use of a majority
filter (Rosenfeld and Kak, 1982). Kittler and Foglein (op.cit.)
propose an algorithm which they claim is no more complex in
computational terms than conventional algorithms. The
recursive technique that these authors describe is similar to
a probabilistic relaxation algorithm such as that described by
Harris (1985), although Harris's method is a post-classification
technique of the kind described by Kittler and Foglein (op.cit, p.13)
as being handicapped by the fact that an attempt is made to
recover lost information. Other methods of a more experimental
kind are described and illustrated in a comprehensive review
published by the Norwegian Computing Center (Saebo et al.,
1985).

It is clear that no single method or group of methods for
the incorporation of textural information into the classification
of remotely-sensed multispectral images has yet received
widespread acceptance, despite the fact that the inclusion of
textural information can result in more accurate classifications
(Shih and Schowengerdt, 1983). Since reliability (in the eyes
of the ultimate user of the classified image map) is a function
of accuracy it is evident that this problem must receive more
attention from mathematicians and statisticians if significant
progress is to be made. Time does not permit the elaboration
of other important, related problems such as the assessment of

classification accuracy and the selection of training samples.
It should, perhaps, be stressed that the problems involved are
not solely theoretical for, as pointed out several times already,
the major constraint on the processing and analysis of
multispectral remotely-sensed images is the volume of data
involved; we are considering algorithms that are efficient in
both the statistical and computational sense. As hardware costs
fall and memory capacities increase we shall, no doubt, see a
decline in the problems of data handling. Nevertheless,
computational efficiency will remain as perhaps the single most
important characteristic of algorithms for the processing of
remotely-sensed image data.

4.   CONCEPTUAL PROBLEMS

   The problems described in the preceding section were labelled
"practical" for they involve the numerical processing of data
without regard to the difficulties of interpretation of the
results in terms of real-world phenomena. These difficulties
will be considered next. They are called "conceptual" even though
they involve physical processes and interactions. The quantised
counts recorded on a computer-readable magnetic tape have so
far been presented as if they were pure measurements of ground-
leaving radiance, and it has been implied that different
Earth-surface cover types are characterised by distinctive
spectral responses in the 0.4 - 12.5µm region. It is apparent
that no general class of object (such as "granite" or "wheat")
can possess a uniform or invariant spectral characteristic
through space and time because of internal inhomogeneity and
because local environmental conditions will play a part in
determining that response; also, vegetative features will
undergo seasonal alteration. Even so, a general association
between spectral response and character of the object is well-
established. In this section I will skirt the issue of precise
relationships between land-surface cover types and their
spectral reflectance and emittence characteristics and, instead,
concentrate on a more fundamental problem alluded to earlier,
namely, the processes which must be considered when interpreting
measurements of spectral radiance recorded by a sensor located
above the atmosphere.

   Solar radiation passing through the Earth's atmosphere is
subject to scattering and absorption as mentioned above. Fig.
6 shows that a proportion of the incoming radiance is scattered
by the atmosphere into the field of view of the sensor, thus
appearing to the sensor as energy originating from point P on
the ground. This component of the signal is known as
atmospheric path radiance. Energy reflected from regions of the
Earth's surface other than P is also scattered by the atmosphere
into the field of view of the sensor; equally, radiance

originating at P may be scattered out of the sensor's field of
view or absorbed by atmospheric constituents. Thus, the radiance
recorded at the sensor is the sum of several components and the
product of several processes. The magnitude of each component
will vary over time for the same ground area as the proportion
of atmospheric constituents changes in response to meteorological
conditions, even assuming all other factors remain constant.
Furthermore, the contribution of each component will vary with
wavelength. The implications are: (i) the signal recorded at
the sensor is not a true reflection of ground conditions it is
contaminated by scattered energy and (ii) as atmospheric
conditions change so will the degree and spectral distribution
of the contamination. Hence, remotely sensed measurements are
not "pure", and so correlations between characteristics of
reflection/emission spectra and ground phenomena will be
difficult to ascertain accurately and will not be capable of
extrapolation over space or time unless the effects of the
interaction between electromagnetic radiation and the components
of the Earth's atmosphere can be estimated accurately.

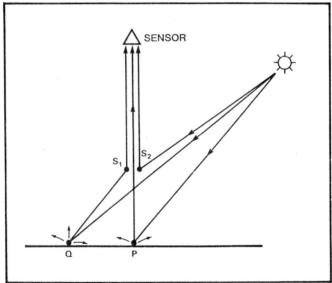

Fig. 6 The signal received at the sensor, apparently from
        point P on the ground, is contaminated by scattered
        energy. The path radiance is scattered before reaching
        the ground (S2) while energy reflected from Q is
        scattered at S1 into the field of view of the sensor.

There is a considerable body of theoretical knowledge
relating to the processes of scattering and absorption of
electromagnetic radiation by the Earth's atmosphere. In
practice, however, the use of such theoretical knowledge is
hindered by the lack of measurements of the state of the

atmosphere at the time of imaging. The collection of
atmospheric data is costly in terms of time, equipment and
manpower given the large area covered by a MSS or TM image.
Atmospheric models are available which allow the use of
standard atmospheres (described in such terms as "mid-latitude
temperature summer") bu these models cannot accurately
evaluate the atmospheric effect at a particular point in space
and time (Kneiszys et al., 1982). An alternative, though still
incomplete, solution is to utilize information present in the
image data themselves. Ahern et al. (1977) suggest that deep
clear lakes be used as standard reflectors, which should
maintain a known spectral response over time and from place
to place. Any change in the expected spectral response of
these standard reflectors is thus attributable to atmospheric
variation. Switzer et al. (1981) describe a multivariable
regression procedure which they call the covariance matrix
method; this method utilises the statistical relationships
between the channels of a multispectral image, and is claimed
to provide a first-order approximation to the path radiance
term. The estimation of atmospheric effects in the thermal
infrared region of the spectrum can be estimated by "split window"
techniques. The TIROS/NOAA AVHRR sensor uses two wavebands in
the thermal region between 10.5 and 12.5μm, and studies have
shown that the effects of the Earth's atmosphere can be estimated
from differences between the two bands. Unfortunately, no single
acceptable algorithm has been developed; Cannizzaro et al. (1985)
compare 26 different split-window methods and report that
significantly different estimates of sea-surface temperatures
are obtained from their use.

The difficulties involved in the separation of the true
ground-leaving radiance from the atmospheric component of the
signal received by a sensor above the atmosphere are thus far
from being resolved. This implies that statistical descriptions
of particular land-cover types are approximate only, and refer
to a specific date and place. They cannot reliably be
extrapolated over space or time. Consequently, fully automated
pattern recognition is not feasible at present, as individual
ground measurements are needed at the date and time of image
collection. This is a major limitation on the use of passive
remote sensing for the routine monitoring of changing
vegetation patterns or states (Duggin, 1984).

In land applications of remote sensing, the allocation of
individual pixels to information classes that are associated
with particular land-cover types is based upon the statistical
properties of the radiance recorded by the sensor for those
pixels in a number of wavebands. As Slater (1980, p. 227) notes,
"The heart of the remote-sensing problem is the identification
of ground features from the radiant flux that emanates from

them and passes through the atmosphere to be recorded by a
remote-sensing system". The distortions introduced by the
passage of electromagnetic radiation through the atmosphere
have already been noted. There are, however, other important
factors, including the illumination geometry of the scene,
which includes relationships between the shape of the ground
surface (the topography), the Solar zenith and azimuth angles
and the viewing angle of the sensor. The first two of these
factors have been discussed by Kimes and Kirchner (1981), who
show that the orientation of slopes with respect to the
direction of illumination, together with the gradient of the
slope, will influence the magnitude, of the radiance recorded
by a nadir-pointing sensor. They conclude that this factor
will be significant whenever classification techniques are
applied to images of mountainous regions. Holben and Justice
(1980) describe a laboratory experiment which showed that
considerable variations in recorded radiance will occur as a
result of variations in slope angle of the ground surface and of
differences in Solar elevation angle. These variations were as
large as 50 quantization levels (out of a range of 128 levels)
when the slope angle was varied from $90°$ to $4.6°$ with a Solar
elevation angle of $62°$. For a lower sun angle ($40°$) the
difference increased to 55 quantization levels.

Relationships with Solar azimuth and zenith angles, the
geometry of the surface of the vegetation canopy and the angle
of view of a sensor has also been the subject of considerable
study. Solar elevation angles vary throughout the year, and
also vary with latitude over the Earth's surface for a given
local time. The magnitude of direct Solar irradiance (and
therefore of reflected radiance) will therefore vary over the
year, adding another variable to the complex of factors which
influence the magnitude of the radiance measured by a remote-
sensing device. It is possible to carry out a correction to
standardize the data for a given date to a chosen Solar
elevation angle, but this assumes that the reflectance of the
target is invariant with respect to incidence and viewing
angles; this is rarely the case, and many natural surfaces
display non-isotropic directional reflectance characteristics.
It is this property of natural surfaces which further complicates
the identification of a relationship between a particular land-
cover type and its spectral reflectance function (Goel and
Thompson, 1985). Finite-element techniques have been used to
model this relationship (Suits, 1972; Kimes, 1984). These
methods are, typically, complex and demanding both in terms of
observational data and computing resources. The problems they
generate will become increasingly important as sensors with an
off-nadir viewing capability are developed. The Landsat MSS,
for example, has a relatively narrow angular field of view, of
the order of $8°$ whereas the French SPOT satellite, launched

on February 22nd 1986 has the ability to point its High
Resolution Visible sensor up to 27° to either side of the
subsatellite track. Aircraft-mounted multispectral scanners
also have wide angular fields of view. The Daedalus scanner
mounted on the N.E.R.C. aircraft has a field of view of 72°.
Relationships between the recorded radiances and Earth-surface
characteristics will vary across the area of an image
generated by a scanner with a wide field of view.

It is evident that the interpretation of remotely-sensed
data is fraught with difficulties, relating to the nature of
the physical interactions between radiant energy and the
Earth's atmosphere and surface. Automatic analysis of remotely-
sensed images will not be a feasible proposition until it
becomes possible to establish reliable and effective mathematical
models of these processes. These models should allow the
effects of interactions between the Earth's atmosphere and
incoming or outgoing energy to be estimated accurately without
the need for extensive sets of observational data, whilst
avoiding the problems of overgeneralisation inherent in the
physically-based models that are currently in use. Quantitative
modelling of relationships between the characteristics of Earth-
surface cover types (particularly the geometry of vegetation
canopies) and the angles of illumination and observation is also
a priority.

5. CONCLUSIONS

It is difficult in a review of this nature to present a
balanced view of the status of a particular discipline. The
natural tendency is to stress those problem areas which are
seen to be outstanding at a particular point in time, and to
ignore the real progress that has been made in a relatively
short period of time. Quantitative remote sensing using data
from orbiting satellite platforms is still a very young subject,
yet considerable technical expertise has been built up, both in
the UK and abroad. Whilst recognising the quality of much of
the work that has been done in the development of numerical
techniques for the processing of remotely-sensed data it is,
nevertheless, important to highlight those problem areas which
are inhibiting the operational use of remotely-sensed data from
the visible and near-infrared regions of the spectrum. In this
paper two such areas have been identified - the practical problem
of data-handling and information extraction and the conceptual
problems of relating remote observations to the real properties
of the materials which cover the Earth's surface. Both of these
areas depend upon the development and application of mathematical
expertise and it is therefore fitting that they should be
highlighted at the outset of this conference.

6. REFERENCES

Ahern, F.J., Goodenough, D.J., Jain, S.C., Rao, V.R. and
   Rochon, G., (1977) Use of clear lakes as standard reflectors
   for atmospheric measurements. Proc. 11th Symposium on
   Remote Sensing of Environment, Environmental Research
   Institute of Michigan, Ann Arbor, Michigan.

Ashkenazi, V., (1984) Satellite positioning techniques. In:
   Steven, Dodson and Mather (eds.) (1984), 5-17.

Bernstein, R., (1976) Digital image processing of Earth
   observation satellite data. *IBM Journal of Research and
   Development,* 1976, 40-57.

Brush, R.J.H., (1985) A method for real-time navigation of
   AVHRR imagery. *IEEE Trans. Geosci. Engng. & Remote Sensing,*
   GE-23, 876-887.

Campbell, J.B., (1981) Spatial correlation effects upon
   accuracy of classification of land cover. *Photo. Engng. &
   Remote Sensing,* **47**, 355-363.

Campbell, N.A., (1980) Robust procedures in multivariate
   analysis. I: Robust covariance estimation. *Appl. Statistics,*
   29, 231-237.

Cannizzaro, G., Ricottilli, M. and Ulivieri, C., (1985) Analysis
   of different algorithms for sea surface temperature retrieval
   from AVHRR data. Presented at 19th Intern. Sympos. on Remote
   Sensing of Environment, Ann Arbor, Mich., October 1985.

Davison, G., (1984) Ground control pointing and geometric
   transformation of satellite imagery. In: Steven, Dodson
   and Mather (eds.) (1984), 45-59.

Duggin, M.J., (1984) Factors limiting the discrimination
   and quantification of terrestrial features using remotely
   sensed radiance. *Int. Jnl. Remote Sensing,* **6**, 3-28.

Goel, N.S. and Thompson, R.S., (1985) Optimal solar/viewing
   geometry for an accurate estimation of leaf area index and
   leaf angle distribution from bidirectional canopy
   reflectance data. *Int. Jnl. Remote Sensing,* **6**, 1493-1520.

Haralick, R.M., Shanmugan, K. and Dinstein, I., (1973) Textural
   features for image classification. *IEEE Trans. Syst.,
   Man & Cybernetics,* 610-621.

Harris, R., (1985) Contextual classification post-processing of Landsat data using a probabilistic relaxation model. *Int. Jnl. Remote Sensing,* **6**, 847-866.

Holben, B.N. and Justice, C.O., (1980) The topographic effect on the spectral response from nadir-pointing sensors. *Photogramm. Engng. & Remote Sensing,* **46**,1191-1200.

Kimes, D.S., (1984) Modelling the directional reflectance from complete homogeneous vegetation canopies with various leaf-orientation distributions. *Opt. Soc. Amer. Ann.,* **1**, 725-

Kimes, D.S. and Kirchner, J.A., (1981) Modelling the effects of various radiant transfers in mountanous terrain. *IEEE Trans. Geoscience and Remote Sensing,* GE-19, 100-108.

Kittler, J. amd Foglein, J., (1984) Contextual classification of multispectral pixel data. *Image and Vision Computing,* **2**, 13-29.

Kneiszys, F.X., Shettle, E.P., Gallery, W.O., Chetwynd, J.H., Abreu, L.W., Selby, J.E.A., Clough, S.A. and Fenn, R.W., (1983) Atmospheric transmittance/radiance: computer code Lowtran-6. Optical Space Div., Air Force Geophysical Laboratory, Massachussets Air Force Command, USAF, AFGL-TR-83-0187.

Labovitz, M.L. and Matsuoko, E.J., (1984) The influence of autocorrelation in signature extraction - an example from a geobotanical investigation of Cotter basin, Montana. *Int. Jnl. Remote Sensing,* **5**, 315-332.

Marble, D.F., (1984) Geographic information systems: an overview. Proc. Pecora 9 Symposium - Spatial information technologies for remote sensing today and tomorrow. Sioux Falls, South Dakota, October 1984. IEEE Catalog No. 84CH2079-2, 18-24.

Mather, P.M., (1985a) A computationally-efficient maximum-likelihood classifier employing prior probabilities for remotely-sensed data. *Int. Jnl. Remote Sensing,* **6**, 369-376.

Mather, P.M., (1985b) Derivation of thematic map information from remotely sensed data. *Land and Minerals Surveying,* **3**, p.628-643.

Rosenfeld, A. and Kak, A.C., (1982) Digital picture processing. Second Edition, volume 1, Academic Press, New York.

Saebo, H.V., Braten, K., Hjort, N.L., Llewellyn, B. and Mohn, E.,
    (1985) Contextual classification of remotely sensed data:
    statistical methods and development of a system. Report no.
    768, Norwegian Computing Center.

Shih, E.H.H. and Schowengerdt, R.A., (1983) Classification of
    arid geomorphic surfaces using Landsat spectral and textural
    features. *Photogramm. Engng. & Remote Sensing,* **49**, 337-347.

Slater, P.N., (1980) Remote sensing: optics and optical
    systems. Addison Wesley, Reading, Mass.

Steven, M.D., Dodson, A. and Mather, P.M., (eds.) (1984):
    Location. Proc. Remote Sensing Workshop 1, Department of
    Geography, University of Nottingham, England.

Suits, G.H., (1972) The calculation of the directional
    reflectance of vegatative canopy. *Remote Sensing of
    Environment,* **2**, 175.

Switzer, P., Kowalik, W.S. and Lyon, R.P., (1981) Estimation
    of atmospheric path-radiance by the covariance method.
    Photogram. Engng. & Remote Sensing, **47**, 1469-1476.

Townshend, J.R.G., (1980) The Spatial resolving power of
    Earth resources satellites: a review. NASA Tech. Memo. 82020,
    Goddard Spaceflight Center, Greenblet, Maryland 20771.

varmuza, K., (1980) Pattern recognition in chemistry. Lecture
    Notes in Chemistry 21, Springer-Verlag, Berlin.

Weszka, J.S., Dyer, C.R. and Rosenfeld, A., (1976) A
    comparative study of texture measures for terrain classification
    IEEE Trans. Syst., Man. & Cybernetics, 269-285.

ERRORS AND UNCERTAINTIES IN FEATURE RECOGNITION

J.C. Simon
(Institut de Programmation, Universite Pierre et Marie Curie,
Paris, France)

1. INTRODUCTION

1.1 Image representations

A pictorial image is always obtained from the outside world through an optical instrument. A "luminance" intensity is recorded on some support, giving a "representation" of the image. A familiar representation is a photograph, in which the level of grey is supposed to be proportional to the luminance intensity recorded during a short time interval.

The optical physicists represent this luminance intensity in a plane as a continuous function of very specific properties. They remark that this intensity is the result of the superposition of plane waves of the same wavelength λ, for monochromatic images of course. Thus the amplitude of the field is $f(x,y,z)$, with:

$$f(x,y,z) = \int F(u,v,w) \exp\left[-2\pi i(ux+vy+wz)\right] \, du \, dv \, dw \tag{1}$$

where F is the intensity of the plane wave of direction cosines u, v, w,

f and F are Fourier Transforms (F.T.) of each other. As the wave propagation equation is for monochromatic waves:

$$\Delta^2 f + \frac{4\pi^2}{\lambda^2} f = 0 \tag{2}$$

Thus

$$\left(u^2 + v^2 + w^2 - \frac{1}{\lambda^2}\right) F(u,v,w) = 0 \tag{3}$$

Let

$$u = 1/\lambda \, \sin\theta \, \cos\phi$$

$$v = 1/\lambda \, \sin\theta \, \sin\phi$$

$$w = 1/\lambda \, \cos\theta$$

when from (3)

$$u^2 + v^2 = \frac{\sin^2\theta}{\lambda^2}$$

and if

$$a = \frac{\sin\theta_{max}}{\lambda}$$

then

$$u^2 + v^2 \leq a^2 \tag{4}$$

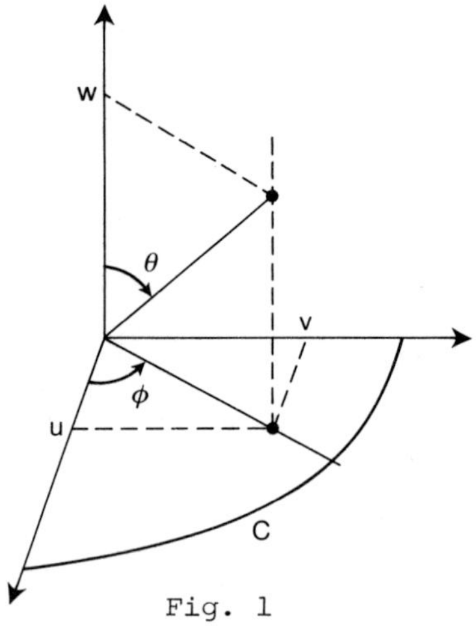

Fig. 1

The F.T. $F(u,v)$ of $f(x,y)$ has a support limited to the circle
C of radius a; and may be assumed as zero outside.

The limitation on the F.T. support has drastic consequences
on the function $f(x,y)$. They are known as Bernstein relations.
As $f(x,y)$ is often taken equal to $h(x)$ by $g(y)$, let us give
these relations for one dimension functions:-

let $|F(u)| = 0$ if $|u| \geq a$, then

$$\text{MAX} \, |f^{(n)}(x)| \leq (2\pi a)^n \, \text{MAX} \, |f(x)| \tag{5}$$

(5) implies limitations on the derivatives of the function $f(x)$,
directly related to the limit of the spectrum.

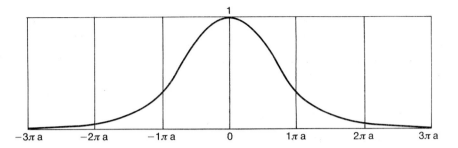

Fig. 2 Shortest non-negative pulse satisfying relation (5)

Fig. 2 shows a non-negative "pulse", as short as possible, satisfying relation (5). For more details see Simon (1984) or (1986,b).

To separate two of these shortest pulses, their spacing δx must be greater than 1/2πa. If δu = 2a is the "bandwidth", we obtain the familiar uncertainty relation:

$$\delta u \ \delta x \ \geq \ 1 \qquad\qquad (6)$$

This relation was pointed by Gabor (1946) under similar arguments. Slepian (1976) and more recently Wilson and Granlund (1984) underlined the importance of uncertainty relations in wave and signal theory, using elaborate arguments, derived from quantum theory, which do not seem fully necessary here.

1.2 Computers and Image Processing

The advent of very fast computing facilities and easy programming allowed treatment of images with a much greater freedom than the "analog" optical physical processes, limited to linear operations, but the use of digital computers introduces limitations. The representation of an image has to be handled as a set of samples, the pixels. Fortunately, and because of the band-limited nature of f(x,y), it is possible to restore without errors f(x,y) from the data of the pixels, but the necessary condition is that the sampling interval α of f should be smaller or equal to 1/2a.

On the reverse, when the pixels are given at a sampling interval, it should be assumed that the former condition is respected, and thus relations (5) and (6) on the continuous representation f, result of the measured luminance of an optical sensor.

In particular, binary images, frequently considered in picture processing because they imply less memory and seem "simpler", are not the result of an optical process alone. Some non-linear process, such as a thresholding, has to be added on the grey level image given by an optical system. Certainly one may consider any binary representation and declare that it represents an image, but can it be obtained in a real situation, with optics and thresholding?

In other words, the problem considered here is to define precisely the properties of a representation given by an optical system, under the data sampled, the grey level pixels, stored in the computer memory. In this paper, we assume that from the pixels, the luminance function f, given by an optical system, can be restored exactly. This cannot be done, of course, from a binary representation. For example, we should remember that in an image representation, a short pulse has at least three non-zero pixels in x and y directions.

1.3 Computing Complexity

The volume of memory necessary to store an image may be quite large, exceeding the capacity even of modern fast computer memories. A TV picture requires 0.4 Mbyte, a LANDSAT image with four wavelengths 60 Mbyte. Image processing and recognition may require from 10 to 10^4 operations by pixels, which saturate readily powerful modern processing units. Parallel processing, proposed to overcome this limitation, is not yet powerful and efficient enough to attain the human vision performances.

Whatever the chosen approach, extremely fast GaAs chips, or parallel processing with hundreds of thousands of conventional chips, complexity problems may be overcome only by multilevel recognition steps.

On the first step elementary features are identified from the numerical pixels; the next step uses the features to identify sub-images; and so on, up to scene analysis.

A very important issue is the nature of recognition or identification operators or programs. Template matching is an exponential algorithm, and soon impossible to use even for a small number of pixels. But identification of an image pixel by pixel is a hopeless effort, as teledetection specialists learned the hard way. In fact there are two possible ways to overcome these difficulties:

- limit the number of pixels on which to take a decision;
- use operators of reduced complexity, polynomial rather than exponential.

1.4 Primitive features and operators

Let us discuss briefly the first level of identification: from the image pixels to the primitive features.

In practice this first step of a Pattern Recognition process conditions the success of the final result, as the following steps rely on the correct identification of these primitive features. Many clever schemes, inspired by A.I., have failed because of the unreliability of first level primitives. Experience has shown that poor phoneme identification could not be saved by semantic information of speech (ARPA project of the late seventies). Poor edge and texture detection do not allow successful scene analysis. Reliable primitive identification is thus of prime interest in Pattern Recognition.

Observing a photograph, one is able to find nearly without error primitive image features, such as edges, contours, lines, uniform regions; but one cannot tell how the vision system is achieving these results. One may give properties of a feature; for example, a straight line is "the shortest path between two points". But what is needed in P.R. is a constructive definition; in other words an operator (algorithm, program) acting upon several pixels (the window) may decide if the corresponding feature exists, and what are its parameters.

In fact, the best definition of a feature is the detecting operator itself. Now of course an edge operator acting on an image may give results different from expected: edges are missing, or found where we do not see any.

The physiology of mammal vision shows that a very large number of neurones (more than 10^9) are devoted to vision in several regions of the cortex. An intricate communication network (up to 10^4 dendrites by neurone) connects these neuronal cells. Nature has provided us with a remarkable parallel system of nearly innumerable operators; today poorly understood. Unfortunately our means are much more limited with our modern computing system. We will examine how to make optimal use of them.

2. VARIABILITY AND ERRORS IN THE DETECTION OF PRIMITIVE
    FEATURES

2.1 Expected Properties of Primitive Features

   Primitive feature identification is the step from the
representation of numerical pixels to the interpretation by
objects having a type, i.e. a name and some properties. What
are these properties?

Reduction of the volume of information. Let $X = (x_1, x_2, \ldots x_n)$
be a list of n pixels x_i, and $W = \{w_1, \ldots w_m\}$ be a set of
feature names. An identification A maps X in W

$$A: X \rightarrow W \qquad\qquad (7)$$

To be implemented in a constructive way, operators (programs)
A_j have to be defined, such that $A_j : (x_1, \ldots x_n) \rightarrow w_j$ (7').
Such operations reduce the volume of information. To n numerical
pixels of q bits correspond one bit, the presence or absence
of the feature w_j, or a few bits, giving some parameters of
the feature.

   The representation complexity is thus reduced. But the
reverse is not at all always true. Any reduction of
representation does not result in a meaningful, or rather
useful, feature identification.

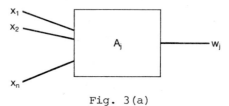

Fig. 3(a)

Template matching, the training set. A theoretical solution of
feature identification would be to keep in a memory the mapping
defined by (7). In other words, to each representation X of n
pixels a feature w_j corresponds in the memory. Template matching
would thus define (7) extensively by giving a priori all the
set of couples (X,w). By looking in the memory for the
identification of an unknown X, an exact answer would be
obtained. But such a goal is unrealizable in practice, as the
number q^n of X to be registered grows exponentially with n.

In a real situation a limited number of couples (X,w) is given. The set of these samples of (7) is called the training set T.

Windows and sub-images. It has been stated already that the need to escape intractability implies several levels of recognition. But the same necessity limits the number n of pixels, on which an operator acts. Sub-images are thus considered; most frequently the n samples are on a compact support, a window. If the window is small, one or a few pixels, a first type of error creeps in. Let P_n be the probability of this error due to noise. It decreases with n. But if n increases, template matching cannot be achieved and an extensive definition of a feature becomes untractable. We have to resort to polynomial operators of low degree (example of linear filters or classifiers). An error may result, the probability of which P_c is the result of the problem complexity. It increases with n. Fig. 3(b) gives the total probability of error $P_e = P_n + P_c$. A flat minimum is found, Kovalevsky (1978). For example, windows of 3x3 to 5x5 are usually used for edge detection.

But we should keep in mind that the size and the shape of these windows are not fixed. The variability of a sub-image window feature may be quite great, depending on the problem, as several studies on edge detection have shown, Levialdi (1981).

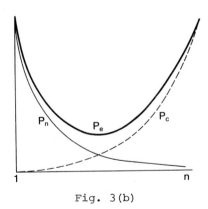

Fig. 3(b)

2.2   The Choice of Features and Corresponding Operators

The choice of features and of feature operators seems to be limited only by the imagination of P.R. specialists, as one may be convinced by examining the proceedings of the International Conferences on Pattern Recognition. A large number of features

and ways of detecting them have been proposed for printed
character recognition, since the first studies on the topic.
In remote sensing, contours, edges, straight segments, textures,
are the usual features looked for, but we may ask: are they
the most useful for the final goals of detection, such as scene
analysis?

The real problem is to find simple operators, which identify
features, whatever they are, with the lowest error. They may
not coincide with what we consider as a meaningful feature such
as an edge, or a phoneme in speech recognition, as long as they
are robust and useful for the final goal. As an example,
research in speech focused first on the detection of "phonemes",
acoustical units defined by the acousticians. If the vowels
were properly identified, the consonants resisted all efforts.
Now in the first level, features are identified with low error,
but which are quite different from the acousticians' phonemes.

On which properties can we rely to find good operators and
robust features? We put them in two classes, properties of
the representation space, properties of the operators or of the
interpretations.

Properties of representation spaces. Such a space X is the set
of possible representations X, with the possible operations on
the X. As X is a list of numerical values, the n pixels x_i,
it is justified to consider X as a multidimensional metric
space, i.e. a n-space with a distance $d(X,X')$. This distance
defines the neighbourhood of an X in a "regular" region. But
if such a representation space is locally regular and compact
in most places, some regions, the transition regions may be
quite irregular. After R. Thom (1972), we may speak of
"catastrophes".

Classification and clustering are well known techniques,
Simon (1978), in digital image processing. Relying on $d(X,X')$,
they find regular regions, which may represent homogeneous
classes. The catastrophe regions are then defined as boundaries
of regular regions.

As an example, the modern techniques of feature detection in
speech find such homogeneous regions by clustering the data X,
using a distance.

Properties of the interpretations; invariant operators. As
stated, several operators A_j may give interpretations of a
representation X. The A_j condition the properties of the
interpretations. The final goal being to identify a pattern,
it is clear that the interpretations should be the same

whatever are the different representations of the pattern.
In other words, invariance of the operator is synonymous with
identification. Looking for a completely invariant operator is
the final goal of a Pattern Recognition problem. We cannot
achieve it in one step; but we can look for operators
invariant to the usual transformations of the representations
X.

Observations shows that human vision and recognition are
quite insensitive to geometrical transformations, which are
well known to be groups G. But unfortunately, they are not the
pure mathematical groups. G is usually defined in a range; for
instance scale invariance is not defined from zero to infinity.
An octopus will recognize correctly some simple shapes in the
range 1 to 12. The displacement group does not suit the
variation of an image; for instance it transforms a square into
a rhomb. But under some limited transformations of this group,
a square is still recognized as a square. The same is true for
printed characters.

It seems then quite essential to find couples of operator-
feature invariant to the usual geometrical transforms in limited
ranges. Let us give examples in the domain of line images.

2.3 Example of Feature Selection in a Line Image

A line image may be defined from the remarks of § 1.1. The
representation of the "shortest black pulse" has at least 3
pixels different of the white level, in any direction. In ˙
memory of Gabor, let us call such a pulse a quantum of information
(a blob). Of course such a quantum may be thicker, having more
than 3 pixels; but it should be in the shape of a "blob", having
approximately the same number on non-white pixels in any
directions. Now in the considered images a line is defined as
a monodimensional list of quanta partially covering each other.

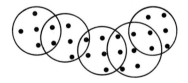

Fig. 4

These feature are invariant for many transformations in the G
groups. All displacements in the plane; plane projections (in
a range; scale transforms (in a range, thickening is allowed
in a certain extent, but not shrinkage under 3 sampling intervals).

Other features, invariant to the former transformations, may
also be chosen:

- end of line; change of direction;

- line crossings or forks;

- closed loops.

   Now of course comes the problem of detecting these features
in the most general images. It is likely that mammal vision
has enough operators (10^9!) to identify directly these shapes
in very various windows; but not us with our still primitive
(?) means. Thus we resort to an indirect approach, which may
be compared to the search of regular regions in the
representation space. Using a robust "line detector", the
features such as end of line and line crossings or forks will
appear naturally as complements of lines detected in a line
image.

2.4 Complementary Identification of Line Image Features

A line detection. According to the former definition, "a line
is a monodimensional list of quanta, partially covering each
other", the first step is to identify quanta in a grey-level
image. It is done with a horizontal scan H.Sc. and a vertical
scan V.Sc. of the image raster

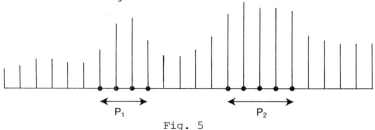

Fig. 5

Fig. 5 shows the result of a H.Sc. By an appropriate set of
rules, line precursors are found on this scan, p_1 and p_2.

Fig. 6 shows on a real example the line precursors (a) of
H.Sc., (b) of V.Sc. Moreau (1985).

   Quanta are found according to the above definition of a
quantum. Fig. 7 gives the quanta found (a) in H.Sc., (b) in
V.Sc. The overlapping regions of two quanta are shown in
black. Lines are easily found. Of course a quantum is an
elementary line. Overlapping is also supposed in regions
which are not quanta themselves but have been detected as line
precursors.

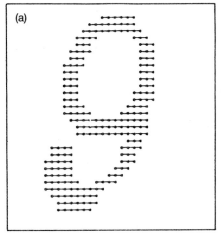
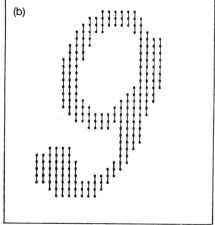

Fig. 6   Line precursors from (a)   Hor. Scan., (b)   Vert. Scan.

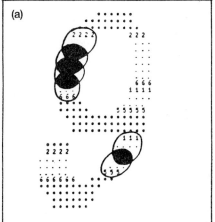
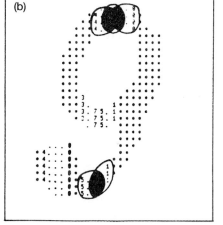

Fig. 7   Quanta and lines found (a)   in Hor. Scan. (b)   in Vert.
         Scan.

Feature detection. Each scan finds lines or quanta in a
sector, for V.Sc. $\pm$ 45°, for H.Sc. 90° $\pm$ 45°. Some results may
overlap. Ambiguities having been solved, lines and quanta are
extracted from the line precursors. What remains are the
primitive features sought: end of line, change of direction,
line crossings or forks. Fig. 8 shows the final results
(a)   with the coding of the lines, (b)   with the interpretations
of the features.

N.B.   Closed loops are found by conventional techniques.

Tested on large number of data of mediocre quality
(distortion and noise), this approach has proved to be quite
robust. Errors are quite small. Inversely a direct approach,
using specialized operators for each feature, would have been
hopeless, due to the very great number of requested operators
and operations. In particular, we should remark on the great
variability of the feature support or "windows", and also on
the need to test the context before making a decision, which
is difficult with a limited size window.

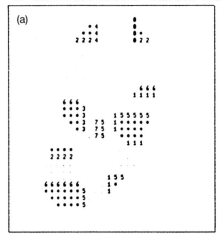 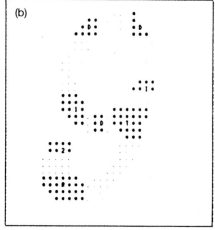

Fig. 8   Final results (a)   with the direction of lines
                        (b)   with the interpretations: T triple
                              point,
        I,D, change of direction, 2 line extremity.

3.   UNCERTAINTIES IN FEATURE DEFINITION

3.1   Variability of features

   We may agree on two points:

i)   The difficulty of the direct determination of features.
Determination of a priori features on windows of fixed and
limited size is prone to errors as the first generation of
edge detectors have demonstrated. Even human beings looking at
a picture in a small window are not able to decide what they
see. On the other hand, several examples, the former one on
line image, show that reliable features may have highly variable
supports.

ii)   The necessity of taking the context into account. Gestalt
physiologists have underlined the importance in human vision of
the "global" organization of a picture. "Dot patterns" are
easily interpreted by human beings. The visual system tries

to find structures in apparently random dots,   Zucker and
Davis (1985).

Thus even if the existence of a feature is highly probable,
it seems difficult to localize it with precision.   In fact,
such a property seems to be the consequence of any identification.
We will express it as an "uncertainty principle", and show its
application.

3.2 Uncertainty Principle

An uncertainty relation has already been expressed by
relation (6). As a consequence, a quantum of information in a
picture has to be represented by at least 3 non-zero level
pixels along 2 orthogonal scans. Once it is found, the
localization of the quantum is undefined inside a window of at
least 3x3 pixels.

The detection of features in line-pictures gives several
robust results in very variable windows. These features
should not be localized in their windows. As it was stated in
2.1, the consequence of feature detection is a reduction of the
volume of representation. The definition "in extension" (with
the pixels) is drastically reduced. Thus we come to the
uncertainty principle, already stated by Aristotle more than
2,300 years ago:

> If definition (representation) in comprehension increases,
>   definition (representation) in extension decreases.

3.3 Applications to higher level identification

Descriptions invariant to transformations. Finding first order
primitive features gives only local information . From this
are needed global information on the object, or part of the
object, to be identified. What is needed is description again
invariant to the possible geometrical transformations of the
object. With such descriptions, identification is a
straightforward comparison between the unknown description and
the reference descriptions stored in the memory.

Again several techniques are possible:
i) A relational data base, i.e. a graph, where the features are
the nodes and the relations between features the arcs of the
graph. To identify, graph comparisons are made.

ii) A set of rules: a description verifying a sub-set of these
rules. Identification may be made either by a "questionnaire",
or more dynamically by an expert-system technique.

iii)   a set of partially ordered lists, as described
later.

   If the relations, rules or lists are invariant to the
transforms, it is clear that the descriptions will also be
invariant.

   Usually, the feature localization is loosely defined.   For
instance A is above B, or A,B,C are in line.   Thus the
"uncertainty principle" is respected, but we are now giving an
example where we had to relax the precision to be able to
identify two descriptions.

Description by partially ordered lists.   The use of lists to
obtain invariant descriptions of the second level was proposed
by Simon et al. (1970), (1971), (1972).   It was thoroughly
studied by Lorette (1978) and Requier (1979), and recently
reported again by Simon (1984), (1986,a).

   Let us consider a set of "points" in a plane A,B,....,H;
see Fig. 9.   Several techniques give descriptions independent
of geometrical transforms.   For example:

   -   Ordering the couples: (BH, GH, BE, AE, ..., BG);
   -   Convex envelopes: (A,B,G,F,E,H) (C,D);
   -   Partition by straight lines;
   -   Scanning by lines: from A a rotating half line: A(H,D,E,C,F,G,B

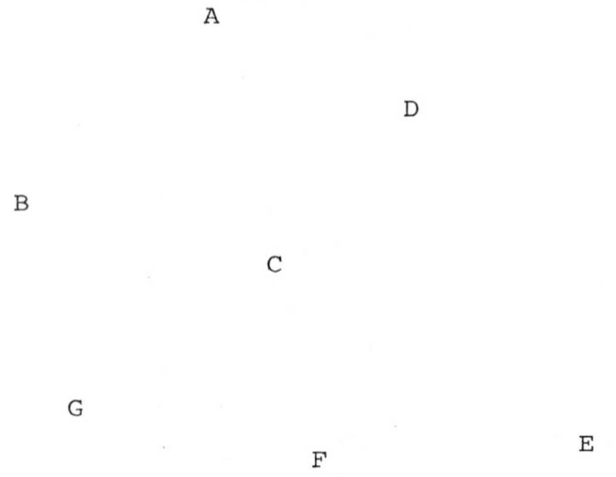

                          A

                                    D

                                                        H

              B

                        C

         G

                            F   E

                       Fig. 9

   Comparing two lists is a well studied problem.   Several
algorithms solve it: Dynamic Programming, steepest descent.
The problem consists in matching an unknown list X to a
reference list Y, in an optimal way.   The result is a distance
D(X,Y).   The best Y corresponds to the smallest D.

If D = O, for a list Y, then the identification is an
excellent one.   But this is rare.   A number of operations due
to distortions or noise on the representation of the unknown
object resulting in X may occur:

- a change of symbol (a different feature);

- a suppression of symbol;

- an insertion of symbol;

- a permutation of two symbols.

Each of these operation implies a cost, which increases
$D(X,Y)$.

The need to relax precision.   In the early 1970s when we
tested these techniques, we found that the attempt to match
two lists X and Y would lead to two types of situation,
according to the precision of ordering e.   If precision was
perfect, e = O, it would never work in practical situations!
However, if the precision is relaxed, e increases, then
impossible comparisons became possible.

We found the following two situations:

a)   Matching is impossible if $e < e_1$;

    Matching is possible if $e_1 \le e \le e_2$;   } matching is possible

    Matching is indeterminate if $e > e_2$.

b)   Matching is impossible if $e < e'_2$;

    Matching is indeterminate if $e \ge e'_2$.   } matching goes from impossible to indeterminate.

In general, e_1 is small and e_2 and e'_2 are close values.   It
may be said that situation (a) the two lists are comparable,
whereas they are not in situation (b).

To mark the relaxation in precision, the exchangeable
symbols are put between parentheses.   For example the list
obtained by scanning with a half-line around A, H,D,E,C,F,G,B
becomes (H,D),E,(C,F),G,B or with a lower angular precision
(H,(D),E),(C,F),(G,B); see Fig. 9.

Such techniques were applied with success to the identification
of printed characters, the primitive features of which being
identified by the method described above.   Thus the precision
of the order of the reference list is given by the fuzzy

localization of the features, undetermined inside their respective
windows. Localization of features by a "point" would lead to
difficulties, avoided by respecting the uncertainty principle,
Simon (1986,a).

   These ideas and approaches may go against some current practices,
such as skeletonization of binary images. In the first place, a
binary image may not be the best data to start with; and in the
second place, the attempts to define a "line" as a chain of pixels
may be real heresy. Undoubtedly difficulties occur at line crossings
or forks, avoided by our techniques. Also too great and unjustified
a precision may lead to real problems in the higher level identifications
So why try to determine a feature with the precision of one pixel,
or less...

Fig. 10   shows the above treatments applied to letters
              a   b   e   (not exactly the same fonts...)
              c   d   g
              f   i

The pixels in "lines" are small dots, features are stars. Some
chains obtained by V.Sc. are compared to 1 or 2 nearest
standard chains, kept in the memory. These results have been
obtained by K. Heng and S. Bouteflika.

4.   CONCLUDING REMARKS

   Summarizing the main ideas, we may say that

i)   Computational complexity compels one to decompose the
identification process in several levels. The first level, from
the extensive representation by pixels to the primitive features,
is essential and most difficult. A low error rate is most
important for the final success of a P.R. process.

ii)   Determination of primitive features should be based upon the
availability and the performances (low complexity, low rate of error)
of the corresponding operators (programs). Rather than use any ad hoc
process, the couples feature-operator should be chosen according
to their quality of invariance to geometrical transformations of
the object to be identified.

iii) Complementary techniques seem to give more robust results than
a direct approach to feature detection. It consists of using operators,
which detect "regular regions" of the picture; in remote sensing,
texture detection plus clustering; in line images, line detection.
The boundary regions (catastrophe regions) are found by subtracting
the regular regions to the total image. Features are thus
complements of the regular regions; in remote sensing, edges,
contours; in line images, end of line, change of direction, forks,
crossings. The main advantages are the drastic reduction in the
number of requested operators, the simplicity and the universality of
the operators detecting regular regions, and, last but not least,
the robustness of the important feature detection.

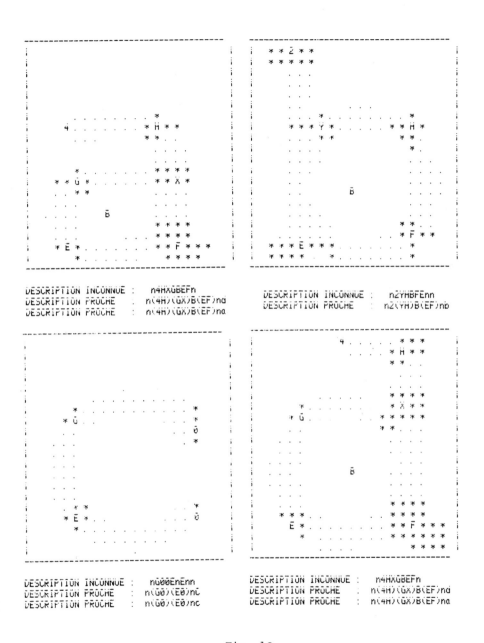

Fig. 10

DESCRIPTION INCONNUE : nGHBFY6En
DESCRIPTION PROCHE : n(GH)B(YF)7ne

DESCRIPTION INCONNUE : nG004004Wn
DESCRIPTION PROCHE : n(G0)(400)(4W0)nt

DESCRIPTION INCONNUE : nHGBEXF4nn
DESCRIPTION PROCHE : n(GH)B(EX)(4F)n9
DESCRIPTION PROCHE : n(GH)B(EX)(4F)n9

DESCRIPTION INCONNUE : nP4HW40nnn
DESCRIPTION PROCHE : nP(4H)(4W0)ni

iv) A contradiction exists between a definition in extension (with the pixels) and the detection of a feature in itself. This may be expressed as an uncertainty relation between a definition in extension and in comprehension. In particular the localization of a feature should be "fuzzy", i.e. undetermined inside its support (window), which may be quite variable, according to the case (example of an edge, a line crossing or fork). An excess of localization may be degrading for the further levels of recognition.

5. REFERENCES

Gabor, D., (1946) "Theory of Communication", *Proc. Inst. Elect. Eng.* **93**, 26, 429-441.

Kovalevski, V.A., (1978) "Recent Advances in Statistical Pattern Recognition". Proc. 4th I.C.P.R. 2-12; Kyoto, Jap.

Levialdi, S., (1981) "Finding the Edge", in "Digital Image Processing" (J.C. Simon & R.M. Haralick, eds.) 105-148. Natō ASI C-77, Reidel.

Lorette, G., (1978) "Operateurs de Description de Figures Planes qui donnent un Résultat invariant après Transformations Géométriques". Proc. Congrès AFCET-IRIA sur RF & IA, 1, 371-377.

Moreau, C., (1985) "Extraction de Formes Caractéristiques dans les Images de Traits". Thèse de 3ème cycle, Univ. Pierre et Marie Curie, Paris.

Requier, J.P., (1979) "Utilisation d'Algorithmes de Comparaison entre Graphes en Reconnaissance des Formes". Thèse d'Etat, Univ. Paris 12.

Simon, J.C., Checroun, A. and Roche, C., (1970) *C.R. Ac. Sc. A* **270**, 1607-1609.

Simon, J.C. and Checroun, A., (1971) I.J.C.A.I. Proceedings, 308-317.

Simon, J.C., Checroun, A. and Roche, C., (1972) Pattern Recognition Jr. **4**, 73.

Simon, J.C., (1978) "Clustering and Digital Image Analysis" in "Machine-aided Image Analysis" (Gardner, W.E., ed.) Inst. of Physics, 20-39.

Simon, J.C., (1984) "La Reconnaissance des Formes par Algorithmes" Masson, Paris.

Simon, J.C., (1986,a) "Invariance in Pattern Recognition; Application to Line Images". Image and Vision Jr. Feb.

Simon, J.C., (1986,b)   "Patterns and Operators; the Foundations
   of Data Classification", North Oxford Academic, London

Slepian, D., (1976)   "On Bandwidth"   Proc. IEEE, **64**, 758-767.

Thom, R., (1972)   "Stabilité Structurelle et Morphogénèse",
   W.A. Benjamin, Reading, Mass. USA.

Wilson, R., Granlund, G.H., (1984)   "The Uncertainty Principle
   in Image Processing".   IEEE Trans. PAMI-6, **6**, 758-767.

Zucker, S.W. and Davis, S., (1985)   "Points and End-points; a
   size/spacing constraint for Dots Grouping", Dept. El. Eng.
   TR-85-3R, McGill Un. Montreal.

THE DYNAMICS OF BRAGG-SCATTERING WAVES ON THE SEA SURFACE

M.S. Longuet-Higgins
(Department of Applied Mathematics and Theoretical Physics,
University of Cambridge)

ABSTRACT

Most of radar backscatter from the sea surface in X- and L-band comes from surface waves of between 2 to 20 cm wavelength. The remote imaging of surface currents, bottom topography, internal waves, and, above all, longer surface waves, depends upon the varying characteristics of the short surface waves when propagated over a non-uniform flow. Here we outline the dynamics of short gravity waves, and show how their wavelength and steepness are affected by their interaction with longer gravity waves of finite amplitude. A mathematical model is constructed for the surface roughness of the longer waves, which takes account of wave breaking and dissipation of the shorter waves, their regeneration by the wind, and the random nature of the longer waves. This leads to a pair of integral equations which are solved by iteration. The solutions are discussed in physical terms, and suggestions are made for further theoretical development.

1.   INTRODUCTION

The scattering properties of a water surface differ basically from those of land. The water surface is nearly always in motion, due to surface waves generated by the action of the wind. Even light winds can raise short waves steep enough to act as Bragg scatterers for electromagnetic radiation at X- and L- band wavelengths. The variable steepness of such short surface waves, interacting with broader features of the ocean such as longer gravity waves, internal waves or steady currents, is at the heart of radar imaging of the sea surface. Fortunately, the propagation of the short waves, and their interaction with larger-scale features, obey the laws of Newtonian dynamics. While some aspects still remain a puzzle,

one can say that, mathematically speaking, the ocean is rather
well behaved.

Typical SAR images of the sea surface (Figures 1 to 5) show
swell being partially refracted by bottom topography; wave
refraction round a headland; large-scale sand-waves on the
bottom in the southern North Sea, as revealed by tidal currents,
and internal waves on the continental shelf.

In this paper we shall not go into details of the SAR
imaging process. Rather, we shall concentrate on the basic
dynamics of the short waves, which renders the imaging
possible in the first place.

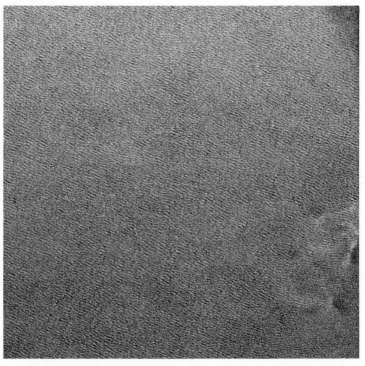

Fig. 1 A SAR image of ocean swell, showing refraction by
        shoaling water (bottom right) in Fair Isle Channel,
        Shetland. SEASAT, 15 Sep. 1978.

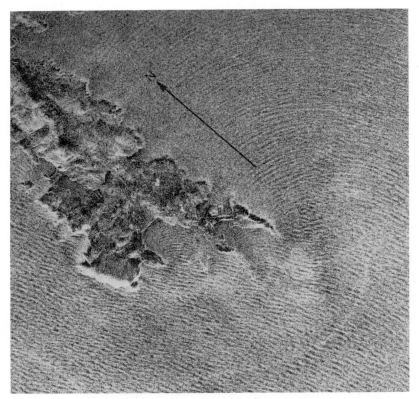

Fig. 2 A SAR image of swell, showing both refraction and
        diffraction by a headland in the Shetland Islands.
        SEASAT, 15 Sep. 1978.

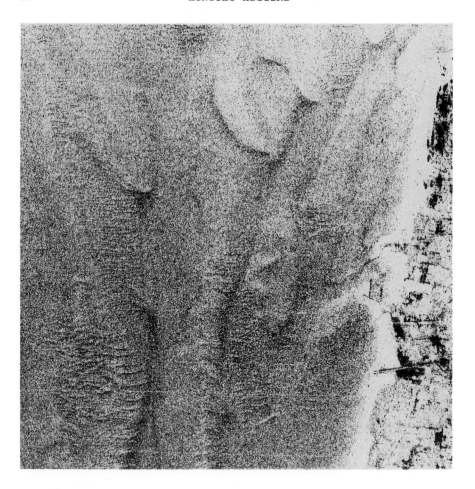

Fig. 3   A SAR image of the southern North Sea, showing sand
        waves on the sea bed near Dunkirk, France.   SEASAT,
        19 Aug. 1978.

Fig. 4 A SAR image showing internal waves in a shallow water
off the west coast of Portugal. SEASAT, 20 Aug. 1978.

L–band   X–Band

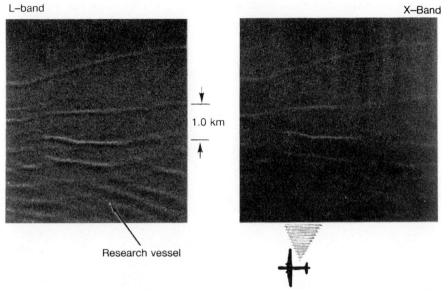

1.0 km

Research vessel

Fig. 5   Two images showing internal waves in shallow water off
        the west coast of Portugal. SEASAT, 20 Aug. 1978.

     In Section 2 below we give a general discussion of the physical
principles, including the principle of action conservation for
the short waves. The most interesting case is the
interaction of short waves with longer gravity waves (see
Section 3). Recent calculations of the short-wave steepening,
which may be strongly influenced by nonlinearities in the
longer waves, are described in Section 4. We then show in
Section 5 how these results can be incorporated in a new model
of sea surface roughness which takes into account both the
dissipation of short-wave energy by breaking, the regeneration
of the short waves by the wind, and the variable height and
group-length of the longer waves. Mathematically, the
application of the model involves the solution of a certain
integral equation. The equation is solved by iteration,
giving physically plausible answers.

     The paper concludes with a discussion (Section 6) suggesting
lines for further theoretical development.

2.   BRAGG SCATTERING

     The basic mechanism is illustrated in Figure 6. Radar waves
of length $\lambda = 2\pi/k$ are incident at an angle θ to the vertical,
on an approximately horizontal surface containing roughnesses
of wavelength $\ell = 2\pi/k$. It is assumed θ is not near θ or $90°$.
Then the condition for first-order backscatter along a path

opposite to the incident ray is that

$$k = 2k \sin \theta, \quad \ell = \frac{1}{2} \lambda \cosec \theta. \qquad (2.1)$$

The intensity of the backscattered radiation is proportional
to the steepness of the surface waves or, in the case of a
continuous wave spectrum, to the spectral density at wavenumber
k.

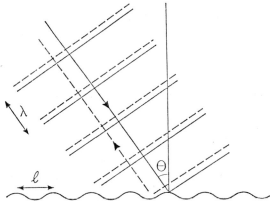

Fig. 6 First-order backscattering of e.m. radiation from a
regular surface wave.

Here we are ignoring other contributory mechanisms such as
Rayleigh scattering from droplets and spray, refraction from
sharp edges and specular reflections from normally oriented
facets. Except possibly at very low or high angles of
incidence the corresponding contributions are usually small.

We have then to consider the dynamics of the short surface
waves. On an otherwise still body of water the radian
frequency σ for free waves is given approximately by the linear
dispersion relation

$$\sigma^2 = gk + Tk^3 \qquad (2.2)$$

where g denotes gravity and T surface tension (see Lamb 1932,
C.9). For simplicity we here discuss the case of pure gravity
waves (applicable to reflections in L-band). Then the
frequency σ and phase-speed c are given simply by

$$\sigma^2 = gk, \quad c = \sigma/k = (g/k)^{\frac{1}{2}}. \qquad (2.3)$$

Consider first the propagation of such waves on a current
U in the direction of wave propagation (Figure 7). In a steady
state the phase of the short waves is conserved, hence

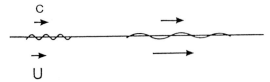

Fig. 7   The propagation of short gravity waves on a non-uniform
         surface current.

$$k(c + U) = \omega, \text{ constant} \qquad (2.4)$$

So from (2.2)

$$c^2 = (g/)(c + U), \qquad (2.5)$$

a quadratic equation for c in terms of U. This has a family
of solutions

$$c = \frac{1}{2}[y \pm (y^2 - 4y\, U)^{\frac{1}{2}}], \quad y = g/\omega, \qquad (2.6)$$

depending on y (see Figure 8). If C and U are in the same
sense (in the first or third quadrant of Figure 8) then as U
increases so does c, hence also the wavelength ℓ. If however
c and U are of opposite sign, as in the second quadrant of
Figure 8, say, then either $0 < |c| < |2U|$, in which case $|c|$
increases with $|U|$, or else $|c| > |2U|$, in which case $|c|$
decreases as $|U|$ increases. The critical case c = - 2U is
indicated by a broken line.

   The amplitude α of the waves is in general, governed by the
principle of action conservation (Bretherton and Garrett 1968).
If E denotes the wave energy density, that is

$$E = \frac{1}{2}\rho\, g\, \alpha^2, \qquad (2.7)$$

then the action-density is defined as E/σ. The action flux
is the action density multiplied by $(c_g + U)$ where c_g denotes
the group-velocity of the waves; in deep water $c_g = \frac{1}{2}c$.
Altogether then, since σ = g/c, we have

$$\tfrac{1}{2}\rho\, \alpha^2\, c(\tfrac{1}{2}c + U) = \text{constant} \qquad (2.8)$$

or

$$\alpha \propto [c(c + 2U)]^{-\frac{1}{2}}, \qquad (2.9)$$

see Longuet-Higgins and Stewart (1961). This equation governs
the wave amplitude as one moves along one of the curves κ =
constant in Figure 8. Consequently, waves encountering a
current increasing in the same direction are generally reduced
in amplitude; waves encountering an increasingly adverse
current are usually shortened and steepened, though the
opposite effect can also occur.

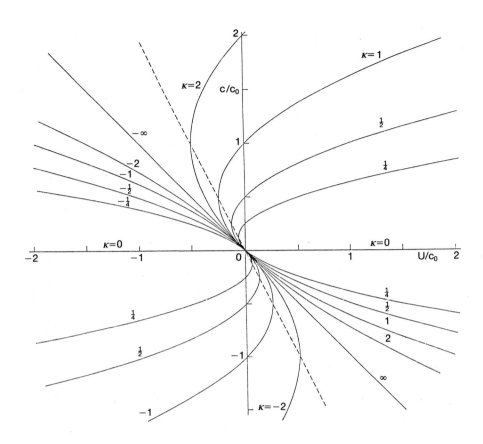

Fig. 3 The relation between the phase-speed c and the surface
current U for short gravity waves; see equation (2.6).
Note that $\kappa = y/c_o = g/\omega c_o$, where c_o denotes the value
of c when $U = 0$.

An exception to the rule (2.9) is in the critical case
$C = -2U$, when the waves encounter a caustic and (2.9) would
yield infinite amplitudes. A linearised theory valid in the
neighbourhood of the caustic has been given by Smith (1975).
In practice, the short waves usually break through before the
caustic is reached, producing a patch of very rough water.

At all events it is clear that an underlying steady current will produce significant changes in the wavelength and amplitude of the short waves. In practice, with a broad spectrum of short waves, the corresponding changes in radar backscatter are very marked.

The radar imaging of underwater sand bars and of internal waves is then easily explained, in principle. Figure 9, for example, illustrates a simple instance of tidal flow over an uneven sea bed. Where the water depth is less, the transverse current tends to increase, by continuity, and where the depth increases again, the current is reduced (we ignore any longitudinal current component). Hence if the waves are in the same direction as the current (Figure 9a) shallow water is associated with a smoothing of the surface, and deeper water with roughening. On the other hand if the current is in the opposite sense to the waves but $|2U| < |c|$ (Figure 9b) shallow water generally produces roughening but, on the far side of the caustic, can produce smoothing. With a reversal of the tide, a "positive" radar image of the bottom topography can be replaced by the corresponding "negative", the wind being unchanged. The case $|2U| > |c|$ is shown in Figure 9(c).

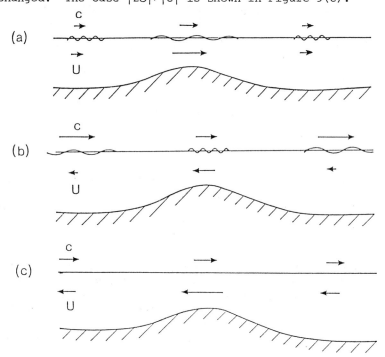

Fig. 9   The effect of uneven depth on a current U in shallow water.
         (a) $c > 0$, $U > 0$.   (b) $c > 0$, $0 > U > -2c$.
         (c) $c > 0$, $U < -2c$.

Similarly in an internal wave. Figure 10 shows a typical
solitary wave on the interface between warm (light) water near
the upper free surface, and cold (denser) water below.
Suppose the phase-speed is to the left as in Figure 10a. Then,
seen by an observer moving with this phase-speed C the solitary
wave appears as a steady current moving to the right. By
conservation of mass, the current is weaker in the trough ₁ of
the wave. So if the short surface waves are moving to the
right (with phase speed c < 2C) they will be <u>smoother</u> over the
trough of the solitary wave than elsewhere. If c > 2C the
solitary wave trough will appear rougher. On the other hand,
if the short surface waves are moving in the opposite sense to
the solitary wave, i.e., if c and C are of opposite sign then
the free surface will tend to be <u>rougher</u> over the solitary
wave tough.

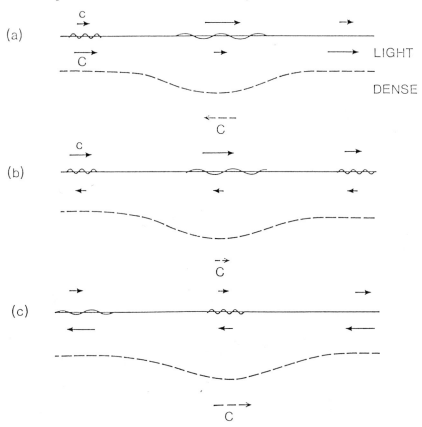

Fig. 10   The effect of an internal solitary wave travelling with
          speed C to the right.
          (a)   c > 0, C < 0.   (b)   c > 0, 0 < C < 2c.
          (c)   c > 0, C > 2c.

In both the above examples we have for simplicity ignored the possible dissipation of the short waves by breaking, and also any regeneration by the wind. These factors will in general produce a shifting of the phase of the roughness relative to the larger-scale features.

3.   LONG SURFACE WAVES OF LOW AMPLITUDE

Consider now the imaging of large-scale waves and swell. Figure 11 illustrates the case of long waves in water of infinite depth, when both short and long waves travel in the same sense. Taking axes moving with the long-wave speed we see that the long waves reduce to a steady, non-uniform current of strength q at the free surface, directed towards the left. The situation is similar to the solitary wave, with the chief difference that the free surface particles now have a significant normal acceleration α say. This means that the short waves are subject to an effective gravity

$$g' = g - \alpha \tag{3.1}$$

instead of g. Because there is no component of the pressure gradient along the surface, it follows that g' is always directed normally to the free surface, and the dispersion relation (2.3) for the short waves is replaced by

$$\sigma^2 = g'k, \tag{3.2}$$

σ and k being the local (intrinsic) radian frequency and the local wavenumber respectively. Moreover, to apply the principle of action conservation we need to replace the expression (2.7) for the energy-density of the short waves by

$$E = \frac{1}{2} \rho g' a^2 \tag{3.3}$$

It is easy to see that because q is greatest in the long-wave troughs and least at the crests, the short waves must always be at their shortest at the long-wave crests, whether they are travelling in the same or the opposite direction to the long waves. Because the short and long waves satisfy the same dispersion relation for gravity waves, the anomalous case $|c| > 2q$ fails to arise, and there are no awkward caustics (the situation is otherwise with gravity-capillary waves, however). Moreover, the short waves are not only shorter but higher, and therefore steeper, at the long-wave crests.

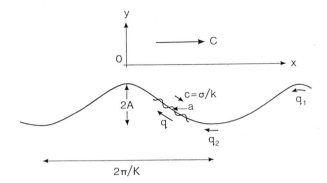

Fig. 11 The propagation of short surface waves on longer
         gravity waves.

The first calculations of this effect, (see Longuet-Higgins
& Stewart, 1960) treated the long waves as linear, so that
quantities of order $(AK)^2$ were neglected. This gave for the
shortening and steepening of the short waves

$$a = \bar{a} \ (1 + AK \cos \psi) \left.\right\}$$

$$k = \bar{k} \ (1 + AK \cos \psi) \left.\right\} \qquad (3.4)$$

$\bar{a}$ and $\bar{k}$ being the values of a and k at the mean level, and ψ
the phase of the long waves (at a long-wave crest, $\psi = 2n\pi$).
Thus, to the same order of approximation

$$ak = \bar{a} \ \bar{k} \ (1 + 2 \ AK \cos \psi) \qquad (3.5)$$

In the two following sections we shall describe how more
general calculations have been performed for finite values of
AK.

4. LONG WAVES OF FINITE AMPLITUDE

For finite values of the long-wave steepness AK a valid
prediction of the behaviour of the short waves depends critically
on the accurate calculation of g', the effective local gravity.
This in turn depends upon evaluating the orbital acceleration
α in the long waves.

If we choose units of length and time so that

$$g = 1, \qquad K = 1, \qquad (4.1)$$

then by taking axes of x and y moving with the phase-speed C
of the long waves, and with a suitably chosen origin of (x,y)
the equation of the free surface may be represented in the
parametric form

$$
\left.
\begin{aligned}
-x &= \phi/C + \sum_{n=1}^{\infty} a_n \sin(n\phi/C) \\
y &= \tfrac{1}{2}a_o + \sum_{n=1}^{\infty} a_n \cos(n\phi/C)
\end{aligned}
\right\}
\qquad (4.2)
$$

where ϕ is the velocity potential at the free surface and a_o,
a_1, a_2, ---- are Fourier coefficients to be determined. As shown
by Longuet-Higgins (1978) the condition of constant pressure
at the free surface is equivalent to the following system of
quadratic equations:

$$
\left.
\begin{aligned}
a_o b_o + a_1 b_1 + a_2 b_2 + a_3 b_3 + ---- &= -c^2 \\
a_1 b_o + a_o b_1 + a_1 b_2 + a_2 b_3 + ---- &= 0 \\
a_2 b_o + a_1 b_1 + a_o b_2 + a_1 b_3 + ---- &= 0 \\
a_3 b_o + a_2 b_1 + a_1 b_2 + a_o b_3 + ---- &= 0
\end{aligned}
\right\}
\qquad (4.3)
$$

----

in which for convenience we have written

$$
b_n = na_n, \ n = 1, 2, ---- ; \quad b_o = 1. \qquad (4.4)
$$

These relations can be quickly and accurately solved to high
order n for a given value of the amplitude

$$
A = a_1 + a_3 + a_5 + ---- \qquad (4.5)
$$

by the procedure described in Longuet-Higgins (1985a). The
particle acceleration α can then be evaluated from the formula

$$
\alpha = u.\nabla u = -q^6 (x_\phi + iy_\phi)^2 (x_{\phi\phi} - iy_{\phi\phi}) \qquad (4.6)
$$

as in Longuet-Higgins (1985b).

Because the evaluation of α involves twice differentiating
the Fourier series (4.2), it is necessary to be sure that these

series have sufficiently converged. This may be done by taking
account of the asymptotic behaviour of the Fourier coefficients
at large values of AK and n, derived in another paper (Longuet-
Higgins 1985c).

The normal component of α and the effective gravity g' for
long waves of various steepness AK including the maximum
value AK = 0.4432, have been calculated as a function of x/L
(Longuet-Higgins (1986a)); some examples are shown in
Figure 12. In the limiting case AK = 0.4432 there is a
singularity at the steep crest, which can be resolved with the
aid of the theory of the "almost-highest wave" (Longuet-Higgins
and Fox, 1977, 1978). For the highest wave, the effective
gravity g' ranges from 0.612g at the wave crest to 1.301g in the
wave trough. The values of the crest and trough accelerations
have also been checked experimentally over practically the
whole range of AK; see Longuet-Higgins (1986).

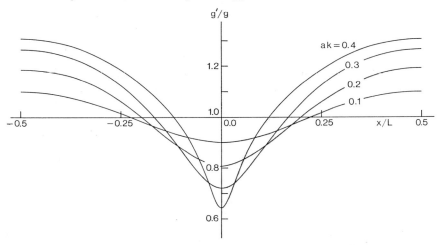

Fig. 12   The effective gravity g' at the surface of deep-water
          gravity waves, for various wave steepnesses AK.

The relative variations in length and steepness of the
short waves have been calculated (Longuet-Higgins 1987a) as
functions of the long-wave steepness AK and of the wavenumber
$\bar{k}$ of the short waves at the mean surface level. The proportional
changes $k/\bar{k}$ and $a/\bar{a}$ are found to be practically independent of
$\bar{k}$. Figure 14, for instance shows the relative values of the
steepness $r = ak/\overline{ak}$ at the crests ($r = r_1$) and troughs ($r = r_2$)
of the long waves, over wide ranges of AK and k. It is
notable that when AK = 0.40, for example, the relative
steepening at the crest is about 8, compared with a value of
less than 2 given by the linear theory, equation (3.6).

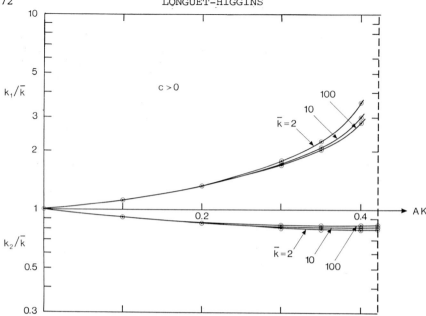

Fig. 13   The relative shortening $k/\bar{k}$ of short gravity waves (1)
at the crests and (2) in the troughs of longer
gravity waves.

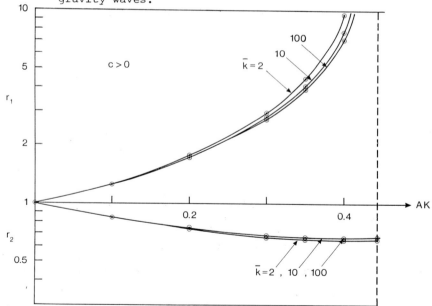

Fig. 14   The relative steepening $r = ak/\bar{a}\bar{k}$ of short gravity
waves at the crests and in the troughs of longer
gravity waves.

Figure 15 shows the steepening r of the short waves as a function of the horizontal distance x/L for a typical value $\bar{k}$ = 8 and for AK = 0.1, 0.2, 0.3 and 0.4. Again the large value at the crest (x=0) will be noticed.

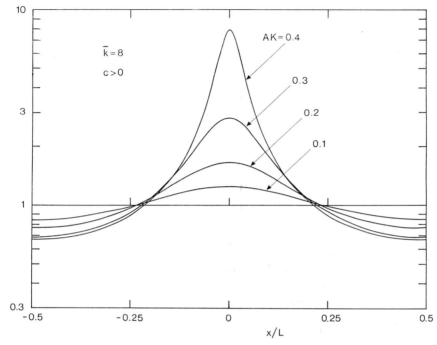

Fig. 15 The steepening r of short gravity waves as a function of the horizontal coordinate x/L, for AK = 0.1, 0.2, 0.3, 0.4 and when $\bar{k}$ = 8.

Figure 16 illustrates the remarkable fact that when the same quantity r is plotted against the vertical coordinate (y - $\bar{y}$)/L (where $\bar{y}$ is the mean surface level) all the values of r for the nonlinear theory, which are shown by full curves, collapse practically onto a single curve, differing significantly from the linear theory (broken curve). This fact will be found very useful in the more general theory described in the following Section.

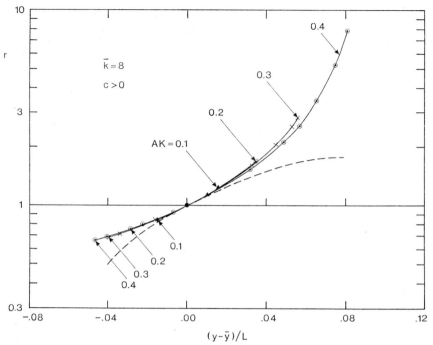

Fig. 16 The steepening r of short gravity waves as a function
of the vertical coordinate $(y - \bar{y})/L$.

5. A STOCHASTIC MODEL OF SURFACE ROUGHNESS

So far we have not included in our discussion any dissipation
of short-wave energy by breaking or any regeneration of the short
waves by the wind. Nor have we taken account of the random
nature both of the long waves and of the short waves riding on
them. We shall now describe a model which includes all of
these features.

We assume that the spectrum of the long, or dominant, waves
is fairly narrow, so that the long-wave amplitude A has a
Rayleigh distribution:

$$p(A) = \frac{A}{\bar{A}^2} e^{-A^2/2\bar{A}^2} \qquad (5.1)$$

Here $\bar{A}$ is a constant denoting the r.m.s. value of the surface
elevation $(y-\bar{y})$. The "significant wave height" is
approximately $2\bar{A}$ (see Longuet-Higgins 1952).

Similarly we assume that the local short-wave steepness
$\varepsilon = ak$ has a Rayleigh distribution:

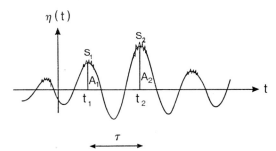

Fig. 17   Propagation of short wave groups at the crests of
longer waves of variable height.

$$p(\varepsilon) = \frac{\varepsilon}{2s^2} e^{-\varepsilon^2/2s^2} \qquad (5.2)$$

where s is the local r.m.s. surface slope: $s^2 = \overline{(\partial \zeta/\partial x)^2}$.
However, s itself will be assumed to be a random variable, with
a certain probability density, depending upon the local state
of the long waves. We shall at first consider the value of
s only at the <u>crests</u> of the long waves, and write $p(s|A)$ for
<u>the probability density of s at the crest of a long wave of
given amplitude A</u>. We propose then to determine $p(s|A)$.

   The key is to consider the value $p(s_2|A_2)$ of $p(s|A)$ at time
$t = t_2$ in relation to the value $p(s_1,|A_1)$ of $p(s|A)$ on the
preceding wave at time $t = t_1$ (see Figure 17). Since the
group-velocity c_g of the short waves is short compared to the
phase-speed C of the longer waves, the time-interval between the
appearance of short-wave energy on successive long-wave crests
is $(t_2-t_1) \approx$ T, the long-wave period. If there were no wind,
the short waves would, in the absence of breaking, be steepened
by a factor

$$s_2/s_1 = f(KA_2)/f(KA_1) \qquad (5.3)$$

where from Figure 16

$$f(KA) = e^{2.08\ KA + 2.94\ (KA)^2}. \qquad (5.4)$$

At first we may retain only the term in KA giving

$$s_2/s_1 = e^{2.08\ K(A_2 - A_1)}. \qquad (5.5)$$

In the presence of wind we assume an additional steepening of
the waves given by

$$s_2/s_1 = e^{\beta T} \tag{5.6}$$

where β is a growth rate found empirically by Plant (1982)
and given by

$$\beta = 3.2 \times 10^{-5} (W/c)^2 \sigma, \tag{5.7}$$

W being the wind-speed at a height of 10 m.   Thus altogether,
if the waves are not breaking, we have

$$s_1 = s_2 F(A_1, A_2) \tag{5.8}$$

where

$$F = e^{B + \gamma(A_1 - A_2)} \tag{5.9}$$

and $B = \beta T$, $\gamma = 2.08K$.   To represent the limitation of the
short-wave steepness by breaking, we suppose that (5.8) is
valid subject to the condition that always

$$s_1 \leqslant s_o, \qquad s_2 \leqslant s_o \tag{5.10}$$

where s_o is a certain steepness which we may take as half the
limiting steepness 0.4432 for regular waves.   According to (5.2)
the probability of ε exceeding s_o is then reasonably small.

The limitation on s implies that $p(s/A)$ must be of the form

$$p(s|A) = \begin{cases} \Phi(s,A) + P(A)\delta(s-s_o), & (s \leqslant s_o) \\ \\ 0, & (s > s_o) \end{cases} \tag{5.11}$$

$\Phi(s,A)$ being a continuous function of s and A, and $\delta(s-s_o)$ a
Dirac delta-function.   In other words, on a crest of given
height A there is a finite probability $P(A)$ of finding the short
waves in a state limited by breaking.   Equation (5.11) is
accompanied by the normalisation condition

$$\int_o^{s_o} \Phi(s,A) \, ds + P(A) = 1 \tag{5.12}$$

for all values of A.

Now the general relation between $p(s_1|A_1)$ and $p(s_2|A_2)$ can be written

$$p(s_2|A_2)\Delta s_2 = \int p(s_1|A_1)\Delta s_1 \, p(A_1|A_2) \, dA_1 \qquad (5.13)$$

where

$$\Delta s_1/\Delta s_2 = ds_1/ds_2 = F(A_1,A_2) \qquad (5.14)$$

by equation (5.8) and $p(A_1|A_2)$ denotes the probability density of A_1, given A_2. But for a fairly narrow-band long-wave spectrum, the joint distribution of A_1 and A_2 is known to be the "bivariate" Rayleigh distribution:

$$p(A_1,A_2) = \frac{A_1 A_2}{(1-K^2)\bar{A}^4} \, e^{-(A_1^2 + A_2^2)/2(1-K^2)\bar{A}^2} \times I_o\left(\frac{KA_1 A_2}{(1-K^2)\bar{A}^2}\right)$$

$$(5.15)$$

where $\bar{A}$ is the same constant as in (5.1) and K is a "groupiness parameter" explicitly related to the spectral density of the long waves (see Kimura 1980; Longuet-Higgins 1984).

For wind-waves K does not exceed 0.5. For a long swell, K may be between 0.7 and 1.0, the latter representing waves with an infinitely long group-length. The general relation of K to the average number of waves in a group is discussed in the reference just given. From (5.15) we have immediately that

$$p(A_1|A_2) = p(A_1,A_2)/p(A_2) \qquad (5.16)$$

where $p(A_2)$ is given by (3.1) with $A = A_2$. Hence all the functions occurring in the kernel of the integral equation (3.13) are in principle known.

However, the condition $s_2 \lesssim s_o$ limits the range of A_2 in (5.13) to $0 < A_2 < H(s_o,A_2)$ where from (5.8)

$$H(s_2,A_2) = A_2 + \gamma^{-1}[B + \ln(s_o/s_1)], \qquad (5.17)$$

and if we separate out the singular part of $p(s|A)$ from the continuous part we find eventually that (3.13) yields

$$\Phi(s_2, A_2) = \int_0^{H(s_o, A_2)} \Phi(s_1, A_1) F(A_1, A_2) p(A_1 | A_2) \, dA_2$$

(5.18)

$$+ \frac{1}{\gamma s_2} P(A_2) p(A_1' | A_2), \quad A_1' = H(s_2, A_2)$$

and

$$P(A_2) = \int_0^{H(s_o, A_2)} \left[\int_{s_o^F}^{s_o} \Phi(s_1, A_1) \, ds_1 + P(A_1) \right] P(A_1 | A_2) \, dA_1,$$

(5.19)

a pair of integral equations for the functions $\Phi(s, A)$ and $P(A)$.

In Longuet-Higgins (1987b) these equations have been solved by iteration. Thus, arbitrary initial functions $\Phi^{(1)}(s, A)$ and $P^{(1)}(A)$ satisfying

$$\int_0^{s_o} \Phi^{(1)}(s, A) \, ds + P^{(1)}(A) = 1$$

(5.20)

were substituted on the right-hand sides of (5.18) and (5.19) and the resulting functions were taken as the next approximations $\Phi^{(2)}(s, A)$ and $P^{(2)}(A)$ respectively. These were again substituted on the right, and the left-hand sides taken as $\Phi^{(3)}(s, A)$ and $P^{(3)}(A)$; and so on. After several (of order 10) iterations the solutions converged to at least four decimal places. This was regardless of the choice of initial conditions (subject to (5.20)). The accuracy of the final solution was checked by how well it satisfied the required normalisation conditions.

A solution with the typical parameter values $\bar{A}K = 0.1$, $B = 0.1$ and $K = 0$ is shown in Figure 18. This shows that for the steeper long waves ($A/\bar{A} \geq 1$, say) the probability density $\Phi(s, A)$ increases monotonically with s, and the probability $P(A)$ of short-wave "breaking" exceeds one-half. For the lower long-waves, however, $\Phi(s, A)$ can have a maximum at values less than s_o, so $\Phi(s, A)$ is actually bimodal.

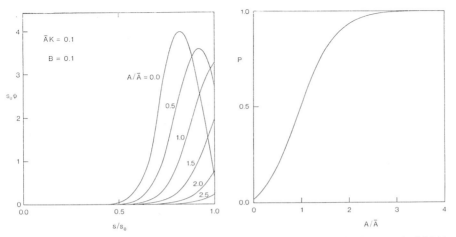

Fig. 18   Graphs of $\Phi(s,A)$, the continuous part of the probability
density $p(s|A)$, and of $P(A)$, the probability of the
short waves being at limiting r.m.s. steepness.
$\bar{A}K = 0.1$, $B = 0.1$, $K = 0$.

The effect of increasing the wind, by raising B to 0.2 can
be seen in Figure 19.   The curves of $\Phi(s,A)$ are pushed to the
right, and the probability $P(A)$ of breaking is increased.

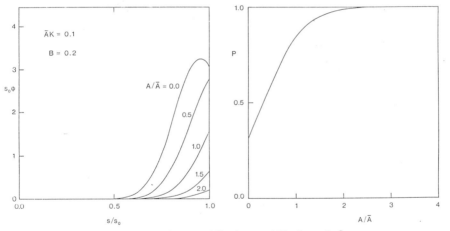

Fig. 19   As figure 18, but with $B = 0.2$.

Increasing $K\bar{A}$, on the other hand, pushes the curves in the
opposite sense (Figure 20).   The physical reason seems to be
that when the straining of the short waves dominates, the short-
wave steepness on the moderate-sized long waves is determined
by the breaking condition on the steep long waves;   the
reduction due to diminished long-wave amplitude overrides the
regenerating effect of the wind.

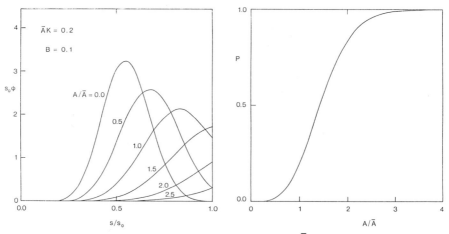

Fig. 20   As Figure 18, but with $\bar{A}K$ = 0.2.

Finally, the effect of increasing the groupiness parameter K from 0 to 0.85 is shown in Figure 22, to be compared with Figure 21. Evidently an increase in K produces an increase in the probability of breaking. This can be understood from the fact that the limiting case K = 1 corresponds to long waves of uniform height. In that case any loss of height from breaking at the crest of one wave would, in a steady state, be make up by energy input from the wind before the next long wave crest. Hence all long waves would support short waves in a breaking condition.

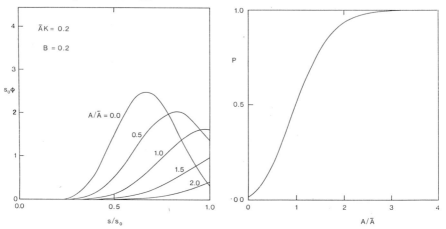

Fig. 21   As Figure 18, but with $\bar{A}K$ = 0.2, B = 0.2.

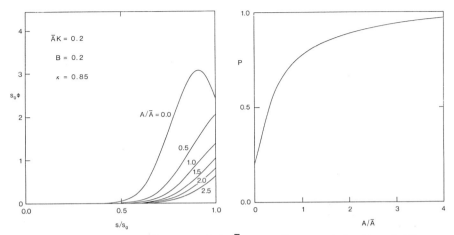

Fig. 22   As Figure 18, but with $\overline{A}K = 0.2$, $B = 0.2$, $K = 0.85$.

6.   DISCUSSION AND SUGGESTIONS FOR FURTHER WORK

The two integral equations (5.18) and (5.19), which we have solved, embody a consistent physical model of the roughening of long-wave crests due to Bragg-scatterers in L-band. Having solved the equations, it is straightforward to calculate the distribution of surface roughness over all phases of the longer waves, and hence to estimate measured parameters of wind-waves: the mean-square slopes (which clearly must depend on both the wind and the longer waves present), the correlation between surface elevation and roughness, and the corresponding phase shifts.

The analysis is clearly capable of extension to arbitrary directions between wind and long waves. Of special interest for X-band scattering is the possible extension of the theory to shorter, capillary-gravity waves.

We have purposely not discussed here the entire problem of SAR imaging, which involves the calculation of the backscattered electromagnetic field, given the form of the sea surface. Rather we have concentrated on understanding the sea surface itself, an essential preliminary to understanding the associated SAR images in terms of "sea truth".

ACKNOWLEDGEMENTS

The work summarised in this paper was carried out partly during visits by the author to the Cal. Tech. Jet Propulsion Laboratory, Pasadena; the Department of Engineering Sciences, University of Florida, Gainesville, and to the Institute of

Hydroengineering of the Polish Academy of Sciences in Gdansk. For the illustrations in Figures 1 to 5 the author is indebted to Dr. T.D. Allan and to Dr. J.R. Apel.

REFERENCES

Bretherton, F.P. and Garrett, C.J.R., (1968) Wavetrains in homogeneous moving media. *Proc. R. Soc. Lond.* A **302**, 529-554.

Kimura, A., (1980) Statistical properties of random wave groups. Proc. 17th Int. Conf. on Coastal Eng., Sydney, pp. 2955-2973. New York, *Amer. Soc. Civil Eng.*

Lamb, H., (1932) Hydrodynamics, 6th ed. Cambridge Univ. Press. 738 pp.

Longuet-Higgins, M.S., (1952) On the statistical distribution of the heights of sea waves. *J. Mar. Res.* **11**, 245-266.

Longuet-Higgins, M.S., (1978) Some new relations between Stokes's coefficients in the theory of gravity waves. *J. Inst. Maths. Applics.* **22**, 261-273.

Longuet-Higgins, M.S., (1984) Statistical properties of wave groups in a random sea state. *Phil. Trans. R. Soc. Lond.* A **312**, 219-250.

Longuet-Higgins, M.S., (1985a) Bifurcation in gravity waves. *J. Fluid Mech.* **151**, 457-475.

Longuet-Higgins, M.S., (1985b) Asymptotic behaviour of the coefficients in Stokes's series for surface gravity waves. *IMA. J. Appl. Math.* **34**, 269-277.

Longuet-Higgins, M.S., (1985c) Accelerations in steep gravity waves. *J. Phys. Oceanogr.* **15**, 1570-1579.

Longuet-Higgins, M.S., (1986) Eulerian and Lagrangian aspects of surface waves. *J. Fluid Mech.* **173**, 683-707.

Longuet-Higgins, M.S., (1987a) The propagation of short surface waves on longer gravity waves. *J. Fluid Mech.* **177**, 293-306.

Longuet-Higgins, M.S., (1987b) A stochastic model of sea surface roughness. *Proc. R. Soc. Lond.* A **410**, 19-34.

Longuet-Higgins, M.S. and Fox, M.J.H., (1978) Theory of the
   almost-highest wave. Part 2. Matching and analytic
   extension. *J. Fluid Mech.,* **85**, 769-786.

Longuet-Higgins, M.S. and Stewart, R.W., (1960) Changes in the
   form of short gravity waves on long waves and tidal currents.
   J. Fluid Mech. **8**, 565-583.

Longuet-Higgins, M.S. and Stewart, R.W., (1961) The Changes in
   amplitude of short gravity waves on steady, non-uniform
   currents. *J. Fluid Mech.* **10**, 529-549.

Smith, R., (1975) The reflection of short gravity waves on a
   non-uniform current. *Math. Proc. Camb. Phil. Soc.* **78**,
   517-525.

Thompson, D.R. and Gasparovic, R.F., (1986) Intensity
   modulation in SAR images of internal waves. *Nature, Lond.*
   320, 345-348.

THE SCATTERING OF ELECTROMAGNETIC RADIATION FROM IRREGULAR SURFACES

P.L.C. JEYNES

(The Blackett Laboratory, Imperial College, London)

ABSTRACT

The scattering of waves by rough surfaces is investigated. Firstly we apply Huygens' principle and find that Crombie's selective grating mechanism is valid only for slight roughness i.e. for $h < \lambda/4\pi\cos\theta$, where h is the roughness height, λ the radiation wavelength and θ the angle of incidence. The Huygens approach is shown to be related to the rigorous integral equations in the tangent plane approximation (TPA) as developed by Isakovich. Although of considerable use, the TPA method predicts only trivial rigid-free surface asymmetry in acoustics and only trivial polarization effects in the electromagnetic case. We devise a technique that accounts for surface curvature and which yields certain non-trivial effects as are observed in experiments and also predicted by more exact, though less versatile, theories.

The implications to the field of radar probing of the ocean surface are considered. Although Crombie's mechanism is applicable for HF radiowave (22m) echo from the sea, it is inappropriate for centimetre microwaves except in conditions of exceptional calm. Instead we propose the Isakovich theory should provide the basis for understanding "wind scatterometry". Likewise the recent developments of Holliday should help explain radar imaging of the ocean. Work remains to be done on the polarization problem.

1. INTRODUCTION

Recent interest in synthetic aperture radar imaging of the ocean has prompted renewed study into the physics of wave scattering by rough surfaces. We shall be concerned mainly with the scatter of plane incident radiation as observed in the

far zone. The most important theories to date are based on
either the Rayleigh (1895) expansion method or on the integral
equation approach. Unfortunately the physics is often
obscured by the complex mathematics. So first we take an
intuitive approach by directly applying Huygens' principle to
the problem. The method is used to illustrate the scattering
mechanism for a slightly rough surface as discovered from
observations by Crombie (1955). Then, our intuitive approach
is related to the integral equation theory in the tangent plane
approximation as developed by Isakovich (1952). We go on to
investigate rigid-free surface asymmetry in acoustics and
polarization effects in the electromagnetic case. A "curved
surface approximation" is developed for this purpose and the
results are compared with various other more exact though less
versatile techniques for slight roughness. The findings are
discussed in the context of microwave scattering from the ocean
and are related to the recent work of Holliday (1985, 1986)
concerning the radar imaging of ocean features and the 2nd
iteration method.

For economy of space, certain details of the work in latter
sections are omitted and may be found in the author's thesis
(1987). Standard notation is used: vectors are underlined,
$\underline{v}$ with magnitude $v=|\underline{v}|$, and with unit direction vector
$\underline{\hat{v}} = \underline{v}/v$.

2.   THE HUYGENS' PRINCIPLE APPROACH

We investigate the manner in which radiation is scattered
by a perfectly reflecting, rough surface. Of particular
interest in many applications is the scattered field at large
distances from the surface. The well-known result of
Fraunhofer diffraction is that "the far-field is the Fourier
transform of the near-field". Consequently the near-field
is the key to the scattering problem. (Fraunhofer diffraction
is the far-zone limit of Fresnel's integral, which itself is
based on Huygens' principle, e.g. Born and Wolf, 1980).

Figure 1 shows the situation. Cartesian coordinates (x,y,z)
are chosen with z vertically upward and the xy-plane defining
the horizontal. The scattering surface is given by the
equation $z=h(x,y)$ where h is the height above the mean
horizontal level, $z=0$.

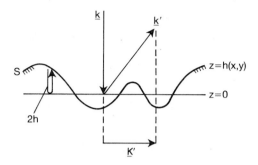

Fig. 1

For simplicity the radiation of wavelength λ is assumed initially to be incident vertically upon the surface. Let us proceed to determine the near field. Evidently, a ray reflected from a surface point with elevation h travels a path shorter by 2h than a ray reflected from the mean horizontal plane. The ray is totally reflected but experiences a relative phase delay of $4\pi h/\lambda$ but is otherwise perfectly reflected. Thus the surface behaves like a "reflecting phase screen". Consequently, the reflected near-field on the xy-plane is effectively $e^{-i4\pi h/\lambda}$. The surface height h plays the role of the phase in an optical phase screen diffraction problem (e.g. Born and Wolf, 1980 pp. 401, 424).

It is instructive to expand the reflected near-field in powers of the roughness scale h/λ:

$$e^{-i4\pi h/\lambda} = 1 - i\frac{4\pi}{\lambda} h - \frac{8\pi^2}{\lambda^2} h^2 + \dots$$

Performing the 2-dimensional (x,y) Fourier transform term by term gives the scattered far-field Fraunhofer diffraction pattern:

$$2\pi\delta - i\frac{4\pi}{\lambda} H - \frac{4\pi}{\lambda^2} H \otimes H + \dots$$

where (i) δ is the Dirac delta function; (ii) H is the Fourier transform of the surface height function h,

$$H(\underline{\kappa}) = \frac{1}{2\pi} \iint dxdy \; e^{-i\underline{\kappa}\cdot\underline{x}} \; h(\underline{x})$$

where $\underline{x}=(x,y)$ is the horizontal position vector and $\underline{\kappa}=(\kappa_x,\kappa_y)$ is a variable horizontal wavevector describing the surface;

and (iii) the symbol $\otimes$ denotes the convolution operation
defined shortly. According to the Fraunhofer diffraction: the
scattered far-field $\underline{k}'=(k'_x, k'_y, k'_z)$ is proportional to the
Fourier integral of the reflected near-field evaluated at
$\underline{K}'=(k'_x, k'_y)$ - the projection of $\underline{k}'$ onto the horizontal xy-plane,
shown in figure 1. Thus each term of the Fourier transformed
expansion must be evaluated at $\underline{K}'$ and to this end we
distinguish a number of cases.

For a perfectly flat surface $(h(x,y)=0)$ only the zeroth (0th)
order term remains, which is unity. The Fourier transform of a
constant gives a Dirac delta function, $\delta(\underline{K}')$. This term
vanishes unless $\underline{K}'=0$, corresponding to a specular reflection
(back into the vertical direction) as is expected for a flat
mirror. We are most interested in the "diffuse" scattering
i.e. into directions other than specular.

For an infinitesimally rough surface $(4\pi h/\lambda \ll 1)$ we see that
only the 1st order term in $H(\underline{K}')$ contributes to the diffuse
reflection wherein the scattered wave $\underline{k}'$ depends on only the
$\underline{K}'$ wavevector component of the surface irregularities. The
strict inequality is the criterion for the 2nd and higher
order terms to be negligible. Nontheless, as long as the
surface is slightly rough $(h \leq \lambda/4\pi)$ then the 1st order term
will dominate the diffuse scattering and provide a reasonable
approximation. In this case the scattering mechanism responds
to (selects) a unique surface wave for each direction of
observation.

As the surface roughness increases $(h > \lambda/4\pi)$ the higher
order terms will dominate the 1st and the scattering can no
longer be attributed to a particular surface wave. E.g., for
$h \approx \lambda/\pi$ the expansion shows the magnitude of the 2nd order term
is twice the 1st, while the 3rd is larger still. The 2nd order
term involves the convolution,

$$H \otimes H \ (\underline{K}') = \int d^2\kappa_1 \ H(\underline{K}'-\underline{\kappa}_1) \ H(\underline{\kappa}_1)$$

$$= \int d^2\kappa_1 \int d^2\kappa_2 \ H(\underline{\kappa}_1) \ H(\underline{\kappa}_2) \ \delta(\underline{\kappa}_2 + \underline{\kappa}_1 - \underline{K}')$$

This is essentially a sum of terms of the form $H(\underline{\kappa}_1)H(\underline{\kappa}_2)$
satisfying the resonance condition $\underline{\kappa}_1 + \underline{\kappa}_2 = \underline{K}'$, wherein pairs
of surface waves contribute to the scattering in a given
direction. Similar statements hold for the 3rd and higher
order terms. The product terms show that waves of large
amplitude must play an important role in these higher terms.

For a very rough surface the scattering process described
by the transformation from h to the Fourier transform of
$e^{-i4\pi h/\lambda}$ is highly non-linear, whereas the diffuse scattering
from a slightly rough surface is linear. In general, if the
surface undulations are scaled by a factor of two, the
scattered field in all direction cannot simply double- by energy
conservation. A slightly rough surface achieves linearity in
the diffuse scattering by having a large specular energy
reflection which upon scaling of the roughness can be linearly
redistributed in diffuse directions. A rough surface with a
weak specular component cannot do this.

The Fraunhofer treatment is readily extended to arbitrary angles
of incidence and the results are similar. It is interesting,
however, to use a slightly different approach based directly
on Huygens' principle from which the result is almost immediately
apparent.

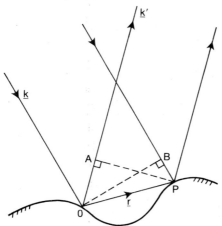

Fig. 2 $OA = \hat{\underline{k}}'.\underline{r}$ $PB = \hat{\underline{k}}.\underline{r}$

Figure 2 shows a scattering point P with position vector
$\underline{r}$ from the origin O. The incoming wavevector is $\underline{k}$ and we are
interested in the contributions to the scattered wave $\underline{k}'$, i.e.
that part of the outgoing "far field" with the wavevector $\underline{k}'$.
The phase delay in the signal from P relative to that from O is
$\underline{q}.\underline{r}$ where $\underline{q} = \underline{k}'-\underline{k}$ is the "change" in the wavevector. Thus
the complex amplitude contribution to $\underline{k}'$ from a reflecting
surface element at $\underline{r}$ is $e^{i\underline{q}.\underline{r}}$. Summing over all points of the
surface gives an integral for the total amplitude of the
scattered wave:

$$\iint dxdy\; e^{-i\underline{q}.\underline{r}}$$

This is just the Fourier integral of $e^{-iq_z h}$ evaluated at $\underline{Q} = (q_x, q_y)$ - the horizontal projection of $\underline{q}$. Expanding in powers of the roughness, h:

$$e^{-iq_z h} = 1 - iq_z h - \tfrac{1}{2} q_z^2 h^2 + \dots$$

and then the Fourier transforming termwise gives the scattered amplitude for $\underline{k}'$

$$2\pi\delta - iq_z H - \frac{1}{4\pi} q_z^2 H \otimes H + \dots$$

when evaluated at $\underline{Q}$. The results are similar to the case of vertical incidence.

For infinitesimal surface roughness ($q_z h \ll 1$) the diffuse scatter is given by the 1st order term in $H(\underline{Q})$, when only the $\underline{Q}$ surface wave (with the wavelength $\Lambda = 2\pi/Q$) contributes to the scatter in the given direction. The 1st order term is expected to be reasonable provided the surface is slightly rough ($q_z h \le 1$).

For the important case of backscattering ($\underline{k}' = -\underline{k}$), the return field strength is proportional to $H(-2k\sin\theta, 0)$ provided that $2kh\cos\theta \le 1$, where θ is the angle of incidence with the vertical. For illustration we refer to figure 3. In terms of $\lambda_x = \lambda/\cos\theta$ and $\lambda_z = \lambda/\sin\theta$, the surface is considered slightly rough if $h \le \lambda_z/4\pi$ when the surface wavelength responsible for backscatter is $\Lambda = \lambda_x/2$ (e.g. for Seasat sar, $\lambda = 23$cm, $\theta = 23°$ then $\Lambda = 33$cm, $h \le 2$cm!).

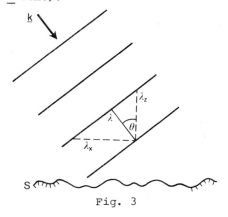

Fig. 3

Let us consider the scattering in the plane of incidence.
Then in the slight roughness regime, the surface wavelength
responsible for scattering, Λ, satisfies

$$\Lambda \mid \sin\theta' - \sin\theta \mid = \lambda$$

where θ,θ' are the angles of incidence and scattering
respectively. For scattering into the direction θ', the surface
is seen as periodic in Λ. This effective surface is familiar
as a reflecting diffraction grating where Λ is the grating
spacing required for constructive interference with a path
difference of one wavelength λ. The above relation is
recognised as the "grating equation" for the principal
diffraction order, e.g. Born and Wolf (1980, p.403, eq.8).

Theoretically, the nature of scattering by a slightly
rough surface has been known for some time, e.g. Rice (1951),
Eckart (1953). Experimentally, Crombie (1955) analysed the
echo of long "HF" radiowaves (λ=22m) from the sea near grazing
incidence and deduced the selective grating mechanism that
was in operation. The return signal was Doppler shifted due
to the motion (crest velocity) of the ocean waves responsible for
backscatter both toward and away from the radar (the shift is
in fact equivalent to the frequency of the backscattering
ocean wave). In recent radio-oceanography literature the 1st
order scattering has been considered analogous to X-ray
diffraction by crystals as studied extensively by Bragg. (e.g
Ashcroft and Mermin, 1976. p. 96). However, there is a formal
equivalence with grating diffraction whereas the analogy with
crystal scattering is heuristic. Thus the terminology "Bragg
scattering" is inappropriate and henceforth we refer to slightly
rough surface scattering as "Crombie's mechanism" or as
"selective grating diffraction".

3.   THE INTEGRAL EQUATION APPROACH

The exact integral equations provide a rigorous formulation
(version) of the intuitive Huygens-Fresnel diffraction theory
of the preceding section. Let us first consider the acoustic
case where the velocity potential ψ satisfies the scalar
Helmholtz wave equation

$$-(\nabla^2 + k^2) \psi = 0$$

Use of Green's 2nd theorem leads to the Helmholtz-Kirchoff (HK)
integral:

$$\psi(\underline{r}') = \oint dS \left(\psi \frac{\partial g}{\partial n} - g \frac{\partial \psi}{\partial n} \right)$$

giving the field at some observation point $\underline{r}'$ in terms of the
field ψ and its normal derivative $\partial\psi/\partial n$ over any surface S
surrounding $\underline{r}'$. The free-space point source Green's function is

$$g = e^{ikR}/4\pi R$$

where $R = |r'-r|$ is the distance from the integration point $\underline{r}$
on the surface to $\underline{r}'$. The derivative normal to S is $\partial/\partial n = \underline{n}.\underline{\nabla}$
where $\underline{n}$ is the unit vector normal to S. The HK theory is
treated by e.g. Born and Wolf (1980) and Baker and Copson
(1950).

In a surface scattering problem, certain exact boundary
conditions must be satisfied at the interface. We consider
perfectly reflecting cases with an incident field ψ and a
scattered field ψ'. For a free surface (soft or pressure
release) the total velocity potential must vanish

$$\psi + \psi' = 0$$

while for a rigid surface (hard) the total normal derivative
must vanish

$$\partial\psi/\partial n + \partial\psi'/\partial n = 0$$

(e.g. Landau and Lifshitz, 1959). In either case, we can
write down the HK integral equation for the scattered field
ψ' and take the integration over the scattering surface. The
scattered field is then given in terms of the fields ψ' and
$\partial\psi'/\partial n$ at the surface, S.

For any approximation of the surface fields ψ', $\partial\psi'/\partial n$ the
HK-integral yields an estimate of the scattered field. This
first estimate may then be used to re-calculate the surface
fields iteratively (the integral equation method).

Maxwell's equations for the vector electromagnetic field
lead to the exact Stratton-Chu integral equations: Stratton
(1941) or Jackson (1976).

For a 2-dimensional problem there is an exact analogy
between the acoustic and electromagnetic cases for a perfectly
reflecting surface. In acoustics the surface may be either
rigid (+) or free (-) which corresponds respectively to
vertically (V) or horizontally (H) polarized EM radiation
incident upon a perfect conductor. The analogy is of frequent
value.

For plane radiation of wavevector $\underline{k}$ incident upon a scattering body, then in a given direction at large distances the outgoing field falls off like e^{ikr}/r. Thus the far-field is conveniently characterised by the scattering amplitude f defined asymptotically by

$$\psi(\underline{r}) = f(\underline{k}',\underline{k}) \; e^{ikr}/r \quad \text{as } r \to \infty$$

where $\underline{k}' = k\hat{\underline{r}}$ is the scattered wavevector. For the EM vector field $\underline{E}'$ we have a vector scattering amplitude, $\underline{f}$.

4. THE TANGENT PLANE APPROXIMATION

For a plane wave $\psi = e^{i\underline{k}\cdot\underline{r}}$ incident upon a plane surface, the exact solution is found by assuming a reflected plane wave in the form $\psi' = Ae^{i\underline{k}'\cdot\underline{r}}$ where A is the amplitude reflection coefficient. It is also assumed that $\nabla\psi = i\underline{k}\psi$, $\nabla\psi' = i\underline{k}'\psi'$ which hold true since ψ,ψ' are plane waves. Using these forms with the exact boundary conditions one finds that $\underline{k}'$ is the specular reflection of $\underline{k}$ and that $A = -1$ for a free surface and $A = +1$ for a rigid surface. It follows that the fields immediately adjacent to the plane reflecting surface are related by:

$$\psi' = A\psi \; , \qquad \partial\psi'/\partial n = - A\partial\psi/\partial n$$

where $\underline{n}$ is the mirror normal.

In the scattering by an arbitrary surface profile a common approximate procedure is the following. At each individual point on the surface, the reflected field is assumed to equal that reflected from a large plane lying tangential to the surface at the point. The surface fields are thus assumed given by the above two equations where $\underline{n}$ is now a variable surface normal. This "tangent plane approximation" (TPA) for the surface fields when used in the HK-integral gives a first iteration to the scattered field. The method is expected to be valid for a surface with radii of curvature, ρ, large in comparison with the wavelength, λ. Brekhovskikh (1952, eq. 45) finds the requirement

$$2k\rho \cos \theta_L \gg 1$$

is necessary for its validity where θ_L is the local angle of incidence. By the integration by parts method of Isakovich (1952), the scattering amplitude in the TPA, neglecting shadowing, multiple scattering and end-effects, is:

$$f = (q^2/4\pi iq_z) \iint dxdy \; e^{-i\underline{q}.\underline{r}}$$

where as before, $\underline{q} = \underline{k}'-\underline{k}$ is the change in wavevector. The same integral arose by our method in section 2 based directly on Huygens' principle. The acoustic rigid (+) and free (-) surface scattering amplitudes f^+ and f^- respectively are given by

$$f^{\pm} = \pm f$$

with only a "trivial" difference in sign. With no further approximations it can be shown (Jeynes, 1987) from the Stratton-Chu integrals that the analogous EM vector scattering amplitude factorises:

$$\underline{f} = \underline{\varepsilon}' \; f$$

where the electric polarisation unit vector of the scattered radiation is

$$\underline{\varepsilon}' = - \; \underline{\varepsilon} + 2(\hat{\underline{q}}.\underline{\varepsilon})\hat{\underline{q}}$$

and $\underline{\varepsilon}$ is the polarisation of the incident wave. Rather surprisingly, $\underline{\varepsilon}'$ is the same as one expects from a plane mirror oriented to specularly reflect the incident radiation into the direction of observation. The result is as simple as possible in view of the transverse and perpendicular nature of an electromagnetic wave. We shall regard such polarisation effects as "trivial". The problem is reduced to evaluation of a scalar integration for f. The polarisation result is well-known in the special case of backscatter and also in the $\lambda \rightarrow 0$ (stationary phase) specular point case (Jackson, 1975 p.450-3).

However, certain exact solutions and experiments show both (i) non-trivial differences between rigid and free surface scattering and (ii) non-trivial polarisation effects in the EM case. These effects are usually small but are often important nevertheless.

As an example consider an infinitesimally rough surface ($q_z h \ll 1$). To first order in h the TPA result is

$$f = - \tfrac{1}{2}q^2 H(\underline{Q})$$

where only the relevant Crombie surface wave $\underline{Q}=(q_x,q_y)$ contributes to the scatter in a given direction. The backscatter amplitude to first order ($2kh \cos\theta \ll 1$) is

$$f = -2k^2 H(-2k\sin\theta \, , \, 0)$$

Now, in some sense the tangent plane approximation "neglects" the surface curvature. Since the curvature (the inverse of the radius of curvature) $1/\rho \simeq \partial^2 h/\partial x^2$ is a quantity of the first order in h, then the TPA is not expected to be exact to the first order (for infinitesimal surface roughness). The TPA is only exact to 0th order (i.e. for a plane mirror). This does not mean the TPA is inappropriate for anything other than a plane. Quite the contrary: the method is an extremely versatile, physics based approximate theory that is not specially suited to any particular surface type. More exact theories tend to be useful only in special cases and are intractable for generally irregular surfaces. However, a major problem with the TPA is its "scalar" nature and we suggest a "physics-based" improvement in the next section.

5. THE CURVED SURFACE APPROXIMATION (details in Jeynes, 1987)

A new method is outlined that takes account of the curvature of the scattering surface, S. In free surface (-) acoustic scattering, the usual tangent plane approximation for the fields on S is

$$\psi' = -\psi \, , \quad \partial\psi'/\partial n = \partial\psi/\partial n$$

The first equation is the exact boundary condition and so the approximation lies in the second which is now scrutinized. Consider a convex point on S, shown P in figure 4.

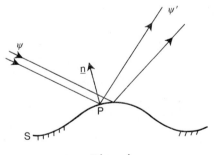

Fig. 4

Parallel incident rays are drawn which upon reflection, diverge. Thus simple ray geometry suggests the intensity in the reflected field diminishes with distance away from the surface

(near P), thus $\partial I'/\partial n < 0$. For incident parallel rays
(a plane-wave) we have $\partial I/\partial n = 0$. For an arbitrary incident
wave, ray geometry shows near a convex point that

$$\partial I'/\partial n < \partial I/\partial n$$

This inequality is considered to be inconsistent with the
second tangent plane approximation, bearing in mind that
$I \propto |\psi|^2$. The cause and the solution of this problem are
outlined below. The argument for a rigid (+) surface and a
partially transmitting surface ($A \neq \pm 1$) is somewhat more subtle
and is not given here.

Previously the plane wave expression $\underline{\nabla}\psi = ik\psi$ and similarly
for ψ' was used with the exact boundary conditions to yield the
TPA. For a curved surface ψ' is not planar. For curved
waveforms generally, we suggest a more appropriate choice is

$$\underline{\nabla}\psi = (ik + \gamma)\ \hat{\underline{k}}\ \psi$$

where the mean wavefront curvature is

$$\gamma = \tfrac{1}{2}(\frac{1}{\rho_a} + \frac{1}{\rho_b})$$

and ρ_a, ρ_b are the principal radii of curvature of the
wavefront surface. Note γ for the incident wave and γ' for
the scattered wave can be found from geometric considerations
since wavefront curvature is equivalent to ray-divergence
(e.g. for the spherical wave $\psi = e^{ikr}/r$ one may readily verify
that $\gamma=1/r$ by evaluating $\underline{\nabla}\psi$). In our case, for the incident
plane wave $\gamma=0$, but the reflected wavefront curvature γ' is
related to the curvature of the scattering surface by
differential geometry and depends in a fairly complicated
manner on the location.

Using the modified expressions for $\underline{\nabla}\psi$ and $\underline{\nabla}\psi'$ we obtain a
new "curved surface approximation" (CSA) relating the fields
on S, which can then be used in the exact HK-integral equation
in place of the TPA. The result is made of two terms. One
is the usual tangent plane approximation, the other is a
"curvature correction term" (CCT), i.e. CSA = TPA + CCT. The
CCT involves the surface curvature ($\gamma_s = 1/\rho$) and is found to
vanish, as it must, when the curvature is small and satisfies
Brekhovskikh's criterion given earlier (section 4). The CSA
is found to predict non-trivial rigid-free surface asymmetry
in acoustic scattering and the corresponding EM polarisation
effects, and is thus an improvement on the TPA in which these
effects are absent.

As an example, let us apply the new method to a problem discussed earlier: the backscatter from a slightly rough surface. A detailed analysis shows that the curved surface approximation backscatter amplitudes for first order ($O \neq 0$) are,

$$f^{\pm} = {}^{-}_{+}2k^2 \; (1 \pm \sin^2\theta) \; H(-2k\sin\theta \; ,O)$$

The electromagnetic cases are related by

$$f_{VV} = f^+, \; f_{HH} = f^-, \; f_{HV} = f_{VH} = O$$

Comparison with the TPA backscatter amplitudes (section 4) shows that the surface curvature, through the CSA, introduces non-trivial acoustic asymmetry and EM polarisation effects by the addition of a $\sin^2\theta$ term in the scattering amplitudes.

That the $\sin^2\theta$ term does in fact arise from the curvature can be seen in the following argument. To first order, the curvature is $\partial^2 h/\partial x^2$. We will use the fact that $\partial/\partial x$ is $i\kappa_x$ in Fourier κ-space. Now the curvature correction term is proportional to the Fourier transform of the curvature $-\kappa_x^2 H(\kappa)$ evaluated at the backscatter Crombie wavevector $\underline{Q}=(-2k\sin\theta, O)$ which gives $-4k^2\sin^2\theta \; H(\underline{Q})$ showing the origin of the extra $\sin^2\theta$ term.

We note that the first order CSA scattering, like the foregoing TPA and intuitive diffraction results, still depends only upon the Crombie component of the surface irregularities. Again the first order result is applicable for very slightly rough surfaces having no justification once the height exceeds $h \simeq \lambda/4\pi\cos\theta$.

It turns out that the CSA results are in fact the exact first order backscatter amplitudes as derived by a variety of other methods, i.e. the Rayleigh (1895) - Rice (1951) small perturbation method (e.g. Ishimaru, 1978) also Maystre (1983) by a new method and Holliday (1986) by a second iteration. Its experimental accuracy has been amply demonstrated, by Crombie (1955), Wright (1966), Bass et al. (1968) and others.

It is not entirely clear at present as to why higher derivatives, $\partial^3 h/\partial x^3$ etc, also of first order in h are important; though it appears to be linked in some manner with the fact that the governing wave equation involves only second derivatives like $\partial^2/\partial x^2$.

6. STATISTICALLY ROUGH SURFACES

Previously the surface $h(x,y)$ was assumed known. Now consider the case where we have only "some idea" of the surface character and we wish to know what the scattered fields are "likely" to be. Let us denote the Fourier transform of the surface on some large area A by $H_A(\underline{\kappa})$. For many natural and artificial "statistical" irregular surfaces, the precise locations of the peaks are unknown (the phases of $H_A(\underline{\kappa})$ are random) and the undulations are variable but with some well defined average (the amplitudes of $H_A(\underline{\kappa})$ are "Rayleigh" distributed about some mean). The statistics of such a surface are fully characterised by the "spectrum"

$$\Phi(\underline{\kappa}) = \lim_{A\to\infty} \frac{1}{A} <|H_A(\underline{\kappa})|^2>$$

where the brackets < > denote the mean average (expectation value).

For the field backscattered diffusely from a slightly rough surface: the random phases of $H_A(\underline{\kappa})$ imply the mean field vanishes $<f>=0$, though the mean intensity does not. For a large horizontal area A of illuminated statistically rough surface the far-zone intensity is characterised by the normalised radar cross-section (NRCS)

$$\sigma = \frac{4\pi}{A} <|f|^2>$$

The exact mean cross-sections to first order are thus

$$\sigma^+ = 16\pi k^4 (1 + \sin^2\theta)^2 \Phi(\underline{Q}_b)$$

$$\sigma^- = 16\pi k^4 \cos^4\theta \ \Phi(\underline{Q}_b)$$

in acoustics, with

$$\sigma_{VV} = \sigma^+, \ \sigma_{HH} = \sigma^-, \ \sigma_{HV} = \sigma_{VH} = 0$$

in the electromagnetic case. The backscatter Crombie wavevector is $\underline{Q}_b = (-2k\sin\theta \ ,0)$. By comparison, the tangent plane approximation leads to

$$\sigma = 16\pi k^4 \ \Phi(\underline{Q}_b)$$

in each case, except for the cross-polarisations which vanish.

For a generally rough statistical surface, the method due to Isakovich (1952) in the tangent plane approximation shows $<f>=0$ again, and the mean cross-section is

$$\sigma = \frac{q^4}{4\pi q_z^2} \; e^{-q_z^2 \eta^2} \iint dx\,dy \; e^{-i\underline{Q}\cdot\underline{x}} \; (e^{q_z^2 \phi(\underline{x})} -1)$$

where the surface autocovariance function $\phi(\underline{x}) = <h(\underline{x}'+\underline{x})\; h(\underline{x}')>$ is related to the spectrum by the Fourier transform (using ergodicity and the convolution theorem):

$$\phi(\underline{x}) = \iint d^2\kappa \; \Phi(\underline{\kappa}) \; e^{i\underline{\kappa}\cdot\underline{x}}$$

Further, the r.m.s. height, η, satisfies $\eta^2 = <h^2> = \phi(0)$. The "Isakovich" cross-section thus depends only on the change in wavevector $\underline{q} = \underline{k}'-\underline{k}$ and the surface spectrum $\Phi(\underline{\kappa})$, properly accounting for diffraction from all wavelengths of the surface. The result is well-known in the texts e.g. Beckmann and Spizzichino (1963), Ishimaru (1978), though the original paper of Isakovich (1952) is of outstanding clarity and has been translated from the Russian to English. Beckmann (pp. 24 and 70) acknowledges that the method is due to Isakovich.

Here we do not wish to evaluate the scattering integral by the method of stationary phase since this yields the $\lambda\to0$ geometric "optics" result that ignores diffraction. Generally, given $\Phi(\underline{\kappa})$ the double 2-d Fourier integrations to give σ will have to be done numerically. Little work has so far been done along these lines though the potential is immense.

It also remains to account for non-trivial acoustic asymmetry and EM polarisation effects for a generally rough surface by use of techniques such as the curved surface approximation or the 2nd iteration in the tangent plane approximation. Encouragingly we have seen these methods work for the case of slight roughness.

7.   MICROWAVE SCATTERING FROM THE OCEAN

In the complete absence of wind the ocean will be mirror smooth and specularly reflect the incident radiation giving zero backscatter. For reference, Phillips (1978, 1985) discusses the generation of a spectrum of ocean waves by the action of the wind. The roughened sea surface causes diffuse scattering, in particular backscattering (echo).

We have seen that Crombie's selective grating mechanism is a first order result in h applicable for a slightly rough surface providing

$$h \leq \lambda/4\pi\cos\theta$$

Thus for long radiowaves, "HF" ($\lambda \simeq 10\text{-}100$ m), particularly near grazing incidence ($\theta \simeq 90^\circ$), the mechanism adequately describes sea echo. However, at microwavelengths (X-band $\simeq 3$ cm/L-band $\simeq 30$ cm) the mean sea surface displacements would need to be less than 2 mm/2 cm (for low to moderate θ) which occurs in open seas only in conditions of exceptional calm (windspeeds less than $0.7/2.3$ ms^{-1}, i.e. Beaufort sea state 0 to 1).

The empirical failure of the first order model is clear in the simple case of scattering from a wind-generated spectrum of ocean waves: a wealth of "wind-scatterometry" data shows a consistent increase in backscatter with windspeed (and "surface roughness") up to 30 or 40 m/s or more (e.g. Schroeder et al., 1982). However the short Crombie waves ($\Lambda \simeq 3$ cm/30 cm) "saturate" (or very nearly so) at very low windspeeds of a few m/s wherein according to just the first order term the backscatter should also saturate rather than increasing, as is observed. Consequently no serious attempt to explain the wind scatterometry data can be made with Crombie's mechanism (the first order term). Nonetheless it has been the major tool in the interpretation of radar imaging of the ocean to date.

To account for the presence of large amplitude long water waves, Wright (1966, 1968) and Bass et al. (1968) use a 2-scale composite surface model suggested previously by other workers and find some success in calm conditions. The model incoherently averages the first order cross-section over the tilts induced by the longer waves. However, the neglect of the phases and the neglect of the finite size of the Crombie patches makes such composite surface models overemphasise the role of Crombie waves and the accuracy of the technique is little understood. Durden and Vesecky (1985) find the ocean wave directional spectrum required by the model to be consistent with wind scatterometry data. The best fit spectrum that results is highly directional near the short Crombie wavelength and is isotropic at longer wavelengths. However, the real ocean is the reverse, becoming more directional for the longer waves near the spectral peak propagating in the downwind direction.

Instead of either of the above models, we suggest the tangent plane approximation as developed by Isakovich (section 6) which accounts for all surface wavelengths in the scattering, will predict the main features of wind scatterometry. Finer details, including polarisation effects could be understood with the curved surface approximation, possibly also with some account of multiple scatter in very high seas. Alternatively the second iteration method could be used.

Certain observed features such as the upwind-downwind asymmetry require a refinement of the surface model as a spectral process. The leaning of the waves toward the downwind direction violates the assumption of random, uncorrelated phases since the harmonic waves are phase locked onto the fundamental.

To a reasonable approximation a radar image is a backscatter cross-section map of the ocean. Surface wavelengths less than the resolution contribute to the roughness responsible for diffuse scattering, while longer waves tilt the mean surface. The wind generated spectrum of ocean waves $\Phi(\kappa)$ is responsible for the "background" cross-section σ (image grey level). Ocean features (e.g. internal waves, swell waves) induce a local perturbation of the spectrum $\delta\Phi(\kappa)$ resulting in a cross-section modulation $\delta\sigma$ enabling the feature to be imaged. Based on the Isakovich approach, Holliday (1985) finds the linear relationship from $\delta\Phi(\kappa)$ to $\delta\sigma$ involving a weighted integration over all surface wavevectors, κ. The weighting kernel is determined by the radar parameters and the wind velocity. All other imaging models to date rely on Crombie scattering and so are severely restricted to calm seas. Comprehensive numerical confirmation of the Isakovich method for wind scatterometry remains to be done and is essential to place Holliday's theory on a firm basis as the first viable ocean imaging model.

8.   CONCLUSIONS

Using Huygens' construction, the selective grating mechanism first observed by Crombie (1955) and erroneously called Bragg scattering, is shown to be a first order effect for slight roughness valid only if $h \leq \lambda/4\pi\cos\theta$, where h is the roughness height, λ the radiation wavelength and θ the angle of incidence. Thus Crombie's mechanism, though applicable for HF radiowave ($\lambda\approx22m$) echo from the sea, is inappropriate for centimetre microwaves except in conditions of exception calm (Beaufort 0 to 1). The integral equation approach in the tangent plane approximation, developed by Isakovich (1952), properly accounts for the diffraction from all wavelengths of the surface spectrum. This approach should provide the first proper understanding of "wind scatterometry", which is the basis of all microwave radar imaging of the ocean. Holliday's (1985) image modulation theory for ocean features (e.g. swell or internal waves) is based on the Isakovich approach.

One drawback is the inherent neglect of polarisation effects important at smaller wavelengths (X-band) and moderate to large angles of incidence. A technique that accounts for surface curvature is devised and found to predict certain non-trivial polarisation effects in addition to those from multiple scattering, as desired. Encouragingly, the "curved surface approximation" in fact yields the exact backscatter for a slightly rough surface including polarisation effects, as shown by various other methods. Work remains to be done on the polarisation problem for a generally rough statistical surface.

9. ACKNOWLEDGEMENTS

It is a pleasure to thank Dr. J.O. Thomas, B.C. Barber, J.R. Perry, Charlie Farrugia and Dennis Holliday for many useful discussions. The work was funded by the Science and Engineering Research Council and The Royal Aircraft Establishment, Farnborough.

10. REFERENCES

Ashcroft, N.W. and Mermin, N.D., (1976) "Solid state physics", Holt-Saunders.

Baker, B.B. and Copson, E.T., (1950) "The Mathematical theory of Huygens' principle", 2nd edn., Clarendon Press, Oxford.

Bass, F.G., Fuks, I.M., Kalmykov, A.I., Ostrovsky, I.E. and Rosenberg, A.D., (1968) "VHF radiowave scattering by a disturbed sea surface, parts I and II" IEEE trans. AP-16, 554-568.

Beckmann, P. and Spizzichino, A., (1963) "The scattering of electromagnetic waves from rough surfaces", Pergamon.

Born, M. and Wolf, E., (1980) "Principles of optics", 6th edn., Pergamon.

Brekhovskikh, L.D., (1952) "Diffraction of waves by irregular surfaces, Pt I" Zh. Eksp. i Teor. Fiz 23, 275-289 (Russian, translated by M.D. Friedman).

Crombie, D.D., (1955) "Doppler spectrum of sea echo at 13.5 Mc/s" Nature 175, 681-2.

Durden, S.L. and Vesecky, J.F., (1985) "Wind speed
    dependence of microwave backscatter from the ocean surface"
    Stanford Center for Radar Astronomy Report (to be published
    in J. Geophys. Res.)

Eckart, C., (1953) "The scattering of sound from the sea
    surface" J. Acoust. Soc. Am. 25, 566-570.

Holliday, D., (1985) "On the interpretation of radar images
    of ocean features" RDA Logicon report Feb. 1985, P.O. Box
    9695, Marina Del Ray, California (to be published in Int.
    J. Rem. Sensing).

Holliday, D., (1986) "Resolution of a controversy surrounding
    the Kirchoff approach and the small perturbation method in
    rough surface scattering" RDA Logicon rep. May 1986, as
    above (to be published in IEEE trans. AP).

Isakovich, M.A., (1952) "Wave dispersion from a randomly
    uneven surface" Zh. Eksp. i Teor. Fiz. 23, 305-314 (in
    Russian, translated by E.R. Hope).

Ishimaru, A., (1978) "Wave propagation and scattering in
    random media" vol. 2, Academic Press.

Jackson, J.D., (1976) "Classical electrodynamics" 2nd edn.,
    J. Wiley, N.Y.

Jeynes, P.L.C., (1987) "On the scattering of radiation by
    surfaces" PhD Thesis, University of London.

Landau, L.D. and Lifshitz, E.M., (1959) "Fluid mechanics",
    Pergamon.

Maystre, D., Mata Mendez, O and Roger, A., (1983) "A new
    electromagnetic theory for scattering from shallow rough
    surfaces" Optica Acta 30, 1707-1723.

Phillips, O.M., (1978) "Dynamics of the upper ocean" 2nd
    edn., C.U.P.

Phillips, O.M., (1985) "Spectral and statistical properties
    of the equilibrium range in wind-generated gravity waves"
    J. Fluid Mech. 156, 505-531.

Rayleigh, Lord, (1895) "Theory of sound", Dover.

Rice, S.O., (1951) "Reflection of electromagnetic waves from
    slightly rough surfaces", Comm. Pure and Appl. Math. 4,
    361-378.

Schroeder, L.W. et al., (1982) "Seasat scatterometer winds"
    J. Geophys. Res. C-87, 3318-3336.

Stratton, J.A., (1941) "Electromagnetic theory", McGraw-Hill,
    N.Y.

Wright, J.W., (1966) "Backscattering from capillary waves with
    application to sea clutter" IEEE trans. AP-14, 749-754.

Wright, J.W., (1968) "A new model for sea clutter" IEEE trans.
    AP-16, 217-223.

Nomenclature/List of Symbols

h	surface height
λ	radiation wavelength
θ	angle of incidence
(x,y,z)	Cartesian coordinates
δ	Dirac delta function
H	Fourier transform of h
$\underline{x}=(x,y)$	horizontal position vector
$\underline{\kappa}=(\kappa_x,\kappa_y)$	horizontal surface wavevector
$\circledX$	the convolution operation
$\underline{k},\underline{k}'$	incident and scattered radiation wavevectors
$\underline{K},\underline{K}'$	horizontal projections of $\underline{k}$ and $\underline{k}'$
$\underline{q}$	change in wavevector, $\underline{k}'-\underline{k}$
$\underline{r}$	position vector.
$\underline{Q}$	horizontal projection of $\underline{q}$ (the Crombie wavevector)
Λ	bistatic Crombie wavelength $(2\pi/Q)$
(λ_x,λ)	respectively $\lambda(1/\sin\theta, 1/\cos\theta)$
ψ	acoustic velocity potential
k	radiation wavenumber, $2\pi/\lambda$
S, dS	surface and surface element
g	Greens function
R	distance $|\underline{r}'-\underline{r}|$
$\partial|\partial n$	normal derivative
$\underline{n}$	surface normal
$\underline{\nabla}$	gradient operator
ψ,ψ'	incident and scattered fields
f	scattering amplitude
$\underline{E}$	electric field
$\underline{f}$	EM vector scattering amplitude
A	amplitude reflection coefficient
ρ	radius of curvature of surface

γ mean wavefront curvature

ρ_a, ρ_b principle radii of curvature

V,H   vertical and horizontal polarisations

+,-   rigid and free surface in acoustics

A   large area

σ normalised radar cross-section

$\underline{Q}_b$ backscatter Crombie wavevector

ϕ surface autocorrelation

Φ surface spectrum

η r.m.s. surface height.

COHERENT AND INCOHERENT COMPONENTS OF SPECULAR SCATTERING
FROM NORMAL AND NON-NORMAL ROUGH SURFACES

C.D. Obray
*(Department of Statistics and Operational Research,
Coventry Polytechnic)*

ABSTRACT

The paper describes the application of statistical optics
models relevant to a simple bench mark assessment of surface
roughness.

The black scattered field reflected from an illuminated
surface consists of:

(1)   a specular coherent component which vanishes everywhere
      except in a precise neighbourhood of the specular direction,
      and

(2)   a diffusely scattered incoherent component.

Broadly speaking, the coherent component depends on the variance
of surface height, while that scattered away from the specular
direction depends additionally on the auto-correlation function
of surface height.

In addition to quantifying the separate coherent and
incoherent components for a variety of statistical distributions
of surface height, the paper also quantifies the percentage of
the scattered field in the specular direction attributable to
surface slopes.

A low-cost comparative optical instrument has been developed
which can be calibrated to give an estimate of the standard
deviation of surface height. The scattering model, developed by
Beckmann[1,2,3] reported here, gives an explanation of the output
of the device, and assists in providing meaningful interpretation
of light scatter data in terms of surface roughness assessment
of the reflecting material, for both normal and non-normal surfaces.

1.   INTRODUCTION

It is well established that the surface finish of a component
could well affect the functional performance of that component
within its destined working environment e.g. a surgical implant
or a piston. Therefore in order to improve the performance and/
or to increase the life of a component greater control over
surface finish is required.

Methods whereby the surface characteristics are measured
directly may have serious disadvantages. For instance the
action of a stylus needle on a fine surface impairs the
surface finish. Indirect, non-contacting methods, whereby for
example an energy stream is projected at the surface and
the interaction of the stream with the rough surface observed,
have much to recommend them. Interpretation of the output in
terms of the surface characteristics is not necessary for some
production control applications, but is necessary for
specification of surface quality in terms of measures used in
current engineering practice.

The interaction of a laser beam and a rough surface has been
widely studied. This paper draws principally on the work of
Beckmann[1,2,3] and focuses on the mean optical intensity of the
reflected beam in the specular direction $<\rho\rho^*>_{spec}$. This
quantity is comprised of a coherent component, which is
dominant for very smooth surfaces or for rougher surfaces
illuminated at grazing incidence, and a diffusely scattered
incoherent component. The coherent component is dependent on
the first order statistics of surface height whilst the
incoherent component depends critically on the second order
statistics of surface height. Furthermore, many ground
surfaces for example have highly skewed distributions of surface
height, the degree of skewness becoming more exaggerated as the
asperity peaks of the surface roughness are removed during the
grinding process. The resultant change of shape from a
symmetrical normal surface height distribution to a highly
skewed non-normal distribution, affects the two components in
a predictable manner.

The theoretical discussion for both normal and non-normal
surfaces is described purely in terms of the first and second
order characteristic functions of the surface height
distribution.

2.   COMPONENTS OF THE SCATTERED FIELD

Suppose that a rough surface can be reasonably modelled by
a stationary, random process. Define Z to be a random deviation
of the surface $Z(x,y)$ from some coordinate xy plane and $p(z)$ to

be the probability density function of Z. A portion of the
surface of area A is illuminated by a monochromatic, coherent
source of electromagnetic radiation of wavelength λ with
$A \gg \lambda^2$. Figure 1 illustrates the three-dimensional scattering
geometry.

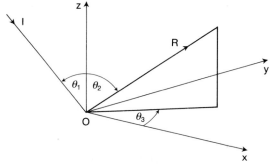

Fig. 1   Scattering Geometry

Suppose further that the geometrical and electrical
properties are such that physical optics can be applied.
(Beckmann and Spizzichino[1] gives a full account of the necessary
surface characteristics and details the results presented in
sections 2 and 3).

It is convenient to introduce a scattering coefficient,
$\rho(\theta_2,\theta_3)$, which is the ratio of the scattered field in the
direction (θ_2,θ_3), to that which would be reflected in the
specular direction $(\theta_2=\theta_1,\theta_3=0)$ by a smooth perfectly conducting
plane of the same dimensions under the same conditions of
illumination.

$$\rho(\theta_2,\theta_3) = \frac{F(\theta_1,\theta_2,\theta_3)}{A} \iint\limits_{A} e^{i\underline{v}\cdot\underline{r}} \, dxdy \qquad (1)$$

where F is a function of the incidence angle θ_1 and the
observer angles θ_2 and θ_3, $\underline{v} = (v_x,v_y,v_z)$ with

$$v_x = \frac{2\pi}{\lambda}(\sin\theta_1-\sin\theta_2\cos\theta_3), \quad v_y = -\frac{2\pi}{\lambda}\sin\theta_2\sin\theta_3,$$

$v_z = \frac{-2\pi}{\lambda}(\cos\theta_1 + \cos\theta_2)$ and $\underline{r} = (x,y,z(x,y))$ is the position
vector describing the surface.

Then $\langle \rho(\theta_2, \theta_3) \rangle = \dfrac{F}{A} \displaystyle\iint_A e^{i(v_x x + v_y y)} \langle e^{iv_z z} \rangle \, dxdy$

$\langle \rho(\theta_2, \theta_3) \rangle = \dfrac{F}{A} \displaystyle\iint_A \chi(v_z) \, e^{i(v_x x + v_y y)} \, dxdy \qquad (2)$

Here $\langle e^{iv_z z} \rangle = \displaystyle\int_{-\infty}^{\infty} p(z) \, e^{iv_z z} \, dz = \chi(V_z)$, is the

characteristic function associated with the p.d.f. $p(z)$ of surface height.

Then $\langle \rho(\theta_2, \theta_3) \rangle = \dfrac{F}{A} \chi(v_z) \displaystyle\int_{-X}^{X}\int_{-Y}^{Y} e^{iv_x x + iv_y y} \, dxdy$

$= \dfrac{F}{A} \chi(v_z) \rho_0 \quad \text{say} \qquad (3)$

where

$$\rho_0 = \frac{\sin v_x X}{v_x X} \cdot \frac{\sin v_y Y}{v_y Y}$$

The mean optical intensity $\langle \rho\rho^* \rangle$ is given by

$\langle \rho\rho^* \rangle = \chi(v_z)\chi^*(V_z)\rho_0^2$

$+ A^{-1}F^2 \displaystyle\iint_A [\chi_2(v_z, v_z; \tau) - \chi(v_z)\chi^*(v_z)] \, dxdy \qquad (4)$

where $\chi_2(v_z, -v_z)$ is the joint characteristic function of surface height and * denotes complex conjugate.

$\langle \rho\rho^* \rangle = \chi(v_z)\chi^*(v_z)\rho_0^2$

$+ 2\pi A^{-1}F^2 \displaystyle\int_0^{\infty} [\chi_2(v_z, -v_z; \tau) - \chi(v_z)\chi^*(v_z)] J_0(v_{xy}\tau)\tau d\tau \qquad (5)$

on transforming to polar coordinates. $(v_{xy} = \sqrt{v_x^2 + v_y^2}$ and $J_0(.)$

is the zero order Bessel function of the first kind.)

Thus the mean optical intensity is described in terms of two components. The first term vanishes except in the close vicinity of the specular direction.

3. NORMAL ROUGH SURFACES

Normal rough surfaces are discussed in Beckmann[1,2]. If Z is a normal random variable $N(0, \sigma^2)$ say, then $\chi(v_z)$ may be written as $e^{-g/2}$ where $g = \sigma^2 v_z^2$. Furthermore, $\chi_2(v_z, -v_z; \tau) = e^{-g[1-c(\tau)]}$ where $C(\tau)$ is the autocorrelation function of surface heights. Substitution into (5) yields

$$<\rho\rho^*> = e^{-g}\rho_o^2 + \frac{2\pi F^2}{A} \int_o^\infty \{e^{-g[1-c(\tau)]} - e^{-g}\} J_o(v_{xy}\tau)\tau d\tau$$

(6)

The mean optical intensity in the specular direction $<\rho\rho^*>_{spec}$ is simply

$$<\rho\rho^*>_{spec} = e^{-g} + \frac{2\pi}{A} \int_o^\infty \{e^{-g[1-c(\tau)]} - e^{-g}\}\tau d\tau$$

$$= e^{-g}\left[1 + \frac{2\pi}{A} \int_o^\infty \{e^{-gc(\tau)} - 1\}\tau d\tau\right]$$

$$= e^{-g}\left[1 + \frac{2\pi}{A} \int_o^\infty \sum_{n=1}^\infty \frac{g^n\{C(\tau)\}^n\tau d\tau}{n!}\right]$$

(7)

If a Gaussian form of autocorrelation function is assumed i.e. $C(\tau) = e^{-\tau^2/T^2}$, and that the order of integration and summation operations can be reversed:

$$<\rho\rho^*>_{spec} = e^{-g}\left[1 + \frac{2\pi}{A} \sum_{n=1}^\infty \frac{g^n}{n!} \int_o^\infty \tau e^{-n\tau^2/T^2} d\tau\right]$$

$$= e^{-g}\left[1 + \frac{\pi T^2}{A} \sum_{n=1}^\infty \frac{g^n}{nn!}\right]$$

(8)

Gaussian forms of autocorrelation are rarely encountered in real life and the negative exponential form $C(\tau) = e^{-|\tau|/T}$ would appear to be more valid (Obray[4]). Whence from (7)

$$\langle \rho\rho^* \rangle_{spec} = e^{-g} \left[1 + \frac{2\pi}{A} \sum_{n=1}^{\infty} \frac{g^n}{n!} \int_{0}^{\infty} \tau e^{-n\tau/T} \, d\tau \right]$$

$$= e^{-g} \left[1 + \frac{2\pi T}{A} \sum_{n=1}^{\infty} \frac{g^n}{nn!} \right] \tag{9}$$

4. NON-NORMAL ROUGH SURFACES

Beckmann[2] details the general procedure for deriving the back scattered field from non-normal surfaces, using expansions for the 2-dimensional p.d.f. in terms of orthogonal polynomials. The example of an exponential surface is described in full.

Elsewhere, Beckmann[3], the case of a gamma surface is discussed. However analysis of these surfaces is particularly easy, since their 2-dimensional density functions with a given correlation and associated characteristic functions have simple forms.

For the exponential surface the bivariate p.d.f.

$$p(z_1, z_2) = \frac{1}{\sigma^2 (1-\rho)} \exp[- \frac{1}{\sigma(1-\rho)} (z_1 + z_2)] I_0 \left(\frac{2\sqrt{\rho}\sqrt{z_1 z_2}}{\sigma(1-\rho)} \right),$$

has characteristic function $\chi_2(v_z, -v_z; \rho) = \frac{1}{1+g(1-\rho)}$.

($I_0(.)$ is the modified Bessel function of the first kind of order zero.) The marginal density function is $\frac{1}{\sigma} e^{-z/\sigma}$ for which $\chi(v_z) \chi^*(v_z) = \frac{1}{1+g}$.

Substituting directly into (5),

$$\langle \rho\rho^* \rangle = \frac{1}{1+g} \rho_0^2$$

$$+ \frac{2\pi F^2}{A} \int_{0}^{\infty} \left\{ \frac{1}{1+g[1-C(\tau)]} - \frac{1}{1+g} \right\} J_0(v_{xy}\tau) \tau \, d\tau \tag{10}$$

and $\langle \rho \rho^* \rangle = \dfrac{\rho_o^2}{1+g} + \dfrac{2\pi F^2}{A} \displaystyle\int_o^\infty \dfrac{1}{1+g} \left\{ \dfrac{1}{1-gC(\tau)/(1+g)} - 1 \right\} J_o(v_{xy}\tau)\,\tau d\tau$

$\qquad = \dfrac{\rho_o^2}{1+g} + \dfrac{2\pi F^2}{A(1+g)} \displaystyle\int_o^\infty \sum_{n=1}^\infty \left(\dfrac{g}{1+g}\right)^n \{C(\tau)\}^n J_o(v_{xy}\tau)\,\tau d\tau$

With a Gaussian autocorrelation function $C(\tau) = e^{-\tau^2/T^2}$,

$\langle \rho \rho^* \rangle = \dfrac{1}{1+g} \left[\rho_o^2 + \dfrac{2\pi F^2}{A} \sum_{n=1}^\infty {}' \left(\dfrac{g}{1+g}\right)^n \displaystyle\int_o^\infty \tau J_o(v_{xy}\tau) e^{-n\tau^2/T^2} d\tau \right]$,

$\qquad = \dfrac{1}{1+g} \left[\rho_o^2 + \dfrac{\pi T^2 F^2}{A} \sum_{n=1}^\infty \dfrac{1}{n} \left(\dfrac{g}{1+g}\right)^n \exp(-v_{xy}^2 T^2/4n) \right]$

$\langle \rho \rho^* \rangle_{spec} = \dfrac{1}{1+g} \left[1 + \dfrac{\pi T^2}{A} \cdot (-1_n \{1-g/(1+g)\}) \right]$

$\qquad = \dfrac{1}{1+g} \left[1 + \dfrac{\pi T^2}{A} 1_n(1+g) \right]$ \qquad (11)

With a negative exponential autocorrelation function,

$\langle \rho \rho^* \rangle_{spec} = \dfrac{1}{1+g} \left[1 + \dfrac{2\pi T}{A} 1_n(1+g) \right]$ \qquad (12)

For a gamma surface, the bivariate p.d.f.

$p(z_1, z_2) = \dfrac{(1-\rho)^p [\,(z_1 z_2)/(\sigma^2(1-\rho)^2)\,]^{p-1}}{\Gamma(p)\,\sigma^2(1-\rho)^2} \exp\left[-\dfrac{(z_1+z_2)}{\sigma(1-\rho)} \right] I_{p-1}\left(\dfrac{2\sqrt{\rho z_1 z_2}}{\sigma(1-\rho)} \right)$

has gamma marginals of order p and correlation ρ. The 2-dimensional characteristic function is

$\chi_2(v_z, -v_z; \rho) = \dfrac{1}{\{1+g/p(1-\rho)\}^p}$, and $\chi(v_z)\chi^*(v_z) = \dfrac{1}{\{1+g/p\}^p}$.

Substituting into (5) as before,

$$\rho\rho^{*}> = \frac{1}{\{1+g/p\}^p}\rho_o^2 + \frac{2\pi F^2}{A} \int_o^\infty \left[\frac{1}{\{1+g/p[1-C(\tau)]\}^p} - \frac{1}{\{1+g/p\}^p} \right] J_o(v_{xy}\tau)\tau d\tau$$

(13)

$$\text{nd } <\rho\rho^{*}>_{spec} = \frac{1}{\{1+g/p\}^p} + \frac{2\pi}{A} \int_o^\infty \frac{1}{\{1+g/p\}^p} \left[\frac{1}{[1-(g/p)C(\tau)/(1+g/p)]^p} -1 \right] \tau d\tau$$

$$= \frac{1}{\{1+g/p\}^p} \left[1 + \frac{2\pi}{A} \int_o^\infty \sum_{n=1}^\infty \frac{p(p+1)\ldots(p+n-1)}{n!} \left(\frac{g/p}{1+g/p}\right)^n \{C(\tau)\}^n \tau d\tau \right]$$

'ith a Gaussian autocorrelation,

$$\rho\rho^{*}>_{spec} = \frac{1}{\{1+g/p\}^p} \left[1 + \frac{2\pi}{A} \sum_{n=1}^\infty \frac{p(p+1)\ldots(p+n-1)}{n!} \left(\frac{g/p}{1+g/p}\right)^n \int_o^\infty \tau e^{-n\tau^2/T^2} d\tau \right]$$

$$= \frac{1}{\{1+g/p\}^p} \left[1 + \frac{\pi T^2}{A} \sum_{n=1}^\infty \frac{p(p+1)\ldots(p+n-1)}{nn!} \left(\frac{g/p}{1+g/p}\right)^n \right]$$

(14)

;pecial cases:-
)=1 reduces to the negative exponential case already
liscussed. The integral form,

$$<\rho\rho^{*}>_{spec} = \frac{1}{\{1+g/p\}^p} \left[1 + \frac{\pi T^2}{A} \int_o^{(g/p)/(1+g/p)} \left\{ \frac{1}{w(1-w)^p} - \frac{1}{w} \right\} dw \right],$$

.s helpful for integer p.

For example for p=2,

$$<\rho\rho^{*}>_{spec} = \frac{1}{\{1+g/2\}^2} \left[1 + \frac{\pi T^2}{A} \int_o^{(g/2)/(1+g/2)} \left\{ \frac{1}{w(1-w)^2} - \frac{1}{w} \right\} dw \right]$$

$$= \frac{1}{\{1+g/2\}^2} \left[1 + \frac{\pi T^2}{A} \{g/2 + 1_n(1+g/2)\} \right]$$

(15)

The limiting case $p \to \infty$ corresponds to the normal surface. Thus if the change of shape of surface height distribution brought about by surface grinding say can be modelled by the transition of p from ∞ to 1, the mean optical intensity in the specular direction varies from (18) to (15) and (11) via (14).

5. GEOMETRICAL SCATTERING

The quantity $g = \sigma^2 v_z^2$ is a discriminating surface roughness parameter since it embodies the ratio σ/λ. For $g \gg 1$, the surface is categorised as being optically rough. The coherent specular component is negligible under this condition. For a one-dimensional rough surface model:

$$<\rho\rho^*> \, \simeq \, \frac{F^2}{2X} \int_{-\infty}^{\infty} e^{iv_x \tau} \chi_2(v_z, -v_z; \tau) \, d\tau.$$

"Saddle point integration" can be used to evaluate $<\rho\rho^*>$: Only in the region of $\tau = 0$ is there any significant contribution to the integral and the Gaussian autocorrelation function $C(\tau) = e^{-\tau^2/T^2}$ may be approximated by $1 - \tau^2/T^2$.

Under these circumstances for each of the model surfaces under consideration, (normal, negative exponential and gamma), $\chi_2(v_z, -v_z; \tau)$ can be expressed as $\phi(\sqrt{g}\tau/T) \cdot \phi(\sqrt{g}\tau/T)$ is the characteristic function of surface slopes and ϕ is a real, symmetric function.

Then, $$<\rho\rho^*> = \frac{F^2}{2X} \int_{-\infty}^{\infty} e^{-iv_x \tau} \phi(\sqrt{2g}\tau/T) d\tau$$

Write $t = \sqrt{2g} \, \tau/T$ it follows that

$$<\rho\rho^*> = \frac{F^2 T}{\sqrt{2g} 2X} \int_{-\infty}^{\infty} \exp(-iv_x Tt/\sqrt{2g}) \phi(t) dt.$$

Then by the inversion theorem,

$$<\rho\rho^*> = \frac{F^2 T}{\sqrt{2g} 2X} \, 2\pi p(v_x T/\sqrt{2g}),$$ where $p(.)$ is the p.d.f. of surface slopes.

Also $$\frac{T v_x}{\sqrt{2g}} = \frac{T v_x}{\sqrt{2} \sigma v_z} = \frac{T}{\sqrt{2}\sigma} \tan\theta/2$$ where $\theta = \theta_1 - \theta_2$.

   The result can be interpreted according to a mirror facet
ray theory model of scattering. According to such a model
surface facets inclined at $\theta/2$ would reflect into θ_2 and thus
the mean optical intensity at receiver angle θ_2 is directly
proportional to the number of facets inclined at the
appropriate angle.

   (In presenting a unified approach to scattering theory the
argument above is again expounded in terms of characteristic
functions. However, without any restrictive mathematical
conditions, the early paper by Cox and Munk[5] on sun glitter
demonstrated that scattering from very rough surfaces is
determined by surface slopes.)

   For a __two-dimensional__ isotropic rough surface model:

$$<\rho\rho^*> \simeq \frac{2\pi F2}{A} \int_0^\infty Jo(v_{xy}\tau)\chi_2(v_z,-v_z;\tau)\tau d\tau.$$

   In the case of a normal surface saddle point integration
yields,

$$<\rho\rho^*> = \frac{2\pi F2}{A} \int_0^\infty \tau e^{-g\tau^2/T^2} J_o(v_{xy}\tau)d\tau$$

$$= \frac{\pi F2 T2}{Ag} \exp\left\{\frac{-T^2 v_{xy}^2}{4\sigma^2 v_z^2}\right\}.$$

   The geometrical optics analogue can be perpetuated. For
if $\tilde\theta$ is the inclination of a facet with respect to the
reference plane then for reflection into (θ_2,θ_3), $\cos\tilde\theta$

$$= \frac{v_z}{[v_x^2 + v_y^2 + v_z^2]^{1/2}}, \text{ and } \tan\tilde\theta = \frac{\sqrt{v_x^2+v_y^2}}{v_z} = \frac{v_{xy}}{v_z}.$$

   For negative exponential and gamma surfaces the saddle point
integration procedure leads to Hankel-Nicholson type integrals
namely:

$$<\rho\rho^*> = \frac{2\pi F^2}{A} \int_0^\infty \frac{\tau J_o(v_{xy}\tau)}{(1+g\tau^2/pT^2)^p} d\tau$$

$$= \frac{2\pi F^2}{A} \frac{pT^2}{g}^{(p+1)/2} \frac{(v_{xy})^{p-1}}{2^{p-1}\Gamma(p-1)} K_-(p-1)(v_{xy}T\sqrt{p/g}).$$

$(K_q(.)$ is the modified Bessel function of the second kind of order q). In particular for the negative exponential surface (p=1) $<\rho\rho*> = \dfrac{2\pi F^2 T^2}{Ag} K_o(v_{xy} T/\sqrt{g})$ which has a singularity at $v_{xy} = 0$, whereas for a one-dimensional model surface the saddle point method for $<\rho\rho*>$ predicted a response directly proportional to the Laplacian distribution of surface slopes.

6. COMPARATIVE INSTRUMENT

Optical scanning devices are appropriate when the requirements of the surface inspection scheme are 100% inspection of bulk material in real time. However there is also an industrial need for a device which provides objective surface assessment, possibly embedded within a framework of a batch sampling plan, for continuous feed-back control of the production process. The device should be flexible, portable, inexpensive, simple to use, and provide readily interpretable information. Such an instrument is the Compari-surf. The Compari-surf measures light intensity reflected from a surface in the specular direction.

Early work by Kluge and Brechmann[5] on specular reflection from ground surfaces showed that this measure could discriminate between ground specimens of varying surface roughness. Whilst from theory the Beckmann model asserts the specimens $\sigma>1\mu m$ ($\lambda=0.914\mu m$) to be optically rough except near grazing incidence ($\theta_1 = 90°$), and the scattering to be virtually diffuse and consequent upon the distribution of surface slopes. For such surfaces $<\rho\rho*>_{spec}$ varies as $1/g$. Whereas for a specimen with $\sigma = 0.5\mu m$ and $K = \pi T^2/A = 0.5$, at $\theta_1 = 75°$ some 30% of the scatter in the specular direction is attributable to the coherent component, and that this component contributes as much as 50% at $\theta_1 = 60°$. At $\theta_1 = 15°$, however the diffuse component is now dominant to the extent of 80% of the total explanation.

Distribution	Specular Component	Diffuse Component	Max	Turning Pt.
Neg.Exp (p=1)	$\dfrac{1}{1+g}$	$\dfrac{Kl_n(1+g)}{1+g}$	K/e $=0.37K$	$g = e-1$ $=1.72$
Gamma (p=2)	$\dfrac{1}{(1+g/2)^2}$	$\dfrac{K[g/2+l_n(1+g/2)]}{(1+g/2)^2}$	$0.43K$	$g =1.6$
Normal (p=∞)	e^{-g}	$Ke^{-g}[E_i(g)-\gamma-l_n g]$ *see eqn. (16)	$0.85K$	$g = 1.5$

Table 1   Components of the specular field

Figure 2 shows the predicted values of $\langle\rho\rho*\rangle_{spec}$ versus angle of incidence for surfaces having negative exponential height and specified RMS roughness. Similar figures are obtained for gamma and normal surfaces with $\langle\rho\rho*\rangle_{spec}$ falling faster as p increases. $\theta_1=75°$ gives good discrimination for RMS σ and is the setting generally adopted when operating the Compari-surf.

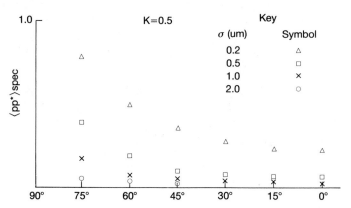

Fig. 2   Specular intensity versus angle of incidence

Calibration plots obtained by the Compari-surf have been published previously[7,8] and are reproduced here.

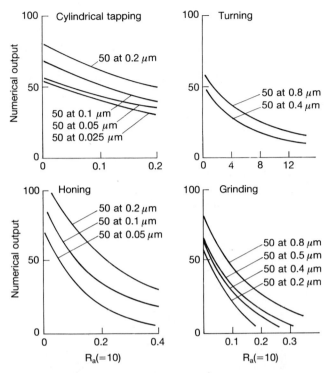

Fig. 3   Calibration Curves obtained for standards representing
four finishing processes (Fig. 4[7]).

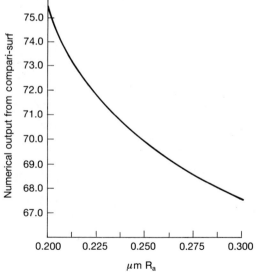

Fig. 4   Calibration curve for a plunge grinding process (Fig. 7[8]).

   Very smooth surfaces fabricated by a cylindrical lapping
process ($0 < R_a < 0.2\mu m, 0 < g < 0.8$) were well fitted by the
Beckmann model based on normal surface statistics. The rate of
fall-off with g was insufficient when fitting non-normal
distributions. At wide angle interrogation ($\theta_1 = 75°$), the
Compari-surf instrument senses only the peaks of the rough
surface. Williamson[9] provides considerable evidence that the
upper-half ($Z > 0$) of the distribution of surface heights is
usually normal.

   For rougher surfaces ($R_a > 1, g > 4$), when scattering is
largely governed by the laws of geometrical optics, log(output)
versus log(R_a) linearised calibration curves.

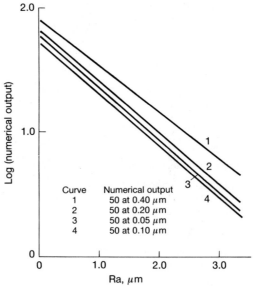

Curve	Numerical output
1	50 at 0.40 μm
2	50 at 0.20 μm
3	50 at 0.05 μm
4	50 at 0.10 μm

Fig. 5 Log calibration graph for ground specimens (Fig. 4[7]).

Finishing Process	Slope	σ_ψ
Turning	-0.8	$\alpha\ \sigma^{0.4}$
Honing	-1.1	$\alpha\ \sigma^{0.55}$
Grinding	-0.9	$\alpha\ \sigma^{0.45}$

Table 2 Variation of σ_ψ with σ for various finishing processes

For isotropic surfaces theory predicts a slope of -2. However, correlation length is not being controlled and σ_ψ, the standard deviation of surfaces slopes, is found to be related to surface roughness height σ according to the above table. Tanner[10] using cylindrical lapped surfaces found $\sigma_\psi \ \sigma^{0.4}$. The surface texture had a definite lay and Tanner used an appropriate one-dimensional theory.

Linear extrapolation into the smoother roughness ranges is not valid. For very smooth surfaces $\log[<\rho\rho*>] \simeq 1-g$ when the coherent component is dominant. A single transformation over the whole range is inappropriate.

A practical option is to use other angles of incidence tailored to the range of RMS anticipated. At normal incidence log-log graphs would again linearise experimental calibration plots for $0.2\mu m-1\mu m$ roughness specimens.

7. CONCLUSIONS

The scattering theory developed by Beckmann supports the empirical calibration curves of section 6 published previously[7,8].

In theory $\log<\rho\rho*>_{spec}$ for very smooth surfaces varies according to $-\sigma^2$. This result is robust in that it is not dependent on the statistical distribution of surface height. Since,

$$<\rho\rho*>_{spec} = \chi(v_z)\chi*(v_z)$$

$$= \exp\left\{2 \sum_{r=1}^{\infty} \frac{(-1)^r \kappa_{2r}(v_z)^{2r}}{(2r)!}\right\}$$

$$\simeq \exp\left\{-g + \frac{2\beta_2 g^2}{4!}\right\}$$

where $\beta_2 = \kappa_4/(\kappa_2)^2$ is the coefficient of kurtosis and κ_{2r} are the even cumulants of surface height.

For very rough surfaces scattering can be wholly attributed to surface slopes. The slope distribution is necessarily symmetrical. Generally, odd moments or odd cumulants play no part in formulating the scatter from a rough surface. In short skewness of the surface height distribution does not appear to be a determinand of scattering. The skewness can only be inferred from variance, kurtosis etc. estimates based on beam scatter information, from the shape of the response curve from an optical scanning device for example, together with prior knowledge of the mechanical profiles of the comparison surfaces. Such prior knowledge needs to be employed in setting suitable ranges for the Compari-surf instrument response or for choosing appropriate angles of incidence for the surface interrogation.

For normal rough surfaces the approach to geometrical optics can be assessed by means of the identity:-

$$E_i(x) - \gamma - l_n x = \sum_{n=1}^{\infty} \frac{x^n}{nn!}, \quad (x > 0). \tag{16}$$

Here $E_i(x)$ is the exponential integral $\int_{-\infty}^{x} \frac{e^t}{t} dt \ (x > 0)$

and $\gamma = 0.57721$ is Euler's constant. The ratio,

$$e^{-g} \frac{\pi T^2}{A} \sum_{n=1}^{\infty} \frac{g^n}{nn!} \bigg/ \frac{\pi T^2}{Ag} = e^{-g} \{E_i(g) - \gamma - l_n g\} / 1/g$$

gives the proportion of the diffuse component of scatter which is attributable to surface slopes, from which the following table can be derived.

g	Prop[n] of $<\rho\rho*>_{spec}$ explained by geometrical optics
4	75%
10	88%
20	94%

Table 3   Variation of $<\rho\rho*>_{spec}$ attributable to geometrical optics with surface roughness.

The linearity of log-log plots for g of the order of 10 or more is again robust regardless of the shape of the surface height distributions considered.

8.   ACKNOWLEDGEMENTS

In the preparation of this paper I am grateful to Professor K.J. Stout (Coventry(Lanchester) Polytechnic) for several fruitful discussions. I am also indebted to Professor M.S. Longuet-Higgins (Cambridge) for drawing my attention to the paper by Cox and Munk.

9. REFERENCES

1. Beckmann, P. and Spizzichino, A., (1963) "The scattering of Electromagnetic Waves from Rough Surfaces", Pergamon, London.

2. Beckmann, P., (1973) *Trans. IEEE Ant. and Prop.*, AP-21, 2, 169.

3. Beckmann, P., (1973) "Orthogonal Polynomials for Engineers and Physicists", Boulder, Colo.: Golem Press.

4. Obray, C.D., (1983) "Application of Scattering Models to Optical Surface Inspection," Ph.D. thesis, The City University, Department of Systems Science.

5. Cox, C.S. and Munk, W.H., (1954) *J. Opt. Soc. Am.* **44**, 838.

6. Kluge, J. and Brechmann, G., (1944) "Comparison of Surface Roughness of Highly Finished Plane Surfaces", *VDI Zeitschrift,* Vol. 88.

7. Stout, K.J., (1984) *Prec. Eng.* 6, 35.

8. Stout, K.J., Obray, C.D. and Jungles, J., (1985) "Specification and Control of Surface Finish: Empiricism versus Dogmatism", *Opt. Eng.* 24, 3.

9. Williamson, J.M.P., (1968) "Properties and Metrology of Surfaces", *Proc. Inst. Mech. Eng. London,* 182, 3K, 21.

10. Tanner, L.H. and Fahoum, M., (1976) *Wear,* 36, 299.

CORRELATED CLUTTER IN COHERENT IMAGING

C.J. Oliver
*(Royal Signals and Radar Establishment, Malvern,
Worcestershire)*

ABSTRACT

   This paper presents a summary of the statistical properties
and simulation of correlated clutter in coherent imaging.
Clutter can often be represented in terms of correlated
K-distributed noise, as indicated by radar and sonar examples.
This paper contains a derivation of a theoretical model which
should describe any coherent scattering and imaging process
applied to clutter surfaces. Finally a simulation technique,
based on linear filtering, which generates such textures is
outlined with examples.

1. INTRODUCTION

   In this paper we set out to present a summary of the
present situation in the interpretation and simulation of
correlated clutter texture in coherent images. The work was
initiated by the demonstration that many clutter textures
could be well represented as K-distributed noise [1-3]. More
recently the great importance of the correlation properties
of this clutter has been demonstrated [4,5] and a number of
theoretical models derived for correlated surfaces [6-9].
This paper examines the manner in which the surface and
imaging properties interact when such surfaces undergo
coherent illumination and imaging and its implications for
image interpretation. We also demonstrate the use of a linear
filter method to simulate image textures which duplicate the
two-point correlation properties of the original image [9,10].

   In this paper we shall omit the derivations of various
theoretical results. We refer interested readers to the
original publications for such detail and concentrate on the
physical reasoning underlying the approach.

2. PROPERTIES OF NATURAL TEXTURES

In Figure 1 we show a section of high resolution Synthetic-
Aperture Radar (SAR) imagery obtained by the RSRE airborne
X-band system. The feature enclosed by the full line
represents the image of a field. The image is reminiscent of
laser speckle, which we know arises from the random
interference between the many scatterers within the
illuminating beam. This suggests a related model for SAR
images of fields - namely that we have a large number of
random scattering centres per resolution cell. The random
interference between these generates a complex Gaussian field
with Rayleigh magnitude and negative-exponential intensity
distributions.

Fig. 1 SAR image of fields

In Figure 2 we show a section of SAR imagery of a wooded
area for comparison. It is apparent that there is additional
bunching within the image than was present for the field area.
It is tempting to associate this with the presence of the
trees and the spacing between them. Similar excess
fluctuations have been observed in laser scattering
experiments where they have been attributed to fluctuations in
the number of scatterers illuminated [2,11]. An alternative
explanation is to attribute the result to fluctuations in some
surface cross-section variable chosen to represent the wood

[1,6-9]. The two effects are indistinguishable on a scale smaller than the resolution cell. In both cases a coherent imaging process would give rise to K-distributed intensity statistics.

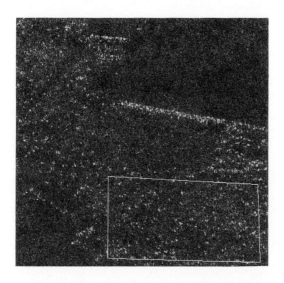

Fig. 2   SAR image of woods

Let us first establish whether the observed textures in a variety of coherent imaging situations can indeed be expressed in terms of K-distributed intensity distribution given by

$$p(I) = \frac{2}{<I>\Gamma(\nu)} \left(\frac{I}{<I>} \right)^{(\nu-1)/2} K_{\nu-1}\left[2 \left(\frac{I}{<I>} \right)^{\frac{1}{2}} \right] \qquad (1)$$

where $\Gamma(\nu)$ is the gamma function and $K_{\nu-1}$ is a modified Bessel function of order $\nu-1$. As already noted such a distribution would arise from coherent imaging of a gamma-distributed surface whose cross-section is given by

$$p(\sigma) = \frac{1}{<\sigma>\Gamma(\nu)} \left(\frac{\sigma}{<\sigma>} \right)^{\nu-1} \exp\left[-\frac{\sigma}{<\sigma>} \right] \qquad (2)$$

The measured single-point intensity moments for the field and wood regions illustrated in Figures 1 and 2 are summarised in tables 1 and 2 where they are compared with theoretical predictions for K-distributions with the same second moment given by

$$<I^n>/<I>^n = n! \; \Gamma(n+\nu)/\nu^n \; \Gamma(\nu) \qquad\qquad (3)$$

[3,8]. In making the comparison with the theory we assume that the errors in the measured moments are related to the values of the higher-order moments as analysed by Oliver [9]. The results in tables 1 and 2 indicate that K-distributions of order ν=33.3 and 2.55 are consistent with the data for the field and wood respectively.

	MEASURED	PREDICTED SD	THEORY
SECOND	2.060	.001	2.060
THIRD	6.528	.012	6.55
FOURTH (x 10^1)	2.82	.11	2.86
FIFTH (x 10^2)	1.54	–	1.60
SIXTH (x 10^3)	1.01	–	1.10
SEVENTH (x 10^3)	7.5	–	9.1
EIGHTH (x 10^4)	6.2	–	8.8

TABLE 1   Intensity moments for SAR field image

	MEASURED	PREDICTED SD	SIMPLE MODEL	EXACT THEORY
SECOND	2.785	.042	2.785	2.784
THIRD (x 10^1)	1.49	.08	1.49	1.55
FOURTH (x 10^2)	1.23	.15	1.30	1.48
FIFTH (x 10^3)	1.37	–	1.67	–
SIXTH (x 10^4)	1.9	–	3.0	–
SEVENTH (x 10^5)	3.1	–	7.0	–
EIGHTH (x 10^6)	5.5	–	21.0	–

TABLE 2   Intensity moments for SAR woods image

However, the treatment so far in terms merely of statistics has ignored the correlation properties within the image. If we study the woods image (figure 2) then it is obvious that there is an underlying texture which will have characteristic correlation properties. Following previous analysis it is reasonable to assume that the underlying statistics remain gamma-distributed. We have shown that a surface with the required single-point moments and two-point correlation properties can be generated by considering a random walk process with narrow-band Gaussian noise [6]. When such a surface is illuminated coherently and imaged the resulting intensity will combine the effects of the instrument function, which determines the speckle size, with the underlying cross-section fluctuations. Theoretical forms for the received field autocorrelation function (ACF), the detected intensity ACF and the intensity single-point moments have been derived for a variety of surface models assuming a Gaussian instrument function [6-9]. In particular the two-point intensity ACF at lag values of X and Y for a homogeneous texture has been shown [9] to be given by

$$\langle I(0,0)I(X,Y)\rangle/\langle I\rangle^2 = 1 + \exp[-X^2/W_x^2 - Y^2/W_y^2]$$

$$+ \frac{1}{\pi W_x W_y} \int\int_{-\infty}^{\infty} dudv \; [\langle \sigma(0,0)\;\sigma(u,v)\rangle/\langle\sigma\rangle^2 - 1] \qquad (4)$$

$$\{\exp\{-(u-x)^2/W_x^2 - (v-Y)^2/W_y^2] + \exp[-(u^2+x^2)/W_x^2 - (v^2+Y^2)/W_y^2]\}$$

where W_x and W_y are 1/e widths of the Gaussian imaging response. The total surface ACF, $\langle \sigma(0,0)(u,v)\rangle$, can be expressed in terms of its factorisation properties using the random-walk derivation and assuming gamma statistics [8]. In the absence of any cross-section fluctuations only the first two terms of eqn 4 are relevant. These demonstrate the classical speckle intensity ACF with speckle size determined by the instrument function widths W_x and W_y which take the values 0.52 and 1.6 pixels respectively. We might expect the intensity ACF for the field image to be well represented by these two terms alone, as is indeed demonstrated in figure 3 where the data are compared with theory (full line - X axis, dashed line - Y axis). The second two terms of eqn 4 include the interaction between the surface fluctuations and the imaging function. This relationship is discussed in more detail by Miller [12]. In general, however, we find that the third term approximates to a convolution between the surface and the imaging function (and is precisely that for a Gaussian surface). The fourth term has a dependence which is largely dominated by the imaging function. Only the third term, therefore, contains useful information about the underlying surface fluctuations.

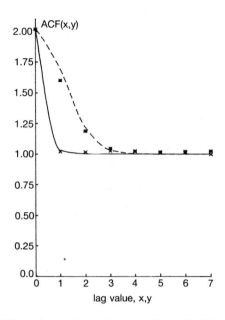

Fig. 3 Intensity ACF for fields

Let us next examine the extent to which these fluctuations
can be represented with simple analytic models. As mentioned
above, we assume a Gaussian instrument function which is close
to that observed experimentally in the RSRE SAR system [9].
Provided that the surface texture is homogeneous the number of
dimensions of each integral can be reduced by one along each
axis, as is assumed in eqn 4. Two basic analytic models have
been considered, each with gamma-distributed statistics. The
first assumes that the spectrum is Lorentzian, corresponding
to a negative-exponential ACF; the second assumes a Gaussian
spectrum and, hence, ACF. The gamma-distributed noise
contribution may also be mixed with a local oscillator of
constant amplitude but offset frequency or with a similar,
frequency-shifted, noise source to give an oscillatory
component to the surface [8,9]. With the Gaussian spectrum
the contributions along the two axes may be completely
factorised and full analytic solutions are available. With
the Lorentzian spectrum, on the other hand, the contributions
along the two axes cannot be factorised and a numerical
evaluation is required.

In figure 4 we show a comparison of the measured intensity
ACF for the woods image with such a theoretical prediction.
The data are well represented if the surface ACF is assumed
to take a negative-exponential form, i.e

$$<\sigma(0,0)\ \sigma(u,v)>/<\sigma>^2-1\ =\ \exp\ [-(u^2/\ell_x^2+v^2/\ell_y^2)^{\frac{1}{2}}]\ .\qquad(5)$$

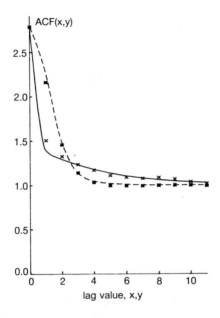

Fig. 4   Intensity ACF for woods

No oscillatory component is required and the correlation lengths, ℓ_x and ℓ_y, take values of 3.3 and 1.2 pixels respectively. Substituting this form of surface ACF into eqn (4) the predicted intensity ACF can then be derived numerically, as already noted. The agreement between the simple model and the data indicates the validity of the approach.

In addition to the two-point statistics the higher-order single-point statistics also depend on the interaction between the surface and the imaging function. Analytic results for these moments have been derived previously [6-9]. Predicted values are compared with those for the original woods image in the last column of table 2. An order parameter of $\nu=1.32$ was used for the surface fluctuations as compared with the value $\nu=2.55$ required by the simple model. Thus the imaging process has essentially halved the spikiness of the observed distribution by averaging out the surface fluctuations. It is instructive to note that the moments of the data are fitted

equally well by the predictions of both the simple model and
the exact theory. A filtered gamma-distribution is very
similar, though not identical unless the filter is
rectangular, to a gamma-distribution with a different order
parameter [6,9]. It is in the dependence of these statistics
on the imaging resolution that it is essential to use the
exact theory.

As a further comparison we demonstrate the application of
the model to a sonar image , illustrated by figure 5, of sand
ripples on the sea bed. In this case an oscillatory
component to the surface is required, introduced by a
frequency-shifted local oscillator, and the spectrum appears
well-respresented by a Gaussian form. Thus the surface ACF is
given by

$$<\sigma(0,0)\sigma(u,v)>/<\sigma>^2 - 1 = 2ab \exp[-u^2/2\ell_x^2 - v^2/2\ell_y^2] \cos\Omega_x u/(a+b)^2$$

$$\tag{6}$$

$$+ b^2 \exp[-u^2/\ell_x^2 - v^2/\ell_y^2]/(a+b)^2$$

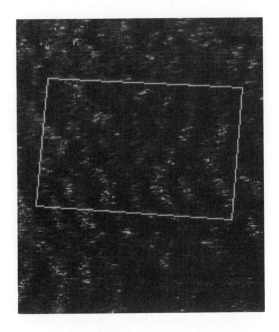

Fig. 5   Sonar image of sand ripples

The local oscillator strength (a) is twice the noise term strength (b), while the surface correlation lengths are 16 (ℓ_x) and 6.5 (ℓ_y) pixels respectively. The period of the oscillation along the x axis ($2\pi/\Omega_x$) is 24 pixels. For a Gaussian spectrum the contributions along the two axes may be factorised and evaluated analytically [6,9] leading to predicted ACFs shown in figure 6. Agreement is adequate over the first 40 pixels but fails to represent the surface fluctuations at large lag values in the x direction. A comparison of the measured and predicted intensity moments is given in table 3. For the simple model we ignore the presence of the local oscillator and assume that the intensity is K-distributed with the correct second moment. This yields an order parameter of 1.86 compared with the value of 0.938 required for the exact theory. Both models give adequate agreement with the measured data.

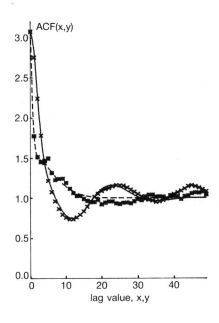

Fig. 6   Intensity ACF for sand ripples

	MEASURED	PREDICTED SD	SIMPLE MODEL	EXACT THEORY
SECOND	3.08	.13	3.08	3.08
THIRD (x 10^1)	1.97	.26	1.92	1.75
FOURTH (x 10^2)	1.97	.52	2.01	1.92
FIFTH (x 10^3)	2.6	-	3.2	-
SIXTH (x 10^4)	4.1	-	7.0	-
SEVENTH (x 10^5)	7.0	-	21.0	-
EIGHTH (x 10^7)	1.3	-	8.0	-

TABLE 3   Intensity moments for sonar sand ripple image

3.   TEXTURE SIMULATION

As a corollary to the determination of the statistical
properties of clutter textures, it would be useful to be
capable of simulating such textures with defined properties
to resemble genuine radar images. These images could be
simulated based on maps, which would define the boundaries of
the different textures, together with statistical correlation
properties which define the individual textures. The general
problem of generating noise with defined statistics for a
two-dimensional image is complex. However, Oliver and Tough
[10] have demonstrated a linear filter method which
approximates to what is required, reproducing the two-point
statistics precisely. The higher-order statistics are then
determined by the filter process and will not, in general,
exhibit the required properties. However, as we shall
demonstrate, reproducing the second-order texture of the
surface gives rise to very realistic images.

The different stages in this method [9] are illustrated in
figure 7 where the wood texture is chosen as an example.
First we select an appropriate analytic model with associated
parameter values from those outlined in the previous section.
Based on this model we may define the form of the surface ACF.
We next require to deduce that linear weighting function that
will yield the correct surface ACF when uncorrelated noise (x)

is passed through it. We have shown that the ACF of the
filter output at lag value r will then be given by [10] .

$$<\sigma(0)\sigma(r)>/<\sigma>^2 - 1 = \frac{\text{Var } x}{<x>^2} \sum_{j=1}^{N} \omega_j \omega_{j\cdots r} / \left(\sum_{j=1}^{N} \omega_j \right)^2 \qquad (7)$$

The ACF of the filter output thus depends on the second-
order properties of both the input noise and the filter. We
assume that x is gamma-distributed so that eqn (7) thus
determines the order parameter of x and the ACF of the filter
weight, from which we need to establish the weight itself.
For a homogeneous, ergodic surface the ACF will be real and
symmetric while the surface power spectrum will be real,
symmetric and positive. We may deduce an amplitude spectrum
by taking the square root of the power spectrum. We choose
the positive value of this root; an arbitrary decision which
is justified since only the ACF of the weight is defined. On
taking the inverse Fourier Transform of the amplitude spectrum
a real, symmetric, weighting function is derived. For the
wood texture this has the form shown on the top left of
figure 7. Such a derivation is only applicable to homogeneous
random textures.

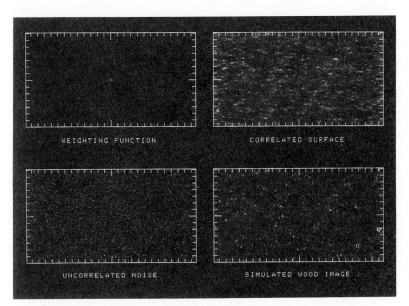

Fig. 7   Stages in simulation of image of woods

A correlated random variable which describes the local
cross-section, as shown in the top right of figure 7, is thus
generated by convolving white, gamma-distributed, noise of

appropriate order, shown in the bottom left of figure 7, with
this weighting function. The scattered electro-magnetic field
is then simulated by generating real and imaginary Gaussian
random components from a distribution with variance
proportional to the cross-section at that pixel. This
simulates the effect of the random interference between many
scatterers within the resolution cell giving rise to speckle
in the image. Finally, this scattered field is then imaged by
convolving it with the instrument function to give an image
such as that on the bottom right of figure 7.

As an example of the results of this process the azimuthal
components of the simulated, and original, intensity ACFs for
the wood are shown in figure 8 together with the theory.
Agreement is seen to be close. A more sensitive test of the
simulation method, however, is to examine the higher-order
statistics. The discrepancy between those for an exact gamma
distribution and the results of simulation are demonstrated in
table 4. The statistics are measured for a set of ten
simulated surfaces and then compared with the exact theory and
that for the filter technique [9,11]. It is apparent that the
simulated distribution is more spikey than a true gamma-
distribution with moments close to the linear filter theory.
On generating the intensity following scattering and imaging
speckle is found to dominate the statistics. The difference
and indeed the simulation results lie between the two limits.
A detailed description of these errors is given elsewhere [9].

	SIMULATION			THEORY	
	MEAN	SD	PREDICTED SD	EXACT GAMMA	LINEAR FILTER
SECOND	1.75	.10	.28	1.76	1.76
THIRD	7.4	1.9	8.0	4.4	7.7
FOURTH (x 10^1)	8.5	5.0	26.0	1.4	9.4
FIFTH (x 10^3)	1.8	1.6	–	.058	–
SIXTH (x 10^4)	4.8	6.0	–	.028	–
SEVENTH (x 10^6)	1.5	2.3	–	.0015	–
EIGHTH (x 10^7)	5.1	8.7	–	.00098	–

TABLE 4 Cross-section moments for the simulated wood texture

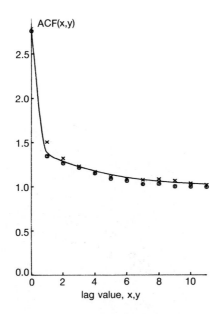

Fig. 8 Comparison of theory, data and simulation of
azimuthal intensity ACF for woods

We asserted earlier that real textures and those simulated
with the correct two-point correlations appeared very similar.
Let us compare the simulated texture in figure 7 with the real
woods texture in figure 2. To the eye the results are almost
indistinguishable. Obviously the detail is not expected to be
identical; it is the overall correlation statistics that are
duplicated. In figure 9 we make the equivalent comparison for
the sonar images of sand ripples analysed earlier. Again, the
results are closely similar. It is instructive to remark that
the theoretical model does not, in itself, contain any
textural information that relates directly to the long ridges
that appear in the image which are significantly longer than
any correlation length in the intensity ACF. They are
constructed by the observer joining the randomly-positioned
contributions and interpreting the results as continuous
ridges.

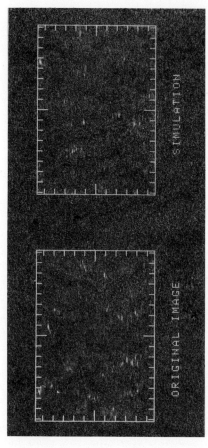

Fig. 9   Comparison of real and simulated images of sand ripples

4. LIMITATIONS OF THE MODEL

There are two main areas of limitation in the type of
approach adopted here; the first relates to problems with
random textures, the second to the completely different class
of deterministic textures which largely corresponds to man-
made objects, eg urban areas.

Let us initially indicate the problems associated with
random textures. The theoretical models we have discussed
have simple analytic forms. However, even the examples
chosen do not fit these forms exactly. For example, the sonar
image of sand ripples failed to reproduce the correlation
properties at lag values greater than about 40 pixels. It is
important to be capable of simulating textures which do not
closely match these models. Work is at present in progress on
an approach which treats the observed intensity spectrum as a
perturbed version of the simple analytic model and then
derives the corrected surface ACF and weighting function and
simulates the surface in the same way as before.

One of the drawbacks of the simulation method is that the
higher-order statistics of the simulated image are not K
distributed because of the action of the imaging filter on the
original gamma-distributed white noise. Miller [13] has
proposed an approach which allows us to modify the form of
this original distribution so as to reproduce the desired
single-point statistics following the filter process. Not
only should this allow us to achieve correlated K-distributed
clutter for arbitrary spectra, when combined with the previous
modification, but would also, in principle, enable arbitrary
statistics to be duplicated. Further work is in progress on
this development of the linear filter model for simulation.

A more fundamental limitation stems from the choice of the
filter approach. This obviously restricts the class of
textures that can be reproduced to those which are consistent
with a simple filter process. Arbitrary higher-order
correlation properties could not be satisfied in general. In
principle we require to be able to define an arbitrary set of
statistical properties for a texture and then simulate a
random texture that meets those conditions. An approach based
on Markov Random Fields may offer a route for future research.

The discussion above has considered the problems associated
with random textures. Scenes containing man-made objects such
as towns are not generally random. This problem raises the
whole question of how deterministic information could be
included in texture interpretation and simulation. One result
of the existence of man-made objects in a scene is that the

statistics are no longer K-distributed but show much more violent fluctuations than the model would suggest. Since the model is based on ergodic noise processes it is not surprising that it should fail to represent this increase which denotes an excess bunching of cross-section beyond a random process. Furthermore, the intensity spectrum of a town shows periodicities which are explicitly related to the spacing of streets and buildings, ie deterministic information. A simulation attempting to reproduce this spectrum from random fluctuations will not achieve sufficient regularity of the dominant scatterers to duplicate the image structures. This deterministic property of such scenes also evidences itself on the scale of the individual bunch of scatterers comprising a building. Each building is made up of a series of fundamental scatterer elements, such as corner reflectors, separated by distances determined by the building dimensions. Again it is apparent that a random texture approach will not be appropriate. The requirement for combining both deterministic and random textures is a key research topic in coherent image interpretation and simulation.

5. CONCLUSION

The purpose of this paper was to indicate the physical concepts that underlie the approach we adopt to the interpretation and simulation of natural textures in coherent images. Correlated K-distributed noise was shown to represent many natural clutter images very closely. Further, a simulation method has been proposed, based on linear filtering, which produces textures with identical second-order statistical properties to the original image. The fact that the same approach could be applied to both sonar and radar images indicates its generality. It would be expected to be relevant to all types of coherent image of random textures. The direction of further research along the same basic route has also been indicated and the important of combining such textures with deterministic information stressed.

6. REFERENCES

[1]   Jakeman, E. and Pusey, P.N., 1976, *IEEE Trans*, AP-24, 806-814.

[2]   Jakeman, E. and Pusey, P.N., 1977, Radar 77, IEE Conf Publ. 155 (London: Institution of Electrical Engineers) pp 105-109.

[3]   Jakeman, E. and Pusey P.N., 1978, *Phys Rev Letts*, **40**, 546-550.

[4]   Ward, K.D., 1981, *Electron Lett,* **171**, 561-565.

[5]   Ward, K.D., 1982, Radar 82, IEE Conf Publ 216 (London: Institution of Electrical Engineers) pp 203-207.

[6]   Oliver, C.J., 1984, *Optica Acta,* **31**, 701-722.

[7]   Oliver, C.J., 1986, Proc IMA Conf on Wave Propagation and Scattering (Oxford University Press).

[8]   Oliver, C.J., 1985, *Optica Acta,* **32**, 1515-1547.

[9]   Oliver, C.J., 1986, Inverse Problems, **2**, 481-518.

[10]   Oliver, C.J. and Tough, R.J.A., 1986, 33, 223-250.

[11]   Jakeman, E., 1980, *J Phys A,* **13**, 31-48.

[12]   Miller, R.J., 1986, Proc IMA Conf on Mathematics and its Applications in Remote Sensing, to be published.

[13]   Miller, R.J., 1986, to be published.

CORRELATION PROPERTIES OF COHERENTLY IMAGED NON-UNIFORM SURFACES

R.J. Miller

(GEC Research Laboratories, Marconi Research Centre, Chelmsford)

ABSTRACT

A simple but powerful way of representing coherent images of non-uniform surfaces is proposed, from which it is possible to derive both coherent and incoherent intensity correlation functions. The representation of the image may be expressed in terms of an equivalent surface, which exhibits arbitrary spatial variation of cross-section to which the imaging process introduces multiplicative noise or speckle. Both Gaussian and non-Gaussian speckle may be considered. The intensity correlation function may be decomposed into contributions arising from coherent and incoherent mechanisms, and into contributions arising entirely from lack of uniformity in the surface being imaged. Explicit convolutional forms of the correlation functions are given for shift-invariant systems and the corresponding spectral representations are also given.

1. INTRODUCTION

The images obtained from coherent imaging systems, such as synthetic and real aperture radars and sonar, exhibit pronounced fluctuations in intensity often referred to as speckle. The presence of speckle tends to obscure the underlying variations of intensity which convey information on the nature of the surface being imaged. Useful information may be contained in the spatial spectrum of the intensity fluctuations, or equivalently, the spatial autocorrelation function. Thus, one aim in work on coherent imagery is to be able to deduce the spectrum of the underlying cross-section fluctuations. In order to be able to do this, it is necessary to have some theoretical understanding of how images are affected by variations in surface cross-section. This paper considers the effects of cross-section fluctuations on the

correlation properties of the image using a multiplicative
noise model described below.

2.   IMAGING MODEL

A classical model of surface backscatter considers the
illumination of a surface of uniform cross-section containing
many independent scatterers of roughly equal amplitude, such
that their phases are uniformly distributed and the sum of the
separate contributions tends to a limiting complex circular
Gaussian distribution in accordance with the central limit
theorem.   Hence, the resulting intensity distribution has
negative exponential statistics.   This notion is readily
generalised to consider a surface described by a patch-work of
areas of differing cross-section, each of which contains
sufficient independent scatterers for the central limit
theorem to apply.   The complex return from a patch located at
some point r on the surface may therefore be described by the
following multiplicative noise model:

$$x(r) = a(r)\nu(r) \tag{2.1}$$

where $\nu(r)$ is a complex, circular Gaussian random field and
$a(r)$ is the magnitude of the reflectivity at point r.   Rather
than considering $a(r)$ as some explicit function for a given
scene, it is convenient to model it as a real, non-negative
random field; this gives rise in a natural fashion to a large
class of texture models.   Furthermore, the requirement that
$\nu(r)$ is Gaussian may also be relaxed, so that non-Gaussian
speckle may be considered.   The random field $a(r)$ may be
arbitrarily correlated, but a central assumption of the model
adopted here is that the noise $\nu(r)$ is independent from one
point to another.   This assumption may be justified both on
theoretical and experimental grounds, and gives the
multiplicative noise model its essential character.   A more
complete discussion of the basis of the model and the
conditions required on its components is given in (Miller, 1986).

So far, only the return from a small illuminated patch has
been considered; the system response function simply forms a
weighted sum over a number of such patches, which may be
expressed in convolutional form as:

$$y(r) = T(r)*x(r) \tag{2.2}$$

where $T(r)$ is the point spread function of the imaging system,
and $y(r)$ is the complex signal obtained at the output of the
system.   This completes the formulation of the model of the
imaging of the surface.

3. NOTATION

The following sections will consider the correlation properties of the multiplicative noise model outlined above. A derivation of these properties is given in (Miller, 1986). Two correlation functions are of interest, the simplest of which is the coherent correlation function defined by:

$$\Gamma_c(w) \; \hat{=} \; <y(r)\,y^*(r+w)> \qquad\qquad (3.1)$$

where the angle brackets denote an ensemble average. In the optical literature, this is also known as the mutual coherence function, and gives a measure of the coherence properties of the complex field $y(r)$. Its main interest here lies in its relationship to the intensity correlation function defined below. Some elementary properties of the intensity are considered first of all. The intensity is defined by:

$$I(r) \; \hat{=} \; \left| y(r) \right|^2 \qquad\qquad (3.2)$$

From (2.2), the mean intensity is therefore given by:

$$<I(r)> \; = \; <\sigma(r)> \; <\left| v(r) \right|^2> \int dr \; \left| T(r) \right|^2 \qquad (3.3)$$

where the cross-section $\sigma(r)$ is defined as the square of the reflectivity $a(r)$. Without loss of generality, all three of the factors on the right in (3.3) may be normalised to unity; since this allows a much more compact presentation of results, this normalisation will be used throughout the rest of the paper. Hence:

$$I(r) \; = \; <\sigma(r)> \; = \; <\left| v(r) \right|^2> \; = \; \int dr \; \left| T(r) \right|^2 \; = \; 1 \qquad (3.4)$$

Use of this normalisation is equivalent to normalising the coherent correlation function (3.1) by the mean intensity. The intensity correlation function (ICF) is defined by:

$$\Gamma_I(w) \; \hat{=} \; <I(r)\,I(r+w)> \qquad\qquad (3.5)$$

Use of the normalisation (3.4) is equivalent to normalising the ICF by the square of the mean intensity. The form taken by the ICF is the central concern of the present paper, and is discussed in the following section. It transpires that the ICF may be expressed entirely in terms of the system response function $T(r)$ and of a function describing the correlation properties of the surface cross-section. This latter surface correlation function is defined by:

$$\Gamma_\sigma(w) \; \hat{=} \; <\sigma(r)\,\sigma(r+w)>$$ (3.6)

4. INTENSITY CORRELATION FUNCTION

The form of the intensity correlation function may be derived from the multiplicative noise model (Miller, 1986) to yield the following expression:

$$\Gamma_I(w) = \Gamma_\sigma(w) * \left| T(w) \right|^2 * \left| T(-w) \right|^2$$

$$+ \lim_{\varepsilon \to 0} \Gamma_\sigma(\varepsilon) * \left[T(\varepsilon+w/2)\,T^*(\varepsilon-w/2) \right] \overset{\varepsilon}{*} \left[T(-\varepsilon+w/2)\,T^*(-\varepsilon-w/2) \right]^*$$

$$+ \quad <\sigma^2> \; \{<|\nu|^4>-2\}\left| T(w) \right|^2 * \left| T(-w) \right|^2$$ (4.1)

where $\overset{\varepsilon}{*}$ denotes convolution with respect to ε. The normalisations specified by (3.4) are in force. The three terms on the right of (4.1) all have identifiable origins. The first term is entirely equivalent to the intensity correlation function produced by imaging a real surface described by $\sigma(r)$ with an instrument having response function $\left| T(w) \right|^2$; hence, this term is referred to as the incoherent contribution to the ICF. By contrast, the second term is a wholly coherent phenomenon, and arises as the result of interaction of the complex instrument response with the multiplicative noise. It is referred to as the coherent contribution. The third term of (4.1) is similar in form to the first, and is the same as if a delta-function were added to the origin of the surface correlation function. However, this contribution is again due solely to the presence of speckle. This may be seen by noting that, for Gaussian speckle, the fourth moment of the noise is equal to two, so that the third term becomes zero. Hence, this term is due to the presence of non-Gaussian speckle, and is referred to as the non-Gaussian contribution.

A further useful decomposition of the ICF may be made by expressing (4.1) in terms of the surface covariance function $K_\sigma(r)$ rather than its correlation function $\Gamma_\sigma(r)$. In normalised units, the two are related by:

$$K_\sigma(w) \; \hat{=} \; <I(r)\,I(r+w)> \; - \; <I(r)> <I(r+w)> \; = \; \Gamma_\sigma(w) - 1$$

(4.2)

The covariance function is more natural in application, since it tends asymptotically to zero, in contrast with the correlation function, which tends to one.

Substitution of (4.2) into (4.1) results in an expression
for the ICF consisting of the sum of three contributions; for
reasons which will become evident, these are referred to as
non-uniform, uniform and non-Gaussian contributions. The full
expression for the ICF is then given by the following four
equations:

$$\Gamma_I(w) = \Gamma_I^{(NU)}(w) + \Gamma_I^{(UN)}(w) + \Gamma_I^{(NG)}(w) \tag{4.3}$$

$$\Gamma_I^{(NU)}(w) = K_\sigma(w) * |T(w)|^2 * |T(-w)|^2 \tag{4.4}$$

$$+ \lim_{\varepsilon \to 0} K_\sigma(\varepsilon) * [T(\varepsilon+w/2)T^*(\varepsilon-w/2)] * [T(-\varepsilon+w/2)T^*(-\varepsilon-w/2)]^* \tag{4.5}$$

$$\Gamma_I^{(UN)}(w) = 1 + |T(w) * T^*(-w)|^2 \tag{4.6}$$

$$\Gamma_I^{(NG)}(w) = \langle\sigma^2\rangle\{\langle|\nu|^4\rangle - 2\} \, |T(w)|^2 * |T(-w)|^2 \tag{4.6}$$

Equation (4.6) is identical to that for the non-Gaussian
contribution in (4.1), and will not be discussed further.
Equation (4.5) corresponds to the ICF of a surface of uniform
cross-section to which speckle is introduced via coherent
imaging (Goodman, 1984). Furthermore, this may be expressed
entirely in terms of the coherent correlation function (CCF).
It may be shown that the CCF is given by

$$\Gamma_c(w) = T(w) * T^*(-w) \tag{4.7}$$

This result holds for surfaces with arbitrary variations of
cross-section, and is of interest in its own right. Comparison
with (4.6) shows that the uniform contribution is equal to
one plus the squared magnitude of the CCF, a relationship
attributed to Siegert. This relationship is of considerable
convenience for the following reasons. The CCF may be readily
estimated from complex data, from which the uniform contribution
may be derived as above. In Gaussian speckle, the contribution
given by (4.6) is absent, so that the non-uniform contribution
may be directly deduced by subtracting the uniform contribution
from the total ICF. Since the non-uniform contribution contains
all the information relating to the correlation properties
of the surface, this is a valuable and precise step in deducing
surface properties from coherent imagery. Due to the
availability of this reduction process, the non-uniform
contribution given by (4.4) is also referred to as the
correlation residue. Comparison with (4.1) shows that the

residue contains two contributions, one being incoherent, the
other being coherent. As before, only the incoherent
contribution relates to the surface.

5. ILLUSTRATIONS

   The decomposition outlined above is illustrated in Figures
1 to 4. Fig. 1 shows an example of a system response function
$T(r)$ and a surface covariance function $K_\sigma(w)$. The response
function (a) is Gaussian having poorer resolution in one
direction than the other. Gaussian response functions have been
found to be adequate approximate representations of response
functions in both SAR and sonar images since errors involved
in the estimation of correlation functions were larger than
sidelobe levels. The surface covariance function (b) is also
Gaussian, but has its axes rotated 45 degrees wrt the response
function. There is a large degree of correlation along one axis,
but little along the other. Fig. 2 shows the corresponding total
ICF $\Gamma_I(w)$; its decomposition is shown in Figures 3 and 4. The
uniform contribution (Fig. 3a) consists of the squared magnitude
of the coherent correlation function resting on a platform of
unit height. For Gaussian speckle, the residue obtained by
subtracting the uniform contribution from the total correlation
function constitutes the non-uniform contribution (Fig. 3b).
Only the correlation residue bears information about the surface.
The decomposition of the residue is illustrated in Fig. 4. For
Gaussian speckle, a contribution to the intensity correlation
function identical to the covariance function resulting from an
incoherent system having system response function $|T(r)|^2$ is
produced (Fig. 4a). The remaining contribution (Fig. 4b)
results from coherent imaging processes, and is generally a
complicated function of both system response and surface
covariance functions. The widths of the peak are governed by
the system resolution.

   Figures 5 and 6 illustrate total correlation functions and
correlation residues for two different types of surface. Fig. 5
shows correlation functions of an arable field imaged by SAR, in
which there is no significant correlation residue, indicating
that the surface is of uniform cross-section. Fig. 6 shows the
correlation functions of woodland imaged by SAR; the correlation
residue is distinct and contains all of the information
present in the ICF relating to the surface cross-section.
Not all woodland behaves in the way illustrated.

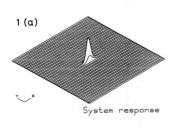

1 (a)

System response

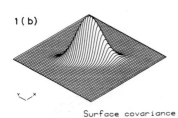

1 (b)

Surface covariance

Fig. 1 System response and
        surface covariance
        functions.

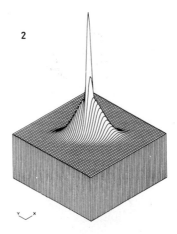

2

Fig. 2 Total intensity
        correlation function
        (ICF).

3 (a)

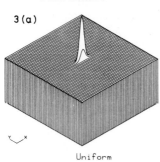

Uniform

3 (b)

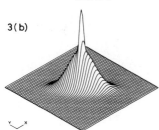

Non-uniform

Fig. 3 Uniform and non-uniform
        contributions to ICF.

4 (a)

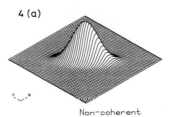

Non-coherent

4 (b)

Coherent

Fig. 4 Non-coherent and
        coherent contributions
        to non-uniform
        contribution.

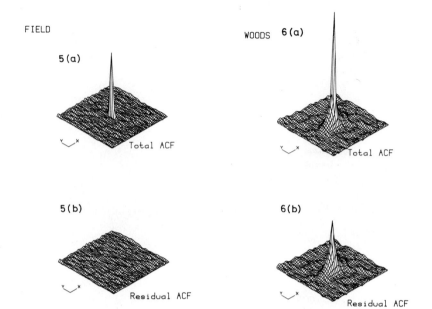

Fig. 5 Total and residual (non-uniform) correlation functions for a field.

Fig. 6 Total and residual correlation functions for a wood.

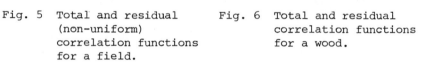

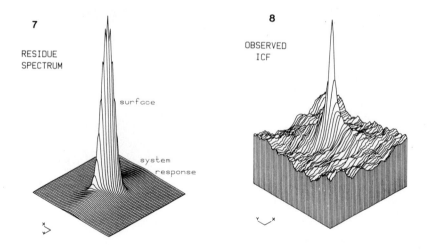

Fig. 7 Spectrum corresponding to a correlation residue.

Fig. 8 Observed ICF from sonar data, analysed in Figures 1-4.

Fig. 7 shows the spectral contribution corresponding to the correlation residue shown in Fig. 3b. The figure is rotated by 90 degrees to show the width of the central peak, which is due to correlation in the imaged surface. The broad contribution due to the system transfer function is much reduced relative to the surface spectrum. In practice, noise spikes present in estimated spectra prevent the situation being so clear-cut as that illustrated.

Finally, Fig. 8 illustrates an example of an intensity correlation function obtained from sonar data. The structure of the correlation function can be made more obvious by use of contour diagrams. The central portion of this ICF is well-modelled by the ICF shown in Fig. 2; hence, the decomposition of Figures 1-4 applies, and the covariance function of the surface being imaged is therefore given by that shown in Fig. 1. In this example, no attempt has been made to model the side-band structures also present in Fig. 8.

6. SPECTRAL REPRESENTATIONS

Given the convolutional form of equation (4.3) to (4.6) expressing the decomposition of the intensity correlation function it is comparatively straight-forward to derive the spatial spectrum, which is simply the Fourier transform of the ICF. Due to the convolution theorem of linear transform theory, the convolutional factors become simple multiplicative factors, so that the only problematical term occurring is the second term in (4.4), describing the non-uniform coherent contribution to the ICF.

The required contribution is most readily evaluated by performing a 2N-dimensional Fourier transform, one wrt ε and the other wrt w. Performing the transform wrt to ε, it is apparent that transforms of the following form are required:

$$\int d\varepsilon \left[T(\varepsilon+w/2) T^*(\varepsilon-w/2) \right] \exp(-2\pi i(\varepsilon,\eta)) \qquad (6.1)$$

where (ε,η) represents the inner product of ε and η. Transforms of this type are well-known in signal-processing theory, and are referred to as ambiguity functions (Cook and Bernfeld, 1967). Generally speaking, the non-uniform coherent contribution is more of nuisance value than of practical significance; however, if situations arise in which it adopts some intrinsic value, the natural context in which to analyse it is that of the theory of ambiguity functions.

Performing the remaining transform of the above contribution allows the spectral decomposition to be written in explicit form. In what follows, the tilde $\sim$ is used to indicate the Fourier transform of the corresponding spatial function. (For definition of the transform, see (6.1) above.)

$$\tilde{\Gamma}_I(v) = \tilde{\Gamma}_I^{(NU)}(v) + \tilde{\Gamma}_I^{(UN)}(v) + \tilde{\Gamma}_I^{(NG)}(v) \qquad (6.2)$$

$$\tilde{\Gamma}_I^{(NU)}(v) = \tilde{K}_\sigma(v) \left| \tilde{T}(v) * \tilde{T}^*(-v) \right|^2 \qquad (6.3)$$

$$+ \lim_{\eta \to 0} \tilde{K}_\sigma(\eta) * [\tilde{T}(v+\eta/2)\tilde{T}^*(v-\eta/2)] * [\tilde{T}(-v+\eta/2)\tilde{T}^*(-v-\epsilon/2)]^*$$

$$\tilde{\Gamma}_I^{(UN)}(v) = \delta(v) + \left| \tilde{T}(v) \right|^2 * \left| \tilde{T}(-v) \right|^2 \qquad (6.4)$$

$$\tilde{\Gamma}_I^{(NG)}(v) = <\sigma^2> \{<|v|^4>-2\} \left| \tilde{T}(v) * \tilde{T}^*(-v) \right|^2 \qquad (6.5)$$

The properties of the spectral contributions reflect those of the correlation function in a reasonably obvious manner. In particular, wide contributions to the correlation become narrow spectral contributions with enhanced amplitude, in obedience to the similarity theorem of linear transform theory.

7.   CONCLUSION

A brief account of the theory of the correlation properties of coherently imaged non-uniform surafces has been given. A much fuller account is given in (Miller, 1986). It has been indicated that correlation functions of interest may be completely described in convolutional form. This description is sufficiently explicit to permit direct evaluation of the correlation functions arising from surfaces having arbitrary covariance functions imaged by systems having arbitrary response functions. Such evaluation can be effected with reasonable efficiency using DFT techniques. When the surface covariance and system response functions have reasonably simple forms, the above analysis allows many of the contributions to be written down immediately, and shows which contributions are unlikely to be expressible in closed form. From an analysis of this type, a large number of techniques arise bearing on the analysis of both correlation data and on the problem of determining the surface covariance function given only intensity correlation data. A simple example of this has been given in the illustrations, demonstrating a natural technique for establishing whether or not the underlying surface is of uniform cross-section. Finally, we note that the entire analysis considered here stems from the multiplicative noise model given by equations (2.1) and (2.2); this naturally leads to a consideration of similar analyses devoted to establishing properties of other moments, such as single-point moments of various orders used in describing the univariate distribution of the intensity in a coherent image. A detailed account of the application of this type of model to the synthesis of coherent imagery is given in (Oliver, 1986).

8. ACKNOWLEDGEMENTS

The author wishes to express his gratitude to Dr. C.J. Oliver, of RSRE, Malvern, for his support and encouragement throughout the course of this work. The work was performed during a secondment at RSRE.

9. REFERENCES

Cook, C.E. and Bernfeld, M., (1967) "Radar Signals - an introduction to theory and applications". Academic Press.

Goodman, J.W., (1984) "Statistical Properties of Laser Speckle Patterns". In "Laser Speckle and Related Phenomena", Springer-Verlag.

Miller, R.J., (1986) "Correlation Properties of Coherently Imaged Non-Uniform Surfaces". Optica Acta (submitted)

Oliver, C.J., (1986) "The Interpretation and Simulation of Clutter Textures in Coherent Images". Inverse Problems (to be published).

BLIND DECONVOLUTION USING THE ZEROS OF THE Z-TRANSFORM

R.E. Burge and D.P. Lidiard
(Department of Physics, King's College London)

1. INTRODUCTION

1.1 Application of proposed technique

A significant proportion of recorded images is degraded, often by processes that are not known. Considerable effort has been spent in the last 10 years to devise digital techniques to remove degradations.

In order to remove a degradation something needs to be known about how it occurred. This paper is concerned with determining the degradation for incoherent imaging in the case where very little is known about it. The technique is intended for linear shift (or space) invariant blurs and is most easily implemented for blurs that take place in 1-D e.g. linear camera motion, or for 2-D blurs that can be reduced to a 1-D problem as in the case of circularly symmetric blurring functions e.g. out-of-focus images and images taken by long exposure, through atmospheric turbulence. The technique can also be extended to general 2-D degradations using a number of images similarly blurred.

1.2 Convolution

The recording of an image, be it by camera film or photoelectronic sensor, is, in general, a very complex process and can be represented mathematically by

$$g(x,y) = s(\phi(f(x,y))) \odot n(x,y) \qquad (1.2.1)$$

(see Hunt, 1976) where g is the recorded image, ϕ an operator that maps the radiant energy distribution of the object into an image e.g. an out-of-focus lens, s is a function that

transforms the energies in the image to a response in the
recording device e.g. logarithmic and saturation response of
film to light intensity, n is the noise in the device e.g.
due to film grain, and $\odot$ is an arbitrary operator combining the
noise with the recorded image. Following corrections for
non-linear recording, $g(x,y)$ is usually expressed with the
assumptions that the imaging is space-invariant with additive,
signal-independent noise. Hence in the case of incoherent
imaging (1.2.1) is simplified to,

$$g(x,y) = \int_{-\infty}^{\infty} \int_{-\infty}^{\infty} h(x-x_1, y-y_1) f(x_1, y_1) dx_1 dy_1 + n(x,y) \quad (1.2.2)$$

or in the Fourier domain,

$$G(u,v) = H(u,v) F(u,v) + N(u,v) \quad (1.2.3)$$

where $h(x,y)$ is the point spread function and G,H,F and N are
the 2-D Fourier transforms of g,h,f and n respectively. It
should be noted that whilst eqn. (1.2.2) is not the case in
reality, degradations can successfully be removed on a
pragmatic level by assuming the model of eqn. (1.2.2) with
added constraints on the form of h,f and n.

1.3 Blind deconvolution

Information is required about the nature of h, the P.S.F.
(point spread function), and n, the noise, before the image
can be restored. If these are not known a priori then they
have to be determined from the degraded image itself. Point
objects such as stars, or straight edges in images, are often
used to determine the P.S.F. More complex techniques (Slepian,
1967) look for a pattern of real zeros in the Fourier domain
when the SNR (signal to noise ratio) is very high, to
determine the nature and parameters of the P.S.F. in the few
cases where they can be determined from the real zeros in the
Fourier domain e.g. uniform motion blur and out-of-focus blur.
Since noise can often frustrate this process several techniques
segment the image into blocks about double the extent of the
P.S.F. and so that there are sufficient image blocks to
average out variations in the undegraded image to leave the
signature of the P.S.F., (Cannon, 1976, Stockham et al, 1975).
Essentially the technique suggested here also uses a number of
image subsections in order to pick out the common signature of
the P.S.F. except that the image subsections are chosen to
highlight this common information.

2. DETERMINING THE POINT SPREAD FUNCTION

2.1.1 *z-Transform and its zeros*

Consider a one-dimensional sequence $\{f_k; k = 0, N-1\}$ then the z-transform of f_k can be defined as

$$F(z) = z^p \sum_{k=0}^{N-1} f_k z^k \qquad (2.1.1)$$

where $z = x + iy$ and p is a constant depending upon the origin of the z-transform. The z-transform is essentially a model for analytically continuing the Fourier domain into the complex plane in a digital sense. The Paley-Wiener theorem states that the analytic continuation of a finite Fourier transform into the complex plane is an entire function of exponential type. If the finite Fourier transform is of even support i.e. $[-a,a]$ then the entire function is completely specified by an infinite product of zeros, the Hadamard product, and a multiplicative constant. However in the same way that the digital representation of a Fourier transform, in practice, only has a finite number of terms corresponding to the important frequencies so only a finite number of zeros is used containing the important frequency information. Mathematically then,

$$F(u) = \int_{-a}^{a} f(t) e^{iut} dt \qquad (2.1.2)$$

is digitally approximated by,

$$F(u) = \sum_{k=-[(N+1)/2]}^{[(N+1)/2]-1} f_k e^{2\pi iku/N} \qquad (2.1.3)$$

where square brackets denote *the integer part of* and $f(t)$ has been sampled N times within $[-a,a]$ at regular intervals to give f_k and there are $N/2$ important frequencies with the rest being considered zero. Eqn. (2.1.2) is continued into the complex plane by,

$$F(z) = \sum_{k=-(N+1)/2]}^{[(N+1)/2]-1} f_k z^k = F(0) \prod_{j=1}^{N-1} (1 - \frac{z}{\xi_j}) \qquad (2.1.4)$$

where $\{\xi_j\}$ are the complex zeros of the z-transform.

Comparing equation (2.1.3) and (2.1.4) it is readily seen that the Fourier domain becomes the unit circle in the complex

plane i.e. for $|z| = 1$ or $z = e^{i\theta}(\theta \in R)$. Hence the zeros for
the z-transform of a constant image $\{f_k = c; k = 0, N-1\}$ are
equally spaced around the unit circle corresponding to the
zeros of a sinc function $(\sin(\pi x)/\pi x)$. The zeros as described
by eqn. (2.1.4) are zeros in complex Fourier space and so each
zero encodes complex frequency information. Several concepts
relating to the zeros that shall be used later on are outlined
here for reference.

2.1.2 The fundamental zero lattice

Referring to eqn. (2.1.3) the Fourier transform values of
$\{f_k\}$ are calculated by setting $z = e^{2\pi i j/N}, j = 0, N - 1$. This
corresponds to N points equally spaced around the unit circle.
For the z-transform of a constant signal the zeros lie at each
of these point except for the one at $z = 1$ showing that the only
frequency information present is in the D.C. term. These zero
positions are termed the fundamental zero lattice or just the
lattice positions. If the zeros are now moved off these
positions then they introduce complex frequencies into the
signal. Fig. 1 shows the zero configurations for various types
of sequences $\{f_k\}$. The crosses denote the lattice positions for
a 16 long sequence with the circles denoting the zero positions
for when $\{f_k\}$ is a constant. For $\{f_k \in R; \forall k\}$ the zeros occur in
conjugate pairs i.e. at θ and $-\theta$, and for a symmetric sequence
$\{f_k = f^*_{N-k}; k = 1, N\}$, where (*) denotes complex conjugate,
the zeros occur in reciprocal pairs i.e. at $re^{i\theta}$ and $\frac{1}{r}e^{i\theta}$.

It is often convenient and easier to consider the zeros in
terms of a coordinate transformation $z \rightarrow e^{i\omega}$. If $\omega = u + iv$
and z is written in polar coordinates (r, θ) then $u = \theta$,
$v = -\ln(r)$. This transforms the unit circle into the abscissa,
the inside of the unit circle into the upper half plane, the
outside of the unit circle into the lower half plane, refer to
Fig. 1 noting the changes in the zero configuation for the
aforementioned types of sequences $\{f_k\}$.

2.1.3 Phase and magnitude values in terms of zeros

The magnitude of any frequency component can be found from

$$|F(e^{2\pi ik/N})| = \frac{F(0)}{\prod_{j=1}^{N-1}|\xi_j|} \quad \prod_{j=1}^{N-1}|\xi_j - e^{2\pi ik/N}| \qquad (2.1.5)$$

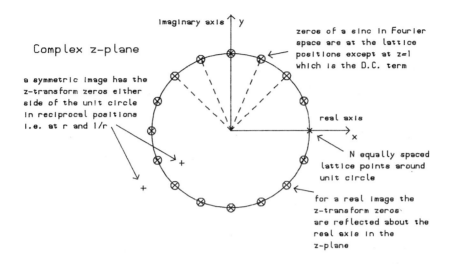

Complex z-plane

a symmetric image has the
z-transform zeros either
side of the unit circle
in reciprocal positions
i.e. at r and 1/r

zeros of a sinc in Fourier
space are at the lattice
positions except at z=1
which is the D.C. term

real axis

N equally spaced
lattice points around
unit circle

for a real image the
z-transform zeros
are reflected about the
real axis in the
z-plane

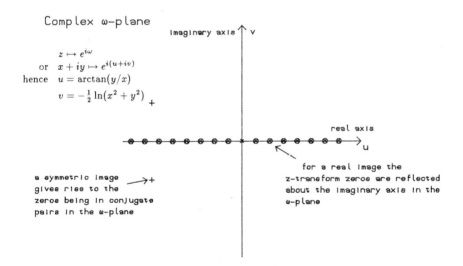

Complex ω-plane

$z \mapsto e^{i\omega}$
or $x + iy \mapsto e^{i(u+iv)}$
hence $u = \arctan(y/x)$
$v = -\frac{1}{2}\ln(x^2 + y^2)$

real axis

a symmetric image
gives rise to the
zeros being in conjugate
pairs in the ω-plane

for a real image the
z-transform zeros are reflected
about the imaginary axis in the
ω-plane

Fig. 1   Positions of fundamental lattice points and of the
         zeros of the z-transform of various types of sequences.

Geometrically this is interpreted as the product of the
magnitudes of the vectors from the kth lattice position to
each of the zeros. Similarly the phase of the kth Fourier
component is geometrically interpreted as the sum of the
angles of the vectors formed from the kth lattice position to

each of the zeros (refer to Fig. 2).

$$\text{phase}(F(e^{2\pi ik/N})) = \psi + \sum_{j=1}^{N-1} \text{phase}(\xi_j - e^{2\pi ik/N}) \qquad (2.1.6)$$

where ψ is a constant phase term depending on the multiplicative constant and the product of the zeros.

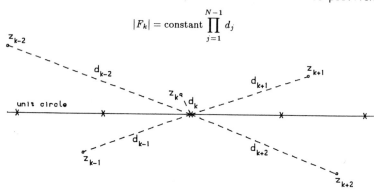

$$|F_k| = \text{constant} \prod_{j=1}^{N-1} d_j$$

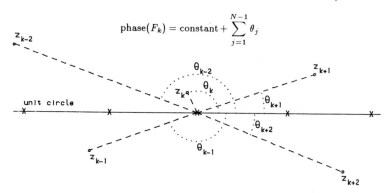

$$\text{phase}(F_k) = \text{constant} + \sum_{j=1}^{N-1} \theta_j$$

Fig. 2 Geometrical interpretation of calculation for the kth fourier magnitude and phase. A section of the unit circle (shown straight for simplicity) around the point for reconstructing the fourier component is shown with vectors to the 5 nearest zeros.

2.2 1-D formulation

The basis behind using the zeros of the z-transform is because, accounting for noise, the set of zeros of a blurred signal contains both those of the undegraded signal and the P.S.F. Rewriting eqn. (1.2.3) in terms of the z-transform,

$$G(z) = F(z)H(z) + N(z) \qquad (2.2.1)$$

Further, in terms of individual elements,

$$(g_0 + g_1 z + \ldots + g_{N+M-2} z^{N+M-2}) =$$

$$(f_0 + f_1 z + \ldots + f_{N-1} z^{N-1})(h_0 + h_1 z + \ldots + h_{M-1} z^{M-1})$$

$$+ (n_0 + n_1 z + \ldots + n_{M+M-2} z^{N+M-2}),$$

$$(2.2.2)$$

which may be written more succinctly as,

$$\sum_{k=0}^{N+M-2} g_k z^k = \sum_{j=0}^{N+M-2} \sum_{k=0}^{k=j} f_k h_{j-k} z^j + \sum_{k=0}^{N+M-2} n_k z^k \qquad (2.2.3)$$

where p is taken as zero in the definition of the z-transform in eqn. (2.1.1).

The convolution of the undegraded image with the P.S.F. is equivalent to multiplication of their z-transforms. If, for the present, the effect of noise is ignored and eqn. (2.2.1) is rewritten in terms of its zeros,

$$G(z) = G(0) \prod_{\ell=1}^{N+M-2} (1 - \frac{z}{\xi_\ell}) = F(0)H(0) \prod_{j=1}^{N-1} (1 - \frac{z}{\eta_j}) \prod_{k=1}^{M-1} (1 - \frac{z}{v_k}),$$

$$(2.2.4)$$

which shows that the zeros of $G(z)$, $\{\xi_\ell\}$ must contain those of $F(z)$, $\{\eta_j\}$ and of $H(z)$, $\{v_k\}$. So by finding the zeros of the z-transform of the degraded image it is possible to pick out a subset of zeros corresponding to the P.S.F., the remainder of zeros corresponding to the undegraded image.

To determine the correct subset of zeros for the point spread function, note that for a 1-D blur such as linear camera motion, each line of the image in the direction of the blur will be convolved with the same function. So the problem

is reduced to detecting a common subset of zeros for every
such line. Reconstruction from the zeros via the Hadamard
product gives the point spread function and the image. In the
case of excellent SNR (signal variance divided by gaussian
white noise variance), of, say, 2000:1 (33db) each set of zeros
can be compared to pick out those that are common and the
image restored. Fig. 3 shows an image which has undergone a
large degree of smear (25% of image width) due to a linear ramp
P.S.F. (uniform deceleration) in the horizontal direction with
the above SNR. The P.S.F. common to every line has been
picked out and is indistinguishable from the original. The
image was reconstructed from the zeros with a good visual
result, an MSE of 0.0376 and a SNR of 700:1. The reason for
the SNR being relatively low is due to the small amount of noise
as discussed in the next section.

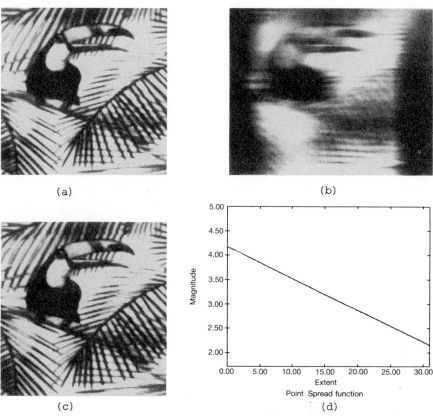

(a)   (b)

(c)   (d)

Fig. 3   Camera motion blur with SNR of 2000:1 and restored
         using the zeros of the horizontal lines. a) Original
         b) Horizontal motion blur c) Restoration from zeros
         d) Comparison of original and calculated P.S.F.s.

2.3 Effect of noise

For images with significant noise the zeros due to the lower frequencies remain close to their original positions but, for the higher frequencies, the zeros are significantly displaced in a random fashion. Three steps are taken to reduce this effect:

1. Blocks of lines are averaged. For large images, say 512 x 512 pixels, every 8 (say) consecutive lines can be averaged to reduce the SNR by about 3, assuming the signal level remains roughly constant.

2. Groups of lines are formed where the lines in each group are distributed evenly over the image. For example, considering the 64 lines that resulted from the averaging procedure above, 4 groups may be formed, with the 16 lines in each group being spaced 4 pixels apart. In this way the undegraded image zeros in the blurred lines should vary whilst those due to the P.S.F. should remain relatively fixed enabling the common zeros to be more easily identified.

For each group of zeros, one particular line of zeros is chosen and for each zero in this chosen line the closest zero in every other line in the group is found and the mean and variance of their distances from the zero in the chosen line found. Those zeros due to the blurring function should have small variances, with the mean position giving an improved estimate of the zero location when noise is present.

3. Once the zeros with the small variances have been picked out for each of the 4 groups then those zeros that are the same in 3 or all 4 groups to within a specified tolerance are chosen as being zeros of the P.S.F. and a finally averaged position for them can be found and the P.S.F. reconstructed. This ensures that if a zero that is not due to the P.S.F. happens to have a small variance in a group and is chosen wrongly as a P.S.F. zero then when compared with other blur zero sets in the other groups it will almost certainly not correlate with other zeros. Similarly if a P.S.F. zero has not been identified in one group it almost certainly will be in several others.

The image may then be restored by deleting those zeros in each line that are closest to the zeros corresponding to the P.S.F. that have been found but in practice this procedure has been found to lead to poor reconstructions. For this reason the proposed technique of using zeros is intended to find

the form and parameters of the P.S.F. to be followed by a
constrained, optimising, inverse method such as MAP (maximum
a posteriori) or MEM (maximum entropy method). Such iterative
methods include information about the type of noise present
in the image (usually an educated guess). Fig. 4 shows the
effect of noise on the variances of the closest zeros as
described above. For a blur with a very good SNR (say 2000:1)
the standard deviations corresponding to the zeros due to the
blurring function are all very small and easily distinguishable.
The lower diagram shows that with a SNR of about 100:1 the dips
in the standard deviations become less pronounced towards
higher frequencies and more difficult to pick out. In the
graph shown there was no averaging of the image lines as in 1
above before the zeros were taken so obviously this would
improve matters. Fig. 5 shows a reconstruction from the
zeros of an image which has been degraded by uniform linear
camera motion with a SNR of 100:1 (20db) showing that all
lines in the reconstructed image are affected by the noise.
The MSE for this image is 0.5131 with a SNR of 20. The low
SNR shows that the zeros reconstruction of the image is not
generally advisable although much of this is due to the fact
that the image was processed in lines and so does not have
much continuity in the vertical direction.

The reason for the reconstruction of the image using the
zeros being unstable in the presence of noise can be seen from
eqn. (2.2.1),

$$G(z) = H(z)F(z) + N(z) = \hat{H}(z)\hat{F}(z) \qquad (2.3.1)$$

where $\hat{H}(z)$ is the nearest approximation to $H(z)$ in terms of the
zeros of the blurred line $G(z)$. Since zeros corresponding to
the P.S.F are being removed rather than $G(z)$ being divided by
$G(z)$ there is no instability caused by $\hat{H}(z) \rightarrow 0$ in the
Fourier plane but obviously the noise is now being included in
both $\hat{H}(z)$ and $\hat{F}(z)$. Hence at high frequency where the noise
will tend to dominate, if $\hat{H}(z)$ is small then a large noise
term will be introduced into the reconstruction, $\hat{F}(z)$. To
avoid this the zeros are searched at high frequency to detect
any positions on the unit circle from which the image is
reconstructed where the nearest zero is an appreciable
distance away compared with neighbouring zeros. This is
because to a first approximation the distance of the nearest
zero from a reconstruction point is proportional to the
magnitude of the Fourier component at that point and hence the
further away the nearest zero is from the reconstruction point
the larger the Fourier magnitude at this frequency. If a
reconstruction point is found with its nearest zero appreciably
further away than its neighbours then this zero is brought

nearer the reconstruction point hence reducing the magnitude to a value comparable with its neighbours but retaining the same phase of the Fourier component. This is a similar scheme to that of the Wiener filter where the inverse filter is attenuated when the power spectrum of the noise becomes larger than that of the undegraded image.

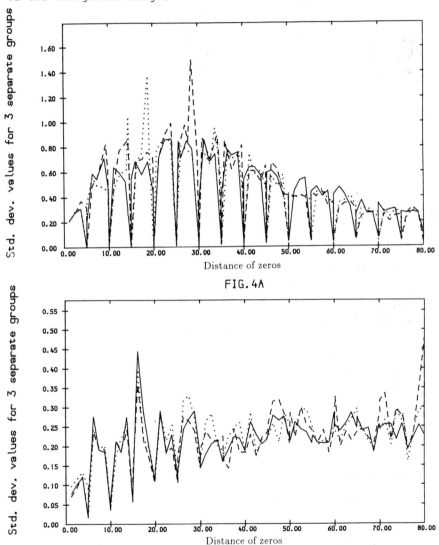

FIG. 4A

FIG. 4B

Fig. 4 The effect of noise on the std. dev. of the closest zeros to the comparison set of zeros for 3 different groups of zeros (solid, dash and dot lines). a) is for an image blurred by a rect with SNR of 2000. b) is the same but with an SNR of 100.

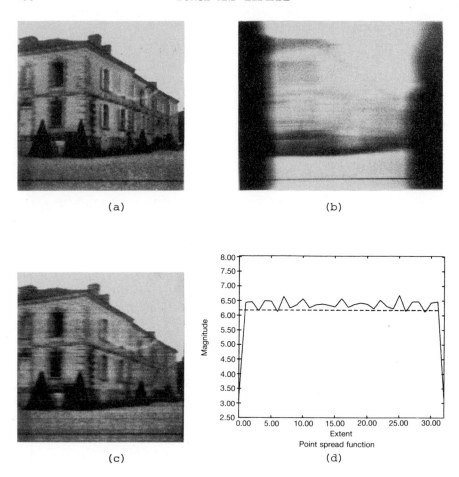

Fig. 5   Camera motion blur with SNR of 200:1 and restored
         using the zeros of the horizontal lines. a) Original
         b) Horizontal motion blur c) Restoration from zeros
         d) Comparison of original and calculated P.S.F.s.

2.4   2-D formulation

2.4.1   Projection sampling and reconstruction

In order to extend this process to 2-D the case of
circularly symmetric P.S.F.s was considered such as out-of-
focus camera blur and long exposure atmospheric turbulence
(assumed to be gaussian shaped). In these cases the projections
at all angles are blurred by the same P.S.F., namely the
projection of the circularly symmetric P.S.F., and hence the
problem can be reduced to 1-D as before.

Since images are normally sampled onto a square array of pixels, it is necessary usually to take projections from the grid sampled image and reconstruct the image in as accurate a manner as possible. Projections were obtained by taking adjacent parallel rays, at a certain angle, of equal thickness through the plane of the image which was itself considered to be made up of square blocks, centred on each grid sample point, of uniform intensity throughout each block and of value given by the intensity of the corresponding sample point (see Fig. 6). The areas of the blocks (or pixels) segmented off by a ray were summed and this was taken as the projection value. Considering all such rays gave the projection of the image at a particular angle. Using this method gave a constant sum of values in each projection, which is desirable. Also the method was found to reproduce the image from its projections, when used with the interpolation technique described below, with an MSE (mean square error) about ten times smaller than when each sample point was projected onto the projection axis and then sinc (squared) interpolated onto the projection sample points as in Fig. 7.

In order to reconstruct the image from projections the direct inverse Fourier transform method due to Stark and Sezan, (1984) was used. This is an exact interpolation of the polar sampled points onto the grid points used by a 2-D FFT in the case where the image is both radially and angularly bandlimited and has been properly sampled. For reference the equations used to interpolate a grid point in Fourier space, $M(\rho,\phi)$, for digital reconstruction from the polar values, $M\left(\dfrac{n}{2A}, \dfrac{\pi k}{K+1}\right)$, of the Fourier transform of the projections are,

$$M(\rho,\theta) = \sum_{n=(n_\rho - L_\rho)}^{n_\rho + L_\rho} \; \sum_{k=(k_\phi - L_\phi)}^{k_\phi + L_\phi} M\left(\frac{n}{2A}, \frac{\pi k}{K+1}\right) \; \mathrm{sinc}\left(2A\left(\rho - \frac{n}{2A}\right)\right) \sigma\left(\theta - \frac{2\pi k}{K+1}\right) t(n)$$

(2.4.1.)

where,

$$\mathrm{sinc}(x) = \frac{\sin(\pi x)}{\pi x} \quad , \quad \sigma(\phi) = \frac{\sin((K+1)\phi)}{(2K+2)\sin(\phi/2)}$$

(2.4.2)

K+1 is the number of projections used between 0 and π, n_ρ = [2Aρ], k_ϕ = [(K+1)ϕ/π] where square brackets denote the nearest integer and are the subscripts of the nearest polar sample point to the grid point being interpolated and 2A is the

diameter that bounds the object in space. $L_\rho = 2$, $L_\phi = 1$
were used so that the nearest 15 polar sample points to the
grid point were used to estimate its value. ρ and θ are the
usual radial and angular distances of a point in Fourier
space. Hence $2A\rho$ is the distance to a grid point in terms of
pixel distances i.e. $1/2A$ is the distance between radial
samples.

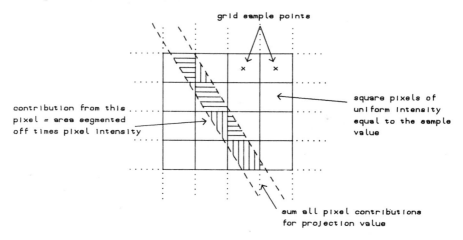

Fig. 6

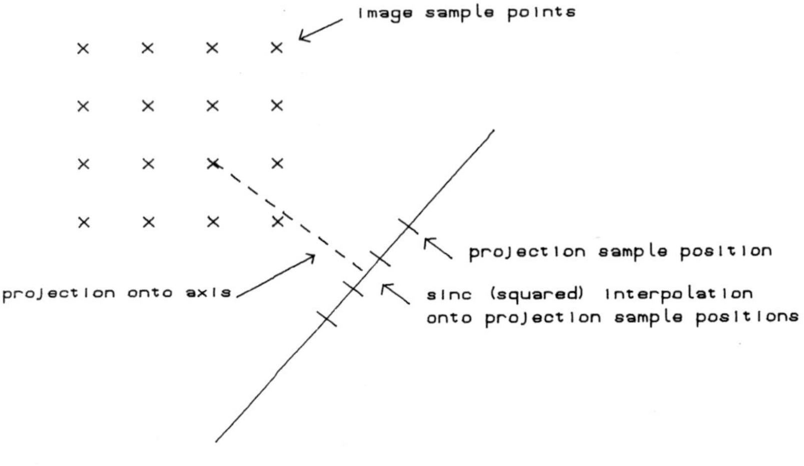

Fig. 7

Figs. 6 and 7   Method of calculation of projections using
                image blocks and sinc squared interpolation from
                a previously grid sampled image.

$$t(n) = \left[\max \left(1 - \frac{|n|}{T}, 0 \right) \right]^2, \; n=0,\pm1,\pm2,\ldots$$

(2.4.3)

which is a radial windowing function included to reduce the
effects of truncating the interpolating function caused by using
a limited number of polar samples. T is taken to be 10.

This interpolation formula is essentially sinc interpolating
along radial lines and using $\sigma(\phi)$ to interpolate along angular
lines, and is very similar to sinc interpolation, in order to
estimate a point not on the polar sampled grid. For
comparison of the method with filtered back projection see
Stark et al., (1981).

2.4.2 Deconvolution in 2-D

As a demonstration images were blurred with various
circularly symmetric P.S.F.s approximated onto a square grid
and then convolved using zero-padded 2-D FFTs in the usual way.
Projections were taken as previously described and the problem
reduced to 1-D. Because the above process does not exactly
model the true projections of an image blurred by a circularly
symmetric P.S.F. then noise was effectively introduced. Fig.
8 shows reconstructions for a couple of P.S.Fs where ringing
has occurred due to the noise.

2.5 General 2-D convolutions

It is well known that a 2-D convolution can be written as
a 1-D convolution, (see MacAdam, 1970 for example), and hence
in theory if a number of different images is degraded through
the same optical system then the problem can again be reduced
to 1-D. The problem in this case is the sheer size of the 1-D
vectors which the zero finding algorithm just cannot cope with.
An alternative approach is to consider the image in Radon space
i.e. in terms of its projections. For N projections the problem
would be reduced to determining the P.S.F. projections by
comparing projections at the same angle in each of the images
and picking out the projection zeros due to the P.S.F. at that
angle and then repeating this procedure N times to get the
projections at other angles.

In the case of atmospheric turbulence usually about 100
images of a stellar object all of short exposure (about 2s) are
summed together to obtain a long exposure image. The P.S.F. for
short exposure images consists of random phases which when
summed over many images tend to a constant giving rise to the
P.S.F. resembling a circularly symmetric gaussian function.
However in the process high frequency detail is often lost

which is of importance to astronomers. An alternative approach
is to process the individual short term exposure images in Radon
space as outlined above except now it is the projections of the
stellar object that remain constant and the P.S.F. projections
that vary from image to image at any given angle. So this is a
case where the zeros of the image rather than the P.S.F. are
picked out and hence the image can be directly reconstructed
without the need for MEM or MAP. Simulations of this nature
are currently being performed.

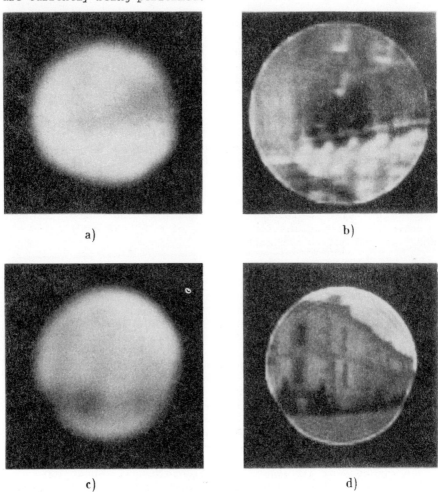

a)   b)

c)   d)

Fig. 8   Images blurred by circularly symmetric point spread
         functions and restored using the zeros of their
         projections. a) Out-of-focus blur b) Restoration from
         the zeros of the projections c) Turbulent atmosphere
         (gaussian shape) blur d) Restoration from the zeros of
         the projections.

3. CONCLUSIONS

A method has been demonstrated, in the context of incoherent imaging, for determining the linear shift-invariant P.S.F. of a degraded image in the case where it is not known. The technique is most readily applied to images with linear 1-D blurs with very good SNR and uses the zeros of the z-transform. An averaging and comparison scheme has been proposed to pick out the common set of zeros due to the P.S.F. when noise of up to about 2% or SNR of 50:1 (17db) is present. Once the P.S.F. has been found it is recommended to use a constrained technique such as MAP or MEM to restore the blurred image since a reconstruction from the remaining zeros can become dominated by noise. The technique also works for 2-D circularly symmetric P.S.F.s degrading images. Finally it was shown how general 2-D P.S.Fs could be tackled with large (~ 100) images and how in the case of short exposure atmospheric turbulent images the zeros yield directly the projections of the required restored stellar image.

4. ACKNOWLEDGEMENTS

D.P.L. gratefully acknowledges an SERC studentship and also support from Perkin-Elmer Ltd. through the CASE scheme.

5. REFERENCES

Hunt, B.R., (1976) "Some remaining mathematical problems in nonlinear image restoration", Image Science Mathematics – Symposium, pp. 123-139.

Slepian, D., (1967) "Restoration of photographs blurred by Image Motion", *Bell Sys. Tech. Jou.*, Vol. 46, No. 2, pg. 2353-2362.

Cannon, M., (1976) "Blind deconvolution of spatially invariant image blurs with phase", *IEEE Trans. on ASSP*, Vol. 24, No. 1, pg. 58-63.

Stockham, Jr. T.G., (1975) "Blind deconvolution through digital signal processing", *Proc. IEEE*, Vol. 63, pg. 678-692.

Sezan, M.I. and Stark, H., (1984) "Tomographic image reconstruction from incomplete view data by convex projections and direct Fourier inversion", *IEEE Trans. on Medical Imaging*, Vol. 3, No. 2, pg. 91-98.

Stark, H. et al., (1981) "An investigation of computerised tomography by direct Fourier inversion and optimum interpolation", *IEEE Trans. on Biomed Eng.*, Vol. 28, No. 7, pg. 496-505.

MacAdam, D.P., (1970) "Digital image restoration by constrained deconvolution", *J. of Opt. Soc. of America,* Vol. 60, No. 12, pg. 1617-1627.

THE RESAMPLING OF SYNTHETIC APERTURE RADAR IMAGES

D. Blacknell

(Marconi Research Centre, Great Baddow, Chelmsford, Essex)

ABSTRACT:

This paper discusses the problems involved in trying to define an optimum procedure for resampling synthetic aperture radar (SAR) images, and suggests ways in which these problems can be resolved. Although described in terms of the SAR image, the results are equally valid for any time series which satisfies the assumed properties. It is shown that the finite nature of the image does not allow the formulation of an exact resampling method. For this reason, assumptions about the nature of the image outside its finite region are considered necessary, (this being shown to be equivalent to estimating the spectrum of the image). Two types of assumption are investigated. The first type involves deterministic assumptions, which, although not robust under small changes, can produce resampling processes which are easily implemented. The second type of assumption uses statistical methods to overcome the problem of robustness under small changes. This provides a valid approach to the problem and leads to a justification of truncated sinc resampling when the assumption of uniformity outside the image applies.

1. INTRODUCTION

A SAR processor can be assumed to act as a bandpass, linear filter, which means that the complex signal output from such a processor can be taken to be, to a good approximation, band-limited. By sampling the complex signal at a grid of sample points (normally a rectangular grid) the complex SAR image is formed. If this complex signal is square-law detected, that is, its in-phase and quadrature components are squared and added, a real, non-negative, band-limited signal results, which can also be sampled at a grid of sample points to form the intensity of (or detected) SAR image. Both these types of

image map, for a given imaging geometry, the radar reflectivity
of a region of the Earth's surface.

It is sometimes required to represent an image with respect
to a grid of sample points which does not coincide with the
grid at which the signal was originally sampled. The process
by which this is achieved is known as the resampling of an
image.

Some examples of situations in which resampling is required
are as follows:-

(i)   Image-Image Matching: To compare two images (perhaps
        obtained by different sensors) of the same region of
        ground it is required that their grids are coincident.
        If this is not the case, then it is necessary to resample
        one of the images onto the grid of the other so that
        image-image matching can be performed.

(ii)   Image-Map Matching: It may be desired to compare an image
        with a map of the imaged region in order to identify
        features in the image. However, in general, the imaging
        geometry will not be the same as the geometry of the
        required map projection and so a grid of sample points
        must be set up which corrects for this. The image can
        then be resampled onto this new grid and comparison made
        between the image and the map.

(iii)   Correction of Geometric Distortion: Extraneous motions of
        the SAR platform can cause geometric distortions in the
        image of a scene. If these distortions can be measured
        then they can be removed by resampling the image onto a
        suitable undistorted grid. (A discussion of geometric
        distortion can be found in [4].)

(iv)   Radargrammetry: The analysis of the vast amounts of SAR
        data which will be produced every day by future satellite
        missions such as ERS-1 involves the transformation of the
        raw SAR imagery into a form compatible with Geographic
        Information Systems and their spatial databases. This
        transformation will produce a final image which is free
        from geometric and radio-metric distortions and is
        located and oriented in a suitable map projection, a
        process which is known as radargrammetry.

        Resampling is an essential part of radargrammetry as it
        is the means by which the raw SAR imagery is interpreted
        in terms of the transformed image space. (The concepts
        of radargrammetry are described in more detail in [5].)

If a band-limited signal is sampled at a high enough rate (the Nyquist sampling rate) then it can be exactly reproduced from its sample values. This is the essential conclusion of the Shannon sampling theorem and is central to much of the analysis to follow. In order for the Shannon sampling theorem to be applied, the Nyquist sampling rate must be maintained over an infinite length of time. For actual complex images this can never be achieved and so the theoretical optimum method cannot be fully implemented in practice, since the original complex signal is not well-defined by the finite set of sample values. In Section 2 the consequences of the finite nature of the image are discussed and it transpires that any resampling process produces resampled values which can be shown to be consistent with the fact that the original complex signal was a finite-energy, band-limited function. Having thus established that consistency with the known information is a property of any resampling process, the paper goes on to consider what assumptions can be made about the form of the image outside the finite region, in order to determine if the optimality of any of the possible resampling procedures can be justified. By making such assumptions prior knowledge is being introduced into the problem. Various forms of prior knowledge are available (see, for example, Bertero et al. [6] and Luttrell [7]) but just two specific forms will be considered in this paper. In Section 3 the deterministic method of assuming values for all the unknown samples is described but it is suggested that this assumtion can be too restrictive. However, the statistical approach of Section 4 proves to be less restrictive and succeeds in producing a justification for using the truncated sinc resampling method when certain assumptions can be made. The main results of the paper are summarised in Section 5.

The properties to be assumed for the continuous SAR signal, which is sampled to form the SAR image, are that it is,

(1)   complex-valued,

(2)   band-limited,

(3)   of finite energy.

This signal will be referred to as the original signal.

The properties to be assumed for the SAR image, which is a sampled version of the original signal, are that it is,

(1)   finite in extent,

(2)   regularly sampled at the Nyquist rate over the finite interval.

The quantization of the SAR image will not be taken into account, so that it will be assumed that sample values can take any complex values, subject to the upper bound on the modulus which is imposed by the finite energy condition.

2. CONSISTENCY WITH THE DATA

An essential requirement of any resampling procedure is that the resampled values it produces are consistent with all the information available. This means that the resampled values must be sample values of a signal which passes through the original sample points and which has the properties required by an original signal as described in the Introduction.

To examine the criterion of consistency with the data the following result will be used.

Theorem

Let S_B be the set of all finite-energy, band-limited, complex-valued function with bandwidth B. Consider a finite time interval $(-T,T)$ which contains R complex sample values taken at the distinct time instants t_r, where R is any finite positive integer and the t_r are not necessarily regularly spaced.

Then there are an infinite number of functions belonging to S_B which pass through the points (a_r, t_r), $r=1, \ldots, R$.

The Shannon sampling theorem [1] will be used in the proof of this theorem and so will be stated here as a reminder (for a discussion of the Shannon sampling theorem see [2]).

The Shannon Sampling Theorem

Let $f(t)$ be a complex-valued function with frequency response $F(\omega)$ and with bandwith $2\omega_o$. Then $f(t)$ is uniquely defined by the values $f(n\pi\omega_o)$, where n is an integer, and can be reconstructed from these sample values by the equation,

$$f(t) = \sum_{-\infty}^{\infty} f(n\pi/\omega_o) \; \text{sinc} \; (\omega_o t/\pi - n) \qquad (1)$$

$$\text{where sinc } \theta = \frac{\sin (\pi\theta)}{(\pi\theta)}$$

A proof of this theorem can be found in [2].

The energy of $f(t)$ is given in terms of the sample values by,

$$E = (\omega_o/\pi) \sum_{-\infty}^{\infty} |f(n\pi/\omega_o)|^2 \qquad (2)$$

Proof of the Theorem

The following is an heuristic proof of the the theorem which illustrates the basic idea. (A rigorous proof is available but has not been included in this paper as it is not essential for understanding what follows.)

For convenience, and without loss of generality, assume the bandwidth to be 2π.

A set of complex numbers associated with the time instants $t = n$, where n is an integer, ($f(n)$ say) which have the property:

$$\sum_{-\infty}^{\infty} |f(n)|^2 < \infty$$

defines a unique member of $S_2\pi$, this being precisely

$$f(t) = \sum_{-\infty}^{\infty} f(n) \ \text{sinc} \ (t-n) \qquad (3)$$

(This is a result of the Shannon sampling theorem.)

Now it is required to find a member of $S_{2\pi}$ which passes through the points (a_r, t_r). Such a member is defined by the set of complex numbers $\{f(n): n \text{ an integer}\}$ provided.

$$a_r = f(t_r) = \sum_{-\infty}^{\infty} f(n) \ \text{sinc} \ (t_r-n), \qquad (4)$$

for $r = 1, \ldots, R$.

These R equations represent R constraints on the possible values of $\{f(n): n \text{ an integer}\}$. Thus all but R of the numbers $f(n)$ can be assigned arbitrarily and it will still be possible to satisfy these constraints by solving the set of R simultaneous equations for the remaining R numbers. The finite energy condition can be easily satisfied by ensuring that the arbitrarily assigned values are modulus squared convergent. This still allows an infinite number of choices for all but R of the $f(n)$, and, since each choice defines a different function with the desired properties, there are infinitely many such functions.

This completes the proof of the theorem.

The problem of ensuring consistency with the data when resampling a complex SAR image can be stated as follows. Given a set of complex values a_n corresponding to times t_n, n=1, ..., N, i.e. a set of points (a_n, t_n), find, or estimate, the values a_m corresponding to the times t_m, m = N+1, ..., N+M subject to the conditions that the points (a_n, t_n), (a_m, t_m) must all lie on a finite energy, band-limited function with bandwith $2\omega_o$. If this condition is achieved then the resampled points are consistent with all the known data, and no error can be calculated. If this condition is not achieved then the resampled points are not consistent with the known data and it is therefore sensible to speak of an error.

Now, in practice, N and M are finite. Thus by enumerating the points (a_n, t_n), (a_m, t_m) (n=1, ..., N; M = N+1, ...N+M) as (a_r, t_r) theorem directly. This means that the points (a_n, t_n), (a_m, t_m) always lie on a member of $S\omega_o$, whatever the values of a_n. This is precisely the consistency condition as defined above, which means that any resampling procedure is consistent with the known data.

Having reached a point where it has been shown that, subject to the known information, any resampling process is valid, it is necessary to attempt to assess which resampling processes are more sensible, or in other words, which resampling processes incorporate a greater physical understanding of the original imaging process. Due to the fact that no more information is available it should be stressed that it is not being stated which processes are better; all that can be done is to assume further constraints on the problem in order to define a smaller set of possible solutions, and to discuss whether these assumptions are sensible or not. In doing this, great care should be taken that the assumptions made are not forcing the resampled image to appear as the resampler thinks it ought to appear rather than how it should in fact appear. This involves imposing the least amount of further structure on the image, i.e. making as few assumptions as possible.

The assumptions made would be concerned with the behaviour
of the undefined sample values. In the time domain the
consequences of these assumptions could be examined by means
of the reconstruction formula

$$f(t) = \sum_{-\infty}^{\infty} f(n) \text{ sinc } (t-n) \text{ (see equation (3)).}$$

This will be done in the next two sections for two types of
assumptions on the nature of the undefined samples.

It should be noted that f(t) has frequency response

$$\hat{f}(\omega) = \sum_{-\infty}^{\infty} f(n) \exp (-in\pi\omega) \quad |\omega| < \pi \qquad (5)$$

$$= 0 \text{ otherwise}$$

which shows that making assumptions concerning the undefined
samples amounts to estimating the frequency spectrum from a
finite time series. This problem is widely encountered in the
physical world and a review of the various methods which have
been developed over the years to solve it is given by
Robinson [3]. Thus, an investigation of these techniques may
provide valuable insight into the problem of resampling,
although the approach has not, as yet, been considered
further.

3.   ASSUMPTIONS - SPECIFIC VALUES

What assumptions on the behaviour of the undefined samples
are possible?

The most obvious answer is the assignment of specific
values to every undefined sample thereby defining a single
band-limited function (of finite energy if the sample values
are chosen suitably) which can be used for resampling. Two
examples of this type of assumption are as follows:-

(a)   Periodic repetition of the sample values.
      A computationally convenient assumption is that of
      periodicity of the sample values over the infinite
      time interval, the spectrum then being given by the
      fast Fourier transform, (FFT)

Example 3.1

Let the bandwidth be 2π $(\omega_o=\pi)$ without loss of generality.
Then the Nyquist sampling rate is unity and the image can

be described as set of values $f(n)$, $|n| < N + \frac{1}{2}$, i.e. $2N + 1$ samples in an interval of length $2N + 1$, a sampling rate of unity. Now assume that the samples repeat periodically with period $T = 2N + 1$, that is

$$f(n + mT) = f(n).$$

Then $f(t + mT) = \sum_{n=-\infty}^{\infty} f(n) \frac{\sin (\pi(t + mT - n))}{\pi (t + mT - n)}$

$$(6)$$

(using the Shannon reconstruction formula).

Let $n' = n - mT$ so that

$$f(t + mT) = \sum_{n'=-\infty}^{\infty} f(n' + mT) \frac{\sin (\pi(t-n'))}{\pi (t - n')}$$

$$= \sum_{n'=-\infty}^{\infty} f(n') \frac{\sin (\pi(t-n'))}{\pi (t - n')}$$

$$= f(t)$$

Thus the function represented by the samples repeats periodically with the same period. If $f(t)$ is expanded as a Fourier series the frequencies present will be of the form $\omega_m = 2\pi m/ (2N + 1)$, i.e. those of period $(2N + 1)/m$. However, since $f(t)$ is band-limited to $[-\pi,\pi]$ it must be of the form

$$f(t) = \sum_{m=-N}^{N} C_m \exp \left(\frac{2\pi mit}{(2N + 1)} \right) \qquad (7)$$

Thus $f(\omega)$ consists of a finite set of delta functions at points $\omega = 2\pi m/(2N + 1)$, $|m| \leqslant N$ with weights C_m.

Now, substituting in the $2N + 1$ known sample values, we get

$$f(n) \sum_{m=-N}^{N} C_m \exp \frac{(2\pi imn)}{(2N+1)} \quad |n| \leqslant N \qquad (8)$$

which gives $2N + 1$ equations for the $2N + 1$ unknowns, C_m. The FFT essentially provides an efficient method for solving these equations and determining C_m.

Once the C_m are known resampling is just a matter of evaluating

$$f(t) = \sum_{m=-N}^{N} C_m \exp \frac{(2\pi mit)}{(2N + 1)}$$

at the relevant time instants.

It should be noted that if the sample values of $f(t)$ are known at any $2N + 1$ points in the interval $[-N - \frac{1}{2} < n < N + \frac{1}{2}]$ a set of simultaneous equations can be set up which completely determine the C_m and thus completely determine $f(t)$. Thus the set of resampled values (providing there are at least $2N + 1$ of them) can be used to reconstruct $f(t)$ exactly. This means that, if the same process is used to resample from the newly resampled points back onto the original sample points, the original sample values will be retrieved.

Not all resampling processes possess this property (e.g. see (b) below) and since the ability to retrieve the original sample value is desirable this gives the method of assuming periodic samples an advantage over those methods which are not reversible.

(b) Zero sample values outside the finite image.
     Let the image be

$$f(n), \quad |n| < N + \frac{1}{2}$$

as before.

Then assume $f(n) = 0 \quad |n| > N$.

This then gives an image reconstruction

$$f(t) = \sum_{-N}^{N} f(n) \frac{(\sin \ (\pi \ (t-n))}{(\pi (t-n))}.$$

Note how the assumption of zeroes has turned an infinite sum, which cannot be practically calculated, into a finite sum which can be evaluated.

However, it should be remembered that having zero sample values outside the image does not imply that the image reconstruction is identically zero outside the image. Thus if resampling is performed onto the points $(n + \alpha)$, $|n| < N + \frac{1}{2}, \alpha < \frac{1}{2}$ then assuming $f(n + \alpha) = 0$ for $|n| > N$ will not produce the same interpolation function and so an attempt to resample back onto

the original sample points will not produce original sample
values. In other words the resampling process is not
reversible in this case.

In Section 2 it was pointed out that the assumptions made
should be as unrestrictive as possible. However, assumptions
(a) and (b), by assuming values for all the unknown samples,
reduce the number of possible original functions from infinity
to one, which seems to show them to be extremely restrictive.
That this is the case can be shown by considering the changes
that can occur within the image if each of the assumed sample
values is allowed to change by an amount δ_n where $|\delta_n| \leqslant \varepsilon$. The
cumulative effect of this change at a point t_o would be an
added contribution of

$$c(t_o) = \sum_{|n| > N} \delta_n \frac{\sin (\pi(t_o - n))}{\pi(t_o - n)} \tag{9}$$

Example 3.2:

$$\text{Let } \delta_n = \varepsilon \ N < n < M \text{ and } n \text{ odd}$$

$$= 0 \text{ otherwise}$$

$$\text{Then } c(t_o) = \sum_{\substack{N<n<M \\ n \text{ odd}}} \varepsilon \frac{\sin (\pi(t_o - n))}{\pi (t_o - n)}$$

$$= \frac{\varepsilon \sin(\pi t_o)}{\pi} \sum_{\substack{N<n<M \\ n \text{ odd}}} \frac{1}{(n - t_o)}$$

Now $\sum_{\substack{N<n \\ n \text{ odd}}} \frac{1}{(n - t_o)}$ is divergent so the contribution can be made

as large as desired by taking M to be sufficiently high, that is
taking a sufficient number of terms. The number of terms
taken will, however, be finite and so the added contribution
to the energy will be finite.

This means that given any particular set of assumed values
for the unknown samples, a slight change in these values at

each point can produce significant changes in the values of
the resampled points. Now, even given that there is reason to
suppose that the undefined samples do take on certain values,
zero say, the evidence cannot be so strong as to exclude the
possibility that these assumed values can be changed by a
small amount, and equally valid, but vastly different results
be obtained. Consequently, the choice of specific values for
the undefined samples, although useful, can be considered
restrictive.

4. ASSUMPTION - STATISTICAL APPROACH

A second approach, which avoids the stringent specification
of values for all the unknown samples, is to assume a
probability distribution for the possible values the unknown
samples can take. The contribution of these samples to be a
resampled value will then also have a probability distribution.

For example,

let $f(n) \sim X_n$, $|n| > N$, where the X_n are independent and
identically distributed (iid).

Then the contribution of these samples to the resampling
value at t_o is,

$$Y_o = \sum_{|n| > N} X_n \frac{\sin (\pi(t_o - n))}{\pi(t_o - n)} \tag{10}$$

where Y_o is a random variable.

This allows a description of the effects of all possible
choices for $f(n)$, $|n| > N$ to be made. The description
incorporates an assumption as to the likelihood of the
different choices, i.e. the assumed distribution of X_n. Thus
the distribution of Y_o describes the differing probabilities
with which different contributions occur when the undefined
samples are generated randomly from the assumed distribution.

If the distribution of Y_o has a narrow peak at y_o then a
small ε can be chosen such that the probability that the
contribution from the undefined samples lies in the interval
$[y_o - \varepsilon, y_o + \varepsilon]$ is great.

Thus by choosing y_o as the resampled value it is possible to say that, under the assumed distribution for the unknown sample values, it is unlikely that the error in the resampling method is greater than ε.

Example 4.1

For the complex image assume that the real and imaginary parts of the undefined samples have zero-mean Gaussian distributions, with identical variances. Consider the real part of the contribution at t_o, which is

$$R_e(Y_o) = \sum_{|n| > N} X_n \frac{\sin(\pi(t_o - n))}{\pi(t_o - n)} \qquad X_n \sim N(0, \delta_o) .$$
$$\text{iid}$$

(11)

The variance of $R_e(Y_o)$ is given by:

$$\text{Var}(R_e(Y_o)) = \text{Var}\left(\sum_{|n| > N} X_n \frac{\sin(\pi(t_o - n))}{\pi(t_o - n)}\right)$$

$$= \sum_{|n| > N} \text{Var}(X_n) \frac{(\sin(\pi(t_o - n)))^2}{\pi(t_o - n)}$$

$$= \delta_o^2 \sum_{|n| > N} \frac{(\sin(\pi(t_o - n)))^2}{\pi(t_o - n)}$$

Now the function $g(t) = \dfrac{\sin(\pi(t_o - t))}{\pi(t_o - t)}$ has bandwith 2π

and so can be written in terms of its sample values as

$$\frac{\sin(\pi(t_o - t))}{\pi(t_o - t)} = \sum_n \frac{(\sin(\pi(t_o - n)))}{\pi(t_o - n)} \frac{\sin(\pi(t - n)))}{\pi(t - n)}$$

Thus putting $t = t_o$ gives:

$$1 = \sum_n \frac{(\sin (\pi(t_o - n)))^2}{\pi (t_o - n)}$$

$$\therefore \text{Var} (R_e(Y_o)) = \delta_o^2 (1 - \sum_{|n| < N} \frac{\sin (\pi(t_o - n)))^2}{\pi(t_o - n)}$$

(12)

Thus for large N this variance will be small and the probability density function of Y_o will be sharply peaked at zero, indicating that by using a zero value for the contribution it is highly probable that the actual contribution will be close to this (provided the distribution assumption is good).

A similar argument holds for the imaginary part and so by this statistical reasoning it is sensible to assume that for the complex image the contribution from the undefined samples is zero.

However, assuming the contribution from the undefined samples to be zero is equivalent to assuming all the undefined samples to be zero, which was earlier stated to be too restrictive. The resolution of this seeming anomaly is as follows. Although the assumption of zeroes gives a zero contribution to t_o, it is also true that the vast majority of other specific assumptions (when weighted by the assumed probability distribution) give contributions which are in the neighbourhood of zero. Thus although it is feasible to produce sets of assumptions which are close to zero at every sample point but which give a large contribution to t_o, the frequency with which these sets occur is much less than the frequency of occurrence of sets which produce a near zero contribution.

Thus, under the assumption that the undefined samples in the complex image are normally distributed with zero mean, the above constitutes a statistical justification for assuming zeroes as in assumption (b).

If the calculation of sinc coefficients for all the sample points in the image is computationally unfeasible then it may be desired to restrict the dependence of the interpolating function to a small number of sample values, e.g. those at the five sample points closest to the point to be resampled. If this is the case then for the purpose of the resampling method, it can be assumed that the values of the samples, other than the chosen five, are unknown.

The problem is then entirely equivalent to resampling an image consisting of five samples with an indication as to the distribution to be assumed being given by the statistics of the unused samples in the image. Thus, if the statistics of the unused samples are zero-mean Gaussian, then, by the reasoning given previously, it is sensible to assume zeroes for all the sample values except for the five being used for resampling. Under this assumption, the optimum method is, therefore, truncated sinc.

It should be noted that this result applies only to the complex image under the given assumption. An attempt to apply the truncated sinc method of resampling to a detected image may produce negative resampled values, which are not permissible.

A homogeneous region when imaged by a SAR system will in theory produce complex sample values which are distributed as zero-mean Gaussians. Thus the assumption in the preceding example amounts to stating that the region outside the imaged area could be assumed uniform. However, it may be more sensible to use the statistics of the image itself to estimate the distribution outside the image. Three possible ways of choosing the assumed distribution are as follows:-

(1)   The distribution can be defined irrespective of the values of samples within the image, as was done in the example.

(2)   The distribution can be entirely specified in terms of its moments estimated from the image.

(3)   A combination of the above two, for example assuming that distribution will be Gaussian and then estimating the mean and variance from the image.

An investigation of these assumptions should lead to the definition of an optimum method, with advantage that the assumptions are less restrictive than those used in the previous section.

5.   SUMMARY

The complex signal produced by a SAR processor is, to a good approximation, band-limited. A sampled version of this signal over a finite interval forms the complex SAR image. The first requirement of any resampling procedure is that it produces resampled values which, together with the image sample values, lie on a band-limited function with the correct bandwidth. This ensures that the resampling process is consistent with the band-limited property of the complex signal. It has been shown in Section 2 that this is true of any resampling procedure. Thus, for a resampling procedure, ensuring consistency with the data is not a problem.

To make progress in producing an optimum resampling procedure it is suggested in Section 2 that the assumptions should be made as to the behaviour of the signal outside the finite image. This was done in Section 3 using the deterministic approach of assuming specific values for all unknown sample points. However, it was found that this approach was too restrictive in the sense that small changes in the assumed values could produce large differences in the resampled values. For this reason a less restrictive approach using statistical methods was considered in Section 4. By assuming the unknown sample values to be described by a probability distribution it was possible to examine the behaviour of all possible choices of the unknown sample values. It was shown that the assumption of a homogeneous region outside the image could provide a justification for using a truncated sinc resampling method.

Thus the main conclusion of this paper is that a valid approach to the determination of an optimum resampling procedure is to consider the assumption of statistical properties for the undefined samples in the complex image. In particular, if it is reasonable to suppose that the region outside the imaged area is homogeneous, then the optimum procedure (as described in Section 4) is to use truncated sinc interpolation. This is, however, only applicable to a complex image for which the assumption is valid. Truncated sinc should not be used on an intensity image as it may produce negative resampled values.

6. REFERENCES

[1] Shannon, C.E., (1949), "Communications in the Presence of Noise", *Proc. IRE*, Vol. 37, pp 10-21,

[2] Jerri, A.J., (1977), "The Shannon Sampling Theorem - Its Various Extensions and Applications : A Tutorial Review", *Proc. IEEE*, Vol. 65, No. 11, pp 1565 - 1596,

[3]   Robinson, E.A., (1982), "A Historical Perspective of
      Spectrum Estimation", *Proc. IEEE,* Vol. 70, pp 885 - 907

[4]   Quegan, S., (1984), "Measurement of Geometric Distortion
      in Airborne SAR Images", *Proc. of IGARSS,* 84 Symposium,
      Strasbourg 27 - 30 August, pp 595 - 599.

[5]   Howard, P.D., (1985), "Radargrammetry : The Link Between
      Radar Imagery and Geographic Information Systems", Proc. of
      the International Conference of the R.S.S. and C.E.R.M.A.
      University of London, September, pp 319 - 329.

[6]   Bertero, M., Brianzi, P., Parker, P. and Pike, E.R.,
      "Resolution in Diffraction - Limited Imaging, a Singular
      Value Analysis III, The Effect of Sampling and truncation
      of the Data", *Opt. Acta,* Vol. 31, pp 181 - 201.

[7]   Luttrell, S.P., (1985), "Prior Knowledge and Object
      Reconstruction Using the Best Linear Estimate Technique",
      Opt. Acta, Vol. 32, pp 703 - 716.

INFORMATION EXTRACTION METHODS

J. Oakley and M. Cunningham
(Department of Electrical Engineering,
University of Manchester)

ABSTRACT

   This paper presents a model, based on functional analysis,
for image sampling, subsampling, resampling and reconstruction.
Some problems with the use of sampling theory for this purpose
are examined and an alternative methodology, involving an
operator model for the digital process, is proposed. This
operator maps a Normed Vector Space (NVS) of functions,
representing the image source, to another function NVS which
represents the image space. The digital error is
characterized by two metrics, a shift variant metric,
corresponding to aliasing error in sampling theory, and an
'inaccuracy' metric, which has no analogue in sampling theory
but gives a 'worst case' error. Both types of error can be
seen as deviations of the implemented operator from the
required operator in a conjugate NVS. An optimal digital
operator is described, which, given a fixed system of sampling
functionals, and given a specification of the required
operator, provides the most accurate image reconstruction. An
example of this operator is given for the case when the source
NVS is a Hilbert space.

1. INTRODUCTION

1.1 Background

   The term "digital image" generally refers to an array of
real or integer numbers which correspond to the grey level of
localised fragments or pixels of an image. In this paper,
digital images are regarded as continuous functions and a model
for digital processing operations is derived using functional
analysis. The basic idea is to treat the sampling,
processing and reconstruction of an image as a single

operation, rather than as simple interfacing steps. Two main
applications of this model are envisaged. The first application
is in the design of subsampling and resampling algorithms for
use in the processing of large pixel arrays. In Remote
Sensing one set of data acquisition hardware often serves diverse
applications, such as agriculture and geology, which have
different feature sizes. Rather than process an overdetailed
discrete image, the image is often subsampled, processed to
extract information or correct degradations, and then resampled.
This paper describes a new class of algorithms for this
purpose. The second application is in the design of 'matching'
image sampling and display hardware. If electronic displays
are designed to complement the characteristics of the image
digitizer, (or vice versa), image quality is improved without
any digital processing. The account here is necessarily brief -
a fuller description will be given in a future publication
(Oakley, 11).

1.2 Operator model

   Recent work on image sampling and reconstruction has been
mainly concerned with the image restoration problem. Given
samples from a blurred, noisy image field it is required to
compute a good estimate $f\hat{}$ of the image source f. One issue
which emerges in the restoration problem is how should $f\hat{}$ best
be represented computationally. In one approach, reported by
Darling et al (Darling, 3), the image source f is modelled as
a point in the Hilbert Space F. Information about f is
available in the form of the values $\{C_n\}$ taken by a finite set
$\{M_n\}$ of N Continuous Linear Functionals (CLF's) on F. By the
well known (Fréchet Riesz) representation theorem there are a
unique set of functions $\{m_n\}$ C F such that:

$$\forall n, \quad C_n = M_n(f) = \ <\ f,\ m_n\ >$$

This set of representers $\{m_n\}$ will be referred to as the
Sampling System. A minimum norm estimate $f\hat{}$ of f is constructed
in terms of the elements of the sampling system:

$$f\hat{} \ = \ \sum_{n=1}^{N} A_n m_n \qquad \text{where } A_n \in R \qquad (1)$$

By the projection theorem, $f\hat{}$ is the projection of f onto the
linear subspace spanned by the sampling system. Prior
knowledge about the form of f can be incorporated in the choice
of the H.S. F. For example the measure on the interval can be
selected to emphasize particularly important regions of the

function. This is equivalent to introducing a weighting
function p^{-1}, leading to an inner product:

$$< f, m > = \int p^{-1}.f.m$$

The function p is chosen to be similar to the expected f, and
this gives useful performance in the presence of noise. This
approach was related by the authors to regularization theory
(Miller, 4), used in ill-posed least squares problems.

Although not stated in (Darling, 3) another way in which
prior knowledge can be used in this formulation is in the
choice of the sampling system. This can be chosen so that, in
F, its linear span is conveniently close to the expected f.
In other words, the sampling system can be chosen to approximate
likely examples of f more closely than, say, the usual
rectangular basis functions. This was recently discussed
by Hanson and Wecksung (Hanson, 5).

The present work deals with a similar, but not necessarily
finite, set of linear measurements derived from a sampling
system $\{m_n\}$ leading to a set of real values $\{c_n\}$. The
sampling system will normally comprise a rectangular, or
perhaps hexagonal, lattice of translates of some prototype
function. Digital image formation is modelled by the operator
T', given by:

$$T'(f) = \sum_{n=-\infty}^{\infty} C_n \phi_n = \sum <f,m_n>\phi_n \qquad (2)$$

The set $\{\phi_n\}$ comprises image basis functions, which are
independent of $\{c_n\}$ and will be referred to as the
Reconstruction System. It is shown (section 3.1) that, with
suitable choice of $\{\phi_n\}$, T' gives the projection of f*K, the
convolution product of f with a spread function K, onto the
linear closure of the sampling system.

$$T(f) = P_H(f*K) \qquad \text{where } H == \overline{< \{m_n\} >}$$

Clearly (1) can be obtained as a limiting case of (2) when K
tends towards the Dirac function. However in the present
model, no change from standard measure on the plane is
proposed, since the numerical conditioning of the problem is
controlled by the choice of K. The function K has an effect
similar to that of a point spread function and so equation (2)
is essentially setting up a discrete simulation of a continuous
image formation process. The usual theory of discrete simulation,
the Sampling Theory for band-limited functions, which was

introduced to engineering by Shannon (Shannon, 1), will not
be used. The motivation for avoiding Sampling Theory is given
in the next section.

1.3 Difficulties with Sampling Theory

   The application of sampling theory to image processing is
based on the proposition that the image field may either be
approximately band-limited, or may be made to be approximately
band-limited, prior to sampling. Sampling theory is generally
taken to imply that this is the only type of signal, which can
be accurately processed by a digital system. The error which
occurs when the image field is not BL is known as aliasing
error and the error resulting from a finite (truncated)
reconstruction summation is known as truncation error. The
total sampling error for a given system is given by the sum
of the aliasing error and the truncation error. Various
recipes for estimating aliasing and truncation error are
available in the literature (Jerri, 2). There are five main
problems with sampling theory in image processing.

1. While images formed by diffraction limited optics are in
principle band limited, image fields produced in real
apparatus contain spatial frequencies well above the
theoretical limit, due to lens aberrations and noise. Because
of this, aliasing error will always be present to some extent
whatever the sample density chosen. This aliasing problem
can become severe when an existing high resolution image must
be digitally subsampled to a lower resolution, since it is
well known that practical digital filtering arrangements cannot
produce a BL image field from arbitrary input.

2. The sin x/x reconstruction algorithm is a badly
conditioned infinite summation and can be only approximately
implemented (with the aid of "Window functions" etc). Simpler
and more efficient algorithms, such as zero order hold (first
order spline) are normally used.

3. In sampling theory the spatial distribution of samples is
not at all significant to the reconstructed image. In practical
applications, where the functions concerned are not entire and
the samples are of finite precision, the opposite has been
found. Regular sampling structures such as the square lattice
are most common in image processing, although there has been
consistent interest in hexagonal sampling lattice (Mersereaux,
6), (Cramblitt, 7). An investigation into variable, image
source dependent, sample distributions has recently been
reported (Luttrell, 8). In all such work on sample distributions,
sampling theory has had to be supplemented with appropriate
models, for example in (Luttrell, 8) information theory is used,

the sample distribution being chosen to maximize
transinformation.

4. Sampling theory leads to a conflict between image quality
and image correctness. This is because of the poor visual and
edge properties of BL images - a consequence of the Paley-
Wiener relation between functions and their Fourier transforms.
Nevertheless, according to sampling theory, it is necessary
to process approximately BL images in order to control sampling
error.

5. The sampling error may be an over-estimate of the error in
a digital image. Since aliasing error is defined with respect
to ideal reconstruction and reconstruction error with respect
to ideal sampling, the combined sampling/reconstruction process
could exhibit lower error.

1.4 Image errors

Pictorial examples of the effects of aliasing error are
available in the literature (eg. Crow, 9). Such erroneous
images feature fringes (Moiré patterns) which are not present
in the image source and which can be thought of as interactions
between the scene and the sampling grid. An alternative source
of intuition is provided by the well known (see for example
Susskind, 10) relationship between sinusoidal functions and
sequences of uniformly spaced samples of those functions. For
typical scenes digital images may be free from artifacts but
lack a basic property of normal images - that of preserving the
form of the same object in different positions and orientations.
This property will be referred to as feature repeatability. The
well known explanation for lack of feature repeatability in
digital images is that scenes contain components of high
spatial frequency and that this leads to aliasing error when
samples are taken at large intervals.

2. DESCRIPTION OF MODEL

2.1 Definitions

The interval E over which the image is formed will be defined
as either the line R, the plane R^2 or the space R^3. The concern
is to provide an argument which is valid for N dimensional
imaging.

The Isometries M(E) of E will be defined as the set of all
tranformations E->E which are bijective and which preserve
distance. Each isometry $f \in M(E)$ has an inverse $f^{-1} \in M(E)$. In
two dimensions the isometries may be classified into translation
(shift), rotation and reflection.

The Orbit of a function $G \in D$ will be the set of functions which are derived from G via an isometry of E, ie. the function whose value is $G(f(t))$, where $f \in M(E)$, is in the orbit of G, and this function will be denoted by $_fG$.

The proposition here is that the image sampling and reconstruction operation is modelled by a Continuous Linear Operator (CLO) T', which maps a Normed Vector Space (NVS) D, representing the image sources, to a NVS. G, which represents the images. The operator T', which is of the form (2), is regarded as an approximation to some error-free operator T, which would usually, but not necessarily, be a convolution – type integral operator. Since both T' and T can be regarded as points in a conjugate NVS (with the usual norm), the distance between them:

$$\| T' - T \| \;=\; \sup_{V \; x \in D} \; \| (T' - T)(x) \| \quad = \sup_{V \; x \in D} \; \| T'(x) - T'(x) \| \tag{3}$$

gives the basic measure of closeness. This will be referred to as the accuracy metric.

As mentioned previously, digital image processing can lead to inconsistent images of the same source. The aim now is to devise a metric which quantifies this type of inconsistency. The repeatability metric will be defined as:

$$\sup_{Vf \in M(E)} \; \| T^1_{f} 1 \; _{-1}T^1_{f} \| \tag{4}$$

where $T'_f(x) \equiv T'(\,_fx)$

2.2 Selection of source and image spaces

Several choices are possible for the input and output NVS's. We choose to investigate the space L_∞, of bounded functions on E, for the image NVS. This is essentially to permit numerical comparisons between the present model and conventional sampling theory (where the uniform norm is used). The NVS for the image source is chosen as the space L_2 of square integrable functions on E. This NVS is chosen here because:

1. L_2 is a Hilbert space. The angle geometry of Hilbert Space has useful consequences.

2. A wide range of linear integral operators can be defined on L_2. These include convolution operators as well as the

Fourier Transform and Discrete Fourier Transform.

3. L_2 is a good model for many physical image sources.

   Having selected the domain and range for the image operator,
it is possible to give the practical meaning of the quality
metrics defined in 2.1. The accuracy metric gives the biggest
pointwise difference between the correct image and the digital
image at any point, for any image source of unit norm. The
repeatability metric gives the biggest pointwise difference
between digital images of an image source and of any function
in its orbit (for example the object in a different orientation)
for any image source of unit norm.

   A precedent for such a repeatability metric exists in
instrumentation engineering where the orbit of an input is
simply the set of all instances of similar waveforms (at
different occasions). It is clear that the repeatability metric
can never be greater than twice the accuracy metric. A given
repeatability metric has, however, no implications for the
accuracy metric. Specification of repeatability is useful
in cases where the accuracy metric would be large and perhaps
irrelevant. A high accuracy (low accuracy metric) would imply
that the digital process is a good approximation to some
specified continuous process. However pattern recognition
systems work on a set of image predicates whose feature
repeatability may be the key performance parameter –
especially for "trained" systems. Knowledge of the repeatability
metric of an image processor enables the specification of image
predicates which are true for the complete orbit of a scene.
For example a set of point intensity predicates could be used
to locate an object. Each predicate could be simply "intensity
at point x,y >= P", or "intensity at x,y <=Q", where x,y are
related to the object reference frame.

3.1 Optimal Reconstruction

   A linear operator model for the digital imaging process
leads to restrictions on the reconstruction system in (2).
These restrictions depend on the nature of the domain and
range spaces. For the spaces D and G, (L_2 and L_∞) selected
in the previous section, we show (appendix 1) that:

$$\{\phi_n(t)\} \in l_2 \quad \forall \ t \in E \qquad\qquad (5)$$

is a sufficient condition for the mapping (2) to be a linear
operator from D to G. We restrict T, the correct or ideal
linear operator, to be of Fredholm type, with kernel K, so
that:

$$(T(x))(t) = \int_E K(t).x \, du(E) \qquad = <K(t),x>$$

(6)

where K is defined on E X E and K(t) is a function on E.

In the case chosen, T is a Continuous Linear Operator (CLO) from $L_2(E)$ to $L_\infty(E)$. This restricts the choice of possible kernels to those with the property that $\parallel K(t) \parallel_2$ is bounded on E. Without loss of generality, it is assumed that:

$$\parallel K(t) \parallel = 1.0 \; \forall \; t \in E.$$

The mapping T', given by (2), may be represented by a similar Fredholm operator with kernel K', given by:

$$K'(t) = \sum_{n=-\infty}^{\infty} \phi_n(t) m_n$$

so that:

$$(T'(x))(t) = \int_E \left(\sum_n \phi_n(t) m_n \right).x \, du(E) = <x, \sum_n \phi_n(t) m_n>$$

The accuracy and repeatability metrics may be written in terms of the operator kernels, rather than the operators. In appendix 2, it is shown that:

$$A = \sup_{\forall \, t \in E} \parallel K(t) - K'(t) \parallel_2 \qquad (7)$$

$$R = \sup_{\substack{\forall \, t \in E \\ \forall \, f \in M(E)}} \parallel K'(t) - {}_f K'({}_f t) \parallel_2 \qquad (8)$$

Given a specified operator T, a digital approximation T' will here be called optimal w.r.t. T if any change to the sampling system or to the reconstruction system leads to an increase in the accuracy metric. The following synthesis problem is now considered. Given some sampling system $\{m_n\}$ and a specified operator T, it is required to find the reconstruction system $\{\phi_n\}$ which leads to the lowest possible accuracy metric for T'.

The solution is found from the projection theorem for Hilbert space, since the accuracy metric is minimised when K'(t) is the projection of K(t) onto the linear closure $<\{m_n\}>$ of the sampling system, i.e:

$$\forall\, n, \quad (\, K(t) - K'(t)\,) \quad \perp \quad m_n \tag{9}$$

An example is given in figure 1. The sampling system comprises a one dimensional system of the well-known cardinal cubic B-splines (fig. 1a) with unit spacing between translates. The specified operator T is set as a convolution operator, with a similar cardinal cubic B-spline as convolution kernel. The matching reconstruction function, fig. 1b, is found via the normal equations (9). If the spacing (translation) between the sampling functions is increased to 2, then the spatial extent of the reconstruction function is reduced (fig. 1c). The accuracy metric for this digital operator depends, critically, on this interfunction spacing. Note that, in this case, the sampling and reconstruction systems can be interchanged, the resulting operator remaining optimal w.r.t. T. This is not true for the general case where K(t) is unrelated to the sampling function.

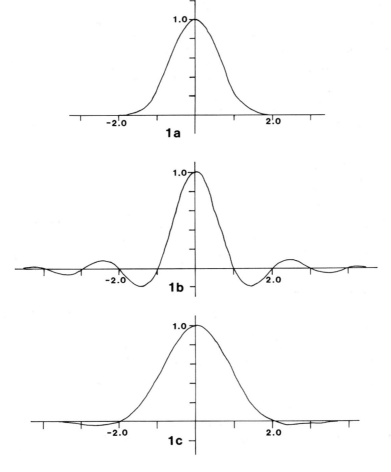

Fig. 1   Sampling and reconstruction functions

3.2　Repeatability metric for optimal operators

As mentioned in section 2, in some applications, the repeatability metric may be of more practical importance than the accuracy metric. The repeatability metric is not necessarily minimized in optimal digital operators. However we cannot directly attempt to construct an image operator with a low repeatability metric (high repeatability), since there are 'trivial' operators, whose repeatability metric is zero, which transmit no information at all from the image source. The approach suggested here is to find the repeatability metric for optimal operators.

In the previous section it was found (appendix 2) that the repeatability metric could be expressed in terms of the kernel K' of the digital operator. It is shown in appendix 3 that, for an operator kernel K', optimal w.r.t. T, the expression $_fK'(_ft)$ is the projection of K(t) onto $<\{_f m_n\}>$ a linear subspace generated by the translated sampling system. It follows that the repeatability metric has the simple interpretation of being the biggest distance between projections of some function K(t) ∈ D onto two linear subspaces of D, the closures $<\{m_n\}>$ and $<\{_f m_n\}>$. The angle geometry of Hilbert space means that the size of R is limited by the norm of K(t) (=1.0 by definition) and by the angle between the two linear subspaces. This can be written as follows:

$$S1 = \overline{<\{m_n\}>} \quad , \quad S2 = \overline{<\{_f m_n\}>}$$

$$P_{S1}(K(T)) = \text{Projection K(t) onto S1}$$

$$P_{S2}(K(t)) = \text{Projection K(t) onto S2}$$

define Q_{max}, the maximum angle between S1 and S2, as follows:

$$\sin Q_{max} = \sup_{\forall g \in D} || P_{S1}(g) - P_{S2}(g) ||_2 / || g ||_2$$

Since g ranges over $L_2(E)$ in this defintion, it is clear that, for any optimum operator R will always be bounded by $\sin Q_{max}$. This bound is probably too weak to be of much practical significance but it has the advantage of not involving K. Q_{max} can be seen as giving a figure of merit for the repeatability performance of a given sampling system.

One case of practical interest is when the required operator of convolution type, with a kernel function which is identical to those of the sampling system. An example of this case, for cubic B splines, was discussed in the previous section. The repeatability metric is then the same as the accuracy metric, and minimising the accuracy metric minimises the repeatability metric.

4. APPLICATION OF MODEL

In the subsampling and resampling problem mentioned in section 1, the practicality of using the optimal image operator defined by (9) depends on the number of terms in the reconstruction algorithm and its numerical characteristics. A well conditioned algorithm with a small number of non zero terms can be efficiently and accurately implemented using standard digital technology.

For the other envisaged application, that of producing matched sampling and display hardware, the problem becomes one of 'tuning' the characteristics of both image digitizer and display, so that the combined effect is that of an operator T' which is an optimal approximation to some desirable operator T.   T might be, for example, a convolution operator with a spline or gaussian PSF. The transfer functions available from conventional vidicon cameras and cathode-ray displays are controllable by aperture adjustment and by optical/electronic defocussing. The reconstruction functions obtainable by this approach may be expected to be positive, but sampling functions with negative segments may be obtained by optical diffraction.

The positivity of the reconstructed image is not assured in this model. However if the required operator T has a non-negative kernel K (for example a B spline), then, given any image source x, the negative amplitude of the image will not be greater than $A.||x||_2$.

5. CONCLUSIONS

1. In the present model, the repeatability metric of an image processing system replaces the aliasing and truncation recipes as a means of predicting digital distortions. Aliasing error essentially gives a figure for the distortion which results when a non-BL source is sampled and ideally reconstructed as a BL image. The repeatability metric can be calculated for a much wider class of images – including those with a B spline type PSF. Although the main interest here is in convolution operators, the model is valid for other types of operators such as the DFT. The accuracy metric does not have an exact analogue in sampling theory since, in sampling theory, the choice of correct images is limited to BL images.

2.   In the model described, there is no contradiction between
obtaining an image with low digital (sampling) errors and
obtaining one with good visual properties.
3.   As general purpose data collection tools, some sampling
systems are better than others, in that they lead to a lower
repeatability metric when used in optimal processing. It is
interesting to note that good sampling systems will consist of
functions of bounded support. Contrast this with sampling theory
where the functions forming the sampling system are ideally
BL and so of unbounded support.

4.   Using this approach, images can be stored and processed at
the most useful resolutions, with consequent economies in
computation. It is anticipated that the number of image
samples required for a given image quality would be much
reduced with an optimal image process. This is to be
experimentally investigated. Also, information extraction
operations, such as the DFT, could be re-designed to work
efficiently with reference to the known image basis $\{\phi_n\}$.

5.   It would be interesting to investiagte other choices for
the source and image NVS. The case where both source and image
spaces are Hilbert may be particularly relevant.

6.   ACKNOWLEDGEMENTS

    The authors wish to thank Graham Little, of the Department
of Mathematics at Manchester University, for his helpful
comments. The support of the S.E.R.C. is gratefully
acknowledged.

7.   APPENDICES

APPENDIX 1 (Conditions for CLO operator model)

    The digital image operator T' can be regarded as the product
of two operators T_1' and T_2'. The sampling operator T_1' maps the
image source to a space V, whose elements are real number
sequences. The operator T_2' maps the space V to the image space
G.

$$T' = T_1' \cdot T_2', \quad \text{where } T_1': D \rightarrow V, \ T_2' : V \rightarrow G$$

For T' to be a CLO from D $\rightarrow$ V it is sufficient that operators
T_1' and T_2' are CLO's. The space V is chosen as l_2. Given a
set of functions $\{\phi_n\}$, with the property that:

$$\forall \ t \in E, \quad \{\phi_n(t)\} \in l_2$$

A linear map of the form:

$$T_2'(a) = \sum_{n=-\infty} a_n \phi_n \qquad \text{where } \{ a_n \} \in l_2$$

provides a well defined continuous linear map from the space l_2 of square summable sequences to the image space $L_\infty(E)$ (the functions $\{ \phi_n \}$ are bounded). From (2) the sampling operator $T_1': D \rightarrow V$ has the form:

$$(T_1'(x))_n = <x, m_n>$$

Provided that the elements of the sampling system are distinct, the resulting number sequences $\{ (T_1')_n \}$ will be square summable as required.

Appendix 2   (Repeatability and accuracy metrics in terms of operator kernels.)

Combining (3) with (7) :

$$A = \sup_{\substack{\forall t \in E \\ \forall x \in D, \text{ s.t. } | \,|x| \,| <= 1.0}} | < x, K(t) > - < x, K'(t) > |$$

$$= \sup_{\forall t \in E} | \,| K(t) - K'(t) | \,|_2$$

From (4) the repeatability metric may be written:

$$R = \sup_{\substack{\forall x \in D \text{ st. } | \,|x| \,| <= 1.0 \\ \forall f \in M(E)}} | \,| T'(x) - {}_f^{-1} T'({}_f x) | \,|_{\infty}$$

using (7) :

$$R = \sup_{\substack{\forall t \in E \\ \forall x \in D \text{ s.t. } | \,|x| \,| <= 1.0 \\ \forall f \in M(E)}} | < K'(t), x > - < K'({}_f^{-1} t), {}_f x > |$$

$$= \sup_{\substack{\forall t \in E \\ \forall f \in M(E)}} | \,| K'(t) - {}_f K'({}_f t) | \,|_2$$

Appendix 3   (Interpretation of $_fK(_ft)$)

Consider the digital operator T' which is optimal w.r.t operator T. By (8), the kernel K' of T' is the best least squares approximation to the kernel K of T. Since D is a Hilbert Space, the normal equations are satisfied,

$$< K(t) - \sum_{n=-\infty}^{\infty} \phi_n(t)m_n , m_j > = 0$$

$$\forall j, \forall t \in E$$

and the digital kernel K'(t) may be regarded as the projection of the __required__ kernel K(t) onto the closure of the sampling system $<\{ m_n \}>$, which is a linear subspace of D. This is true for all $t \in E$ and so $K'(_ft)$ is the projection of the translated kernel $K(_ft)$ onto $\overline{<\{ m_n \}>}$.

$$< K(_ft) - \sum_n \phi_n(_ft)m_n , m_j > = 0$$

$$\forall j, \forall t \in E$$

Applying the isometry f to the inner product terms:

$$< _fK(_ft) - \sum_n \phi_n(_f^T) _fm_n , _fm_j > = 0$$

but $_fK(_ft) = K(t)$ by definition of convolution and so the digital kernel $_fK'(_ft)$ may be regarded as the projection of K(t) onto the closure of $<\{_fm_n \}>$.

7. REFERENCES

1. Shannon, C.E., (1949) "Communications in the presence of noise". _Proc. IRE,_ vol. 37, pp. 10-21.

2. Jerri, A.J., (1977) "The Shannon Sampling Theorem- Its Various Extensions and Applications; A Tutorial Review". _Proc. IEEE,_ vol. 65, no. 11, pp. 1565-1596.

3. Darling, A.M., Hall, T.J. and Fiddy, M.A., (1983) "Stable, noniterative object reconstruction from incomplete data using a priori knowledge". _J. Opt. Soc. Am,_ Vol. 73, No. 11, pp. 1466-1469.

4. Miller, K., (1970) "Least-squares methods for ill-posed problems with a prescribed bound". _S.I.A.M. J. Math. Anal._ No. 1, pp. 52-74.

5. Hanson, M.K. and Wecksung, G.W., (1985) "Local basis-
    function approach to computed tomography". *Applied Optics,*
    Vol. 24, No. 23, pp. 4028-4039.

6. Mersereaux, R.M., (1980) "A Two Dimensional Fast Fourier
    Transform for Hexagonally Sampled Data". In "Digital
    Signal Processing", Cappellini, V. and Constantinides, A.G.,
    (eds.), London, Academic Press.

7. Cramblitt, R.M. and Allebach, J.P., (1983) "Analysis of
    time-sequential sampling with a spatially hexagonal lattice".
    J. Opt. Soc. Am., Vol. 73, No. 11, pp. 1510-1517.

8. Luttrell, S.P., (1985) "The use of transinformation in the
    design of data sampling schemes for inverse problems".
    Inverse Problems, vol. 1, pp. 199-218.

9. Crow, F.C., (1977) "The Aliasing Problem in Computer-
    Generated shaded images". *Comm. ACM,* vol. 20, no. 11,
    pp. 799-805.

10. Susskind, A.K., (1957) "Notes on Analog to Digital
     Conversion Techniques," London, Chapman & Hall Ltd.

11. Oakley, J.P., Cunningham, M.J. and Little, G., "Image
     Sampling, Reconstruction and Display," to be published.

INFERENCE THEORY

S. Luttrell
(RSRE, Malvern, Worcs.)

ABSTRACT

This transcript reviews the need for low-level stochastic
image models, especially for coherent images which are
corrupted by speckle noise. A unified approach to inference
in such models using Bayes' rule and Shannon's information
theory is presented. In the gaussian PDF approximation exact
results for sampling scheme optimisation and for inference from
sample values are derived (super-resolution in particular is
concentrated on). For non-gaussian PDFs the method of graphs
for constructive PDF generation is outlined, and several research
trends are indicated.

1.   INTRODUCTION

I shall assume throughout this paper that the processes that
generate a dataset can be represented as a graph.

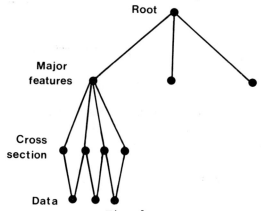

Fig. 1

The essential components of such a representation are

    (1)   Nodes ≡ physical variables (eg. elements of cross
           section), and

    (2)   Links ≡ (mutual) interactions or dependencies amongst
           the variables.

    Such image generating graphs can usually be divided into two
components which are distinguished by the degree of randomness
of their internal interactions: low level components of the
graph (eg. individual pixel variables) typically have noisy
mutual dependencies, whereas high level components (eg. regions
and boundaries) typically have more structured mutual
dependencies.

    Statistical information theory provides a unified framework
for studying systems that are characterised by probability
density functions (PDFs), so I shall use it to analyse low
level graphs. This work is particularly important for the
modelling of coherent images which all suffer from speckle
noise.

2.   TRAINING AND BAYESIAN INFERENCE

    I shall develop information theory by appealing to a standard
experiment as in Figure 2.

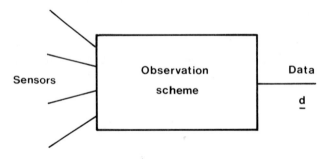

Fig. 2

    Each observation on the system S produces a data vector $\underline{d}^{(j)}$
which is recorded. It is very convenient to gather together all
the $\underline{d}^{(j)}$ into a dataset D, where

$$D \equiv \{\underline{d}^{(1)}, \ldots\ldots, \underline{d}^{(N)}\} \qquad (1)$$

for a set of N observations. Both changes in the state of S
and in the measurement error cause the value of $\underline{d}$ to vary. It
is convenient to plot each $\underline{d}^{(j)}$ in a data space, and it is
appealing to represent the resulting scattergram as in Figure 3.

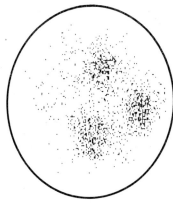

Fig. 3

The scattergram does not record the order in which the $\underline{d}^{(j)}$
were observed, so time dependencies in the variation of $\underline{d}$ have
been lost. This limitation will apply throughout this paper,
although it can be circumvented by regarding the whole set of
$\underline{d}^{(j)}$ as a "super data vector" which is plotted in a "super data
space".

The above procedure is an unsupervised training procedure
where all the $\underline{d}^{(j)}$ are retained but their order is discarded.
I have not yet introduced a separate means whereby the state
of S can be observed directly so as to provide a class label
for the corresponding $\underline{d}^{(j)}$. The scattergram is therefore a
mixture of the scattergrams that would be obtained for the
separate classes.

As the number of observations N increases the "cloud" of
plotted observations becomes more dense, and it provides a
better representation of the distribution of $\underline{d}$. In the limit
N $\rightarrow \infty$ the local density of the scattergram can be used to
define a probability density function (PDF) in the usual way,
and I shall use the generic notation "P(x)" to denote "PDF of x".
Clearly this limit is difficult to realise in practice, so the
scattergram is usually used to estimate only a parametric fit
to the PDF (eg. gaussian PDF). This process of estimating a
PDF from a finite set of observations is not part of information

theory per se; it belongs to approximation theory which I
shall not dwell upon here. The PDF that is obtained from the
mixture scattergram is denoted by $P(\underline{d})$ and is called the
mixture PDF; this is shown schematically in Figure 4.

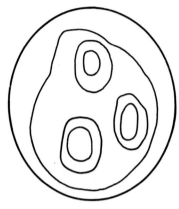

Fig. 4

$P(\underline{d})$ can be decomposed "by eye" into a number of components
(or modes), which can then be arbitrarily assigned class labels.
Because the training procedure is unsupervised these class
labels will not have a known relationship to the state S, and
so it is very difficult to make any further justifiable
analysis of inferences that can be drawn about S given $\underline{d}$.

I shall assume now that the observations are accompanied by
some collateral information (ground truth) derived directly
from S; this is supervised training. This additional
"conditioning signal" comprises a class label $c^{(j)}$ that
accompanies each $\underline{d}^{(j)}$, thus augmenting $\underline{d}$ to $(c,\underline{d})$. Obviously
a scattergram can be constructed from $(c,\underline{d})$ just as it was for
$\underline{d}$; this is equivalent to building a separate scattergram for
each value of c. K datasets $D_1, \ldots, D_K$ (each as in equation
(1)) are constructed when there are K classes, and each dataset
D_j contains only those $\underline{d}$ that derive from class c_j as shown in
Figure 5 for the case K = 3. Each scattergram is used to
estimate a class conditional PDF $P(\underline{d}|c)$, and the total number
of points in each scattergram is used to estimate the a priori
PDF $P(c)$. Obviously

$$P(\underline{d}|c) \; P(c) = P(c,\underline{d}) \qquad\qquad (2)$$

where the joint PDF $P(c,\underline{d})$ is the estimate that would have been
obtained from a scattergram consisting of plots of the $(c,\underline{d})$.

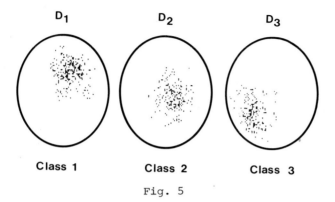

D₁   **D₂**   **D₃**

Class 1   Class 2   Class 3

Fig. 5

The above separation into scattergrams labelled by class c is not the only such separation that can be made. Another separation is into scattergrams labelled by $\underline{d}$ which leads to the data conditional PDF $P(c|\underline{d})$ and the mixture PDF $P(\underline{d})$, with the relationships

$$P(c|\underline{d})\ P(\underline{d}) = P(c,\underline{d}) \qquad (3)$$

Comparing equation (2) and equation (3) leads to Bayes' rule

$$P(c|\underline{d}) = \frac{P(\underline{d}|c)\ P(c)}{P(\underline{d})} \qquad (4)$$

All the quantities in Bayes' rule can be estimated from the various scattergrams that I referred to above.

The data conditional PDF $P(c|\underline{d})$ is also called the a posteriori PDF, because it is the result of combining the a priori PDF $P(c)$ with the class conditional PDF $P(\underline{d}|c)$. Loosely speaking, prior knowledge $P(c)$ that was gleaned from a supervised training procedure is combined with information $P(\underline{d}|c)$ that is derived each time an observation $\underline{d}$ is made. It is very important to distinguish between the original training procedure and the Bayesian inference procedures: training measures PDFs directly by accumulating a large number of class labelled observations, whereas inference from a single unlabelled observation is the act of choosing particular conditional PDF from the trained PDFs (see equations (4)).

3. AN INFORMATION MEASURE

I wish to derive a single measure of the information content of a PDF in order to make clear which properties of a

particular sampling scheme and inference process convey useful
information.   In order to derive a single measure from a
complicated PDF I must first of all reduce the PDF to a
simpler form.   I shall develop the argument in terms of $P(c)$
although it can trivially be carried over to all other PDFs.
Recall that $P(c)$ was the limiting frequency of occurrence of
each class c, so an operational definition of an information
measure must refer to long sequences of class labels
generated by S: this is the key to information theory.

   Consider an N sample sequence (N-sequence) in the K class
case (as above).   Each sample is one of K possible classes, so
the total number of possible N-sequences is given by

$$W_{TOT} = K^N \tag{5}$$

However these sequences do not all have the same probability;
some are overwhelmingly more probable than others when N is
large.   Clearly as $N \to \infty$ the number of times that class c
occurs in the sequence is given by $P(c)N$ - this is the
definition of $P(c)$ after all!   For finite N there is a
distribution about the value $P(c)N$, but this becomes
insignificant for large N.   The effective number of N-sequences
is therefore given by

$$W_{EFF} = \frac{N!}{[P(c_1)N]! \dots [P(c_K)N]!} \tag{6}$$

This expression can be simplified by using Stirling's
approximation $(\log(n!) \sim n\log(n)-n)$ yielding

$$W_{EFF} = \exp[-N \sum_c P(c)\log_e[P(c)]] \tag{7}$$

The argument of the exponential contains the entropy $H(c)$ of
$P(c)$

$$H(c) \equiv - \sum_c P(c)\log_e[P(c)] \tag{8}$$

so that

$$W_{EFF} = \exp[N H(c)] \tag{9}$$

It is easy to verify the inequality

$$0 < W_{EFF} \leqslant W_{TOT} \tag{10}$$

with equality $W_{EFF} = W_{TOT}$ if and only if all classes are equally
likely. The base of the logarithm is unimportant; logarithms
to base x can be used and compensated for by the replacement
exp → x. In particular if logarithms to base K are used then

$$W_{EFF} = K^{NH(c)} \qquad (11)$$

which is compared with equation (5) in order to define

$$K_{EFF} \equiv K^{H(c)} \qquad (12)$$

By the same reasoning that led to equation (10) it follows that

$$0 < K_{EFF} \leq K \qquad (13)$$

The steps of the argument leading to W_{TOT} and W_{EFF} are
presented pictorially in Figure 6.

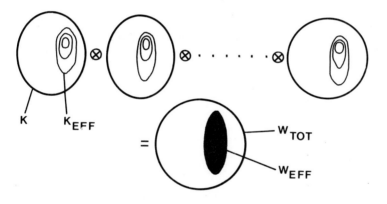

Fig. 6

The effective number of classes K_{EFF} that P(c) generates
when N-sequences are considered is a natural measure to use
when defining the information content of P(c). From equation
(12) it is clear that the entropy H(c) conveys the uncertainty
associated with a PDF P(c). The smaller the entropy, the
smaller the value of K_{EFF}, and therefore the smaller the
the uncertainty about what class c will be sampled next from
P(c).

4. MUTUAL INFORMATION

The operation of Bayes' rule (equation (4)) is shown in Figure 7.

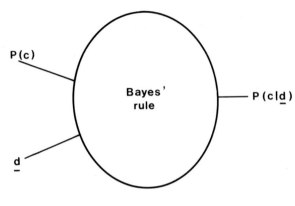

Fig. 7

It is important to define a measure of information which quantifies the usefulness of the data $\underline{d}$ in generating the a posteriori PDF $P(c|\underline{d})$ from the a priori PDF $P(c)$. An obvious measure to use is

$$I(c,\underline{f}) \equiv H(c) - H(c|\underline{d}) \tag{14}$$

where

$$H(c|\underline{d}) \equiv \sum_{\underline{d}} P(\underline{d}) \left[- \sum_{c} P(c|\underline{d}) \log P(c|\underline{d}) \right] \tag{15}$$

and the base of the logarithm used in $H(c)$ and in $H(c|\underline{d})$ must be the same. $I(c,\underline{d})$ measures the reduction in uncertainty associated with the a priori PDF $P(c)$ when data $\underline{d}$ are observed, averaged over observations. $I(c,\underline{d})$ is called transinformation or mutual information, and it provides an information theoretic measure of the usefulness of an observation scheme in collecting data that is suitable for drawing inferences about class membership.

5. GAUSSIAN PROBABILITY DENSITY FUNCTIONS

I shall now derive a closed form expression for $I(c|\underline{d})$ by assuming that $P(c)$ and $P(c|\underline{d})$ are gaussian. The class label c must therefore be continuous, and I shall extend it further to be multi-dimensional (ie. $c \rightarrow \underline{c}$). Defining the a priori

and a posteriori covariance matrices as C_{PRIOR} and C_{POST}
respectively, after some algebra I obtain the result

$$I(\underline{c},\underline{d}) = \log_e \left[\frac{\det[\, C_{PRIOR}\,]}{\det[\, C_{POST}\,]} \right] \qquad (16)$$

A determinant of a covariance matrix is the product of its
eigenvalues, and so it measures the "volume of uncertainty"
that is associated with the corresponding gaussian distribution.
The quantity $\exp[\, I(\underline{c},\underline{d})\,]$ (note the equality of logarithm base
and exponentiating base) therefore measures the ratio of the
effective number of classes associated with $P(\underline{c})$ to the average
of the effective number of classes associated with $P(\underline{c}|\underline{d})$.
Clearly this corresponds to our intuitive notion of the amount
of useful information contained in $\underline{d}$ on average.

An important caveat is that only those class labels which
are of interest should be included in $P(\underline{c})$. All labels
contribute to the value of $I(\underline{c},\underline{d})$, and so to obtain a measure
of the relevant part of the mutual information the irrelevant
class labels should be excluded. This process should be
carried even further so that classes that are equivalent (for
one's particular purposes) are combined into a single class.
Clearly the value of $I(\underline{c},\underline{d})$ that is calculated will depend
strongly on the particular interests and requirements of the
observer, as expected.

6.   SAMPLING SCHEME OPTIMISATION

I shall assume that the classes have been regrouped as I
discussed above. The value of $I(\underline{c},\underline{d})$ then indicates how much
useful information a particular observation (or sampling)
scheme provides on average for resolving the uncertainty about
which class is present. For a particular problem the
sampling scheme can be altered so as to explore variations in
the value of $I(\underline{c},\underline{d})$, and to optimise the sampling scheme by
maximising $I(\underline{c},\underline{d})$.

All the results that I have calculated [1,2] agree with
one's intuition about what a good sampling scheme should look
like. When the number of sample points is fixed the samples
should not be so close together that they measure essentially
the same information, and they should not be so far apart that
they undersample the data space. Mutual information is an
information theoretic measure that allows one to trade off
between these two extreme cases optimally. In particular I
have derived Nyquist sampling of bandlimited functions,
exponential sampling of Laplace transforms, and prior knowledge
dependence of optimal sampling. Mutual information provides a
rigorous means of quantifying the influence of prior knowledge

on the design of sampling schemes, and it produces the expected
results in simple cases.

7.   INDEPENDENT INFORMATION CHANNELS

The expression for $I(\underline{c},\underline{d})$ in equation (16) is simplified
by simultaneous diagonalisation of the two covariance matrices
C_{PRIOR} and C_{POST}. This leads to

$$I(\underline{c},\underline{d}) = \sum_j I_j \qquad (17)$$

where

$$I_j \equiv \log \left[\frac{\lambda_{PRIOR,j}}{\lambda_{POST,j}} \right] \qquad (18)$$

and I use λ generically to denote an eigenvalue. The fact that
in equation (17) the total mutual information $I(\underline{c},\underline{d})$ is
decomposed into a <u>sum</u> over components I_j means that the
eigenvectors which simultaneously diagonalise equation (16)
must correspond to independent information channels, each
carrying an independent piece of useful information about $\underline{c}$.

A simple interpretation of this decomposition is that the
eigenvectors form a feature set for separating the classes $\underline{c}$
optimally. They may also be identified with the Karhunen-Loeve
expansion of a covariance matrix.

8.   SUPER-RESOLUTION

I have related the above decomposition to super-resolution
[3,4] of complex (coherent) images. I model a linear imaging
system of this type by

$$\underline{d} = T\underline{f} + \underline{n} \qquad (19)$$

Defining a gaussian a priori PDF over scattered fields $\underline{f}$

$$P(\underline{f}) \propto \exp[-\underline{f}^+ W^{-1} \underline{f}] \qquad (20)$$

where W is a covariance matrix that expresses the prior
knowledge that $\langle f_j \rangle = 0$ and that $\langle f_j f_k^* \rangle = W_{jk}$ ($\langle .. \rangle$ denotes
ensemble average). There is no factor 1/2 in the argument of
the exponential in equation (20) because the field $\underline{f}$ is complex
(not real). The imaging operator T is linear and the (additive)
noise $\underline{n}$ is modelled also using a gaussian PDF

$$P(\underline{n}) \propto \exp[-\underline{n}^+ N^{-1} \underline{n}] \qquad (21)$$

where usually N is assumed to be proportional to the identity matrix (ie. white noise).

Using the PDFs defined above a little algebra yields

$$P(\underline{f}|\underline{d}) \; \alpha \; \exp[\; -(\underline{f}-\underline{f}_o)^+ M^{-1} (\underline{f}-\underline{f}_o) \;] \qquad (22)$$

where

$$M \equiv T^+ N^{-1} T + W^{-1} \qquad (23)$$

and

$$\underline{f}_o \equiv [\; T^+ N^{-1} T + W^{-1} \;] \; T^+ N^{-1} \underline{d}$$

$$\equiv WT^+ [\; TWT^+ + N \;]^{-1} \underline{d} \qquad (24)$$

Both the a priori PDF $P(\underline{f})$ and the a posteriori PDF $P(\underline{f}|\underline{d})$ are gaussian, so equation (16) may be used with the replacement $\underline{c} \to \underline{f}$.

The simultaneous eigensystem of W and M provides the inputs of the set of independent information channels, which may be transformed using T to produce their outputs in data space; the details of this calculation are contained in [4]. These information channels can be used to invert the data $\underline{d}$, and to obtain an estimate or reconstruction of the scattered field $\underline{f}$. The maximum a posteriori probability (MAP) $\underline{f}$ is clearly $\underline{f}_o$, and I have shown in [4] that $\underline{f}_o$ consists of components that are obtained by independent inversion in each information channel. Note that there is no problem of sensitivity to data noise because such effects have already been modelled in $P(\underline{n})$.

The effect of such an inversion is to reorganise and interpolate the data (by using information channels) in a manner that is dependent on the form of $P(\underline{f})$. In particular, if W expresses the prior knowledge that $\underline{f}$ has large components concentrated only in a small region of order of the resolution area of T, then such an inversion can lead to super-resolution [3,4].

Physically the explanation of super-resolution is simple. With reference to Figure 8 if a small region of the scattered field has larger components than elsewhere, then the fraction of the data space which is dominated by contributions from this region is larger than average; thus a greater than average number of samples contains information pertaining to the form of the large amplitude scattered field, and this leads to the possibility of super-resolution. By a simple extension

of this argument the degree of super-resolution that is
obtainable must decrease either as the size of the large
amplitudes decreases and/or as the competition amongst the parts
of the scattered field for regions of data space increases.

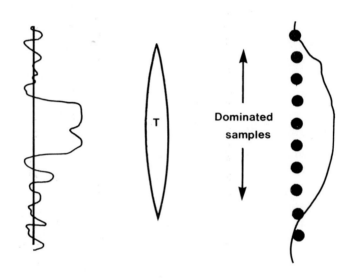

Fig. 8

9.   CONSTRUCTIVE DEFINITION OF NON-GAUSSIAN PDFs

I have presented an extensive analysis of the gaussian PDF
case.   However it is not always realistic to describe prior
knowledge in terms of gaussian PDFs, so I shall now turn to the
more difficult non-gaussian case.

Because analytic results are now difficult to obtain I shall
adopt the constructive approach to defining PDFs.   Thus I will
not write down an analytic expression for $P(\underline{c})$, rather I shall
define a set of rules for generation of $P(\underline{c})$.   I hope that this
approach will permit me to define a richer structure of PDFs
than is possible using the direct analytic approach, and
furthermore that the constructive definition of the PDF will
allow me to generate samples from P(c) with ease.   Both of
these advanatges will turn out to be true.   There are two
essential components involved in constructively defining a PDF
which were defined in the introduction; nodes and links form
the skeletal model of a PDF.

A very simple example of such a model is a first order
Markov chain as shown in Figure 9.

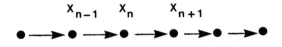

Fig. 9

The joint PDF $P(\underline{x})$ over the variables x_j is generated by developing the chain using the update rule (or stochastic matrix)

$$x_{j+1} = a(x_j)\, x_j + b(x_j)\nu_j \qquad (25)$$

where the ν_j are usually generated by a white (gaussian) noise process. Clearly if $b(x_j)=0\;\forall x_j$ then the relationship between the x_j is deterministic. Usually both terms in equation (25) contribute in which case the x_j have mutual statistical dependency (ie. a PDF $P(\underline{x})$ describes their properties).

Another example that is more closely related to the problem of modelling images is shown in Figure 10.

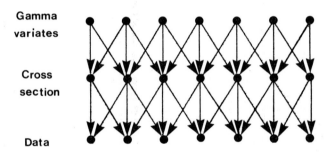

Fig. 10

This depicts (schematically) as nodes and links the clutter model that is discussed by Oliver in his contribution to these proceedings. The meaning of the diagram is self evident; it should be interpreted as an information flow diagram.

10.   GRAPHS, TREES AND TEXTURE MODELS

I shall now analyse more closely the type of information that is encoded into a particular structure of nodes and links. Such structures are called graphs, and their study forms the

field of graph theory. It is not necessary to get involved
with the rigours of graph theory here, because I intend to give
a heuristic treatment of PDF generation models.

I show a simple type of graph in Figure 11 where a parent
node has several children; this is called a tree graph for
obvious reasons. The children naturally have mutual dependencies
because they have a common parent. A graph of this type can
describe very long range mutual dependencies in images. In
particular it can be used to describe a segment of an image,
where the parent node carries information relating to the segment
properties and the children each correspond to a single pixel
value.

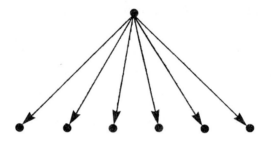

Fig. 11

I show a more sophisticated graph in Figure 12 which has
grandchildren. As in Figure 11 the children have mutually
dependent properties because of the common parent, but now
each set of grandchildren (of one particular child) have
mutually dependent properties because of the common child.
There are thus two levels of mutual dependency; one affects the
mutual dependency amongst a single set of grandchildren (short
range dependency), and the other affects the mutual dependency
amongst different sets of grandchildren (long range dependency).
Loosely speaking there are two correlation scales in the
grandchildren which are naturally described by a tree graph. In
terms of images such as a graph could not only describe a
segment, but also clustering within a segment (sub-segments).

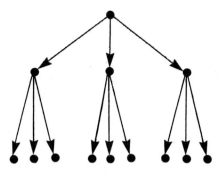

Fig. 12

I show a type of graph that might be used to model texture
in an image in Figure 13.

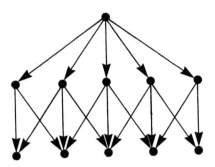

Fig. 13

This differs from Figure 12 only insofar as each grandchild
belongs to more than one family! This is an important
difference for image modelling because it means that the
sub-segments of Figure 12 are now permitted to overlap. This is
necessary if a translation invariant (or homogeneous) texture
model is to be built. Because of this overlap the graph of
Figure 13 is not treelike. Its structure gives rise to two
length scales; segment size and texture coarseness size.

The graphs in Figures 11, 12 and 13 are causal graphs because
the flow of dependency is in one direction only. The overall
dependencies can be written down as nested conditional PDFs by
using the arrows on the links of the graph to identify where there
is dependency. Causal graphs are very simple because there is
no feedback, and therefore there is no danger of any
contradictions arising between the states of the variables at
the nodes. Monte Carlo methods may be used to generate samples

from the PDF that is constructively generated by each causal
graph. This solves the direct problem of generating data from an
an underlying model.

11. BAYESIAN INFERENCE USING GRAPHS

The inverse problem (or inference process) of regenerating
the underlying variables of a model from data can also be
solved within the framework of graphs. This corresponds to
clamping the data variables (eg. grandchildren) with their
observed values, and then deducing the distribution of states
that the remaining variables (eg. parent and children) should
take. However the general causal graph model above does not
provide a simple prescription for inferring the a posteriori
PDF.

A solution to this problem is obtained if a bidirectional
flow of dependency is permitted in the graph model. A simple
way of achieving this is to define the joint PDF over all the
variables of the graph as a Gibbs distribution [5].
Conventionally this is written in the form

$$P(\underline{x}) \propto \exp[-U(\underline{x})] \tag{26}$$

where $U(\underline{x})$ is a sum of terms each of which involves only a few
components of $\underline{x}$. The form of $P(\underline{x})$ is therefore a product of
component factors, which I prefer to emphasise by writing

$$P(\underline{x}) \propto p_1(\underline{x})......P_R(\underline{x}) \tag{27}$$

where each factor depends on only a few components of $\underline{x}$. The
model that is so constructed is difficult to make exactly
equivalent to a causal graph model, because it is necessary
to balance delicately the reverse flow of dependency if the net
forward flow is to be the same as in the causal graph model.
However the Gibbs distribution form of $P(\underline{x})$ (and equation (27))
is very useful because it guarantees that the model has no
internal inconsistencies [6].

The Metropolis algorithm [7] may be used to generate samples
from $P(\underline{x})$, and in particular both the direct and inverse problems
may be solved. The class of the system S is $\underline{c}$, the observed data
is $\underline{d}$, and any other relevant variables are $\underline{h}$, so $P(\underline{x})$ is
replaced by $P(\underline{c},\underline{h},\underline{d})$. The direct problem involves an
unconstrained use of the Metropolis algorithm to generate samples
$(\underline{c},\underline{h},\underline{d})$, where only $\underline{d}$ is of interest for unsupervised training,
and where $(\underline{c},\underline{d})$ is of interest for supervised training. The
inverse problem involves clamping $\underline{d}$ with some data, and then
using the Metropolis algorithm to generate samples $(\underline{c},\underline{h})$; this
generates $P(\underline{c},\underline{h}|\underline{d})$ and hence $P(\underline{c}|\underline{d})$.

The above means of generating samples from the a posteriori PDF can be extended so that the MAP solution is obtained; this is called optimisation by simulated annealing (OSA). The essence of the method is that as samples from the a posteriori PDF are generated by the Metropolis algorithm, the form of the Gibbs distribution should be slowly altered so as to accentuate the differences in probability between different states. In practice this is achieved by the transformation $U(\underline{x}) \rightarrow U(\underline{x})/T$ with $T \rightarrow 0$ in equation (26). The parameter T is analogous to a temperature in statistical thermodynamics, and the way in which it approaches zero is called the annealing schedule. Clearly the MAP solution has the smallest value of $U(\underline{x})$ and so becomes overwhelmingly likely to be sampled as $T \rightarrow 0$. An interesting example of this method is contained in [8], and I recommend that that paper be read by anyone who is interested in OSA in image processing.

There is a pitfall in the above OSA method; the annealing schedule has to be exceedingly slow in order to guarantee convergence to the MAP solution in some cases. Also it can lead to very long computer run times as the Metropolis algorithm has to generate many samples at each stage of the annealing schedule in order to guarantee that the graph has equilibriated. However when OSA is used carefully on a parallel computer architecture it is a powerful tool.

12. THE WAY AHEAD

In my opinion there is a trade-off between the computational expense (in time units) of an inference method and the quality of the inference, where I assume throughout that the degree of parallelism in the method is held constant. The Gibbs distribution method (with or without OSA) has highly desirable properties, but it consumes large resources. I believe that we could speed up the inference process by incorporating heuristics or non-Bayesian methods, although we would probably sacrifice the ability to identify an a posteriori PDF.

A powerful argument against the Bayesian approach is that it relies upon the existence of PDFs. However I introduced these only as a means of representing the form of a scattergram after accumulating a large number (infinite number, strictly) of training sets. In practice the number of training sets available is usually insufficient to produce a densely populated scattergram, and so it is not obvious that a PDF is the best representation.

It seems therefore that the resolution of the computational cost of the inference problem by sacrificing the Bayesian approach and adopting heuristics is intimately bound up with the choice of representation that was chosen for recording the observations during the original training process.   In short, training and inference are not really separate processes at all!

Of course the artificial intelligence (AI) community has adopted a heuristic strategy all along.   However AI has concerned itself principally with the higher level variables and relationships within a graph.   There is no analysis which deals with the whole of a graph including its low level (possibly highly stochastic) variables.   There is a clear opportunity for much original research in this area.

REFERENCES

[1] Luttrell, S.P., (1985) *Optica Acta,* **32**, 255-257.

[2] Luttrell, S.P., (1985) *Inverse Problems,* **1**, 199-218.

[3] Luttrell, S.P., (1985), *Optica Acta,* **32**, 703-716.

[4] Luttrell, S.P., and Oliver, C.J., (1986) *J. Phys. D: Appl. Phys.,* **19**, 333-356.

[5] Kinderman, R. and Snell, J.L., (1980) Contemporary Mathematics, **1**.

[6] Besag, J., (1974) *J. Royal Stat. Soc: series B,* **36**, 192-326.

[7] Metropolis, N., Rosenbluth, A.W., Rosenbluth M.N. and Teller, A.H., (1953) *J. Chem. Phys.,* **21**, 1087-1092.

[8] Geman, S. and Geman, D., (1984) *IEEE PAMI,* **6**, 721-741.

SUPER RESOLUTION BY PROCESSING IN RADON SPACE

J.M. Blackledge and R.E. Burge

(Department of Physics, King's College, London University)

ABSTRACT

There are some digital processes which in 1D are easy to
implement but in 2D become so computationally demanding that
they have little practical use and this can be rather
frustrating when a 1D digital processor is known to yield
important and reliable results. One such example is super
resolution by spectral extrapolation which in 1D requires the
inversion of a square matrix and in 2D, a block matrix. A
common method of overcoming this is to apply parallel strip
processing which reduces the dimensionality of the problem
but is valid if and only if the image is separable. With
non-separable images, we are faced with the problem of
inverting a highly populated matrix. A method of digital
image processing is therefore examined where an image $I(x,y)$
is first converted into a sequence of parallel projections
$p_\theta(x)$ for angles of rotation θ that lie between 0 and π. Each
projection is then super resolved and the image reconstructed
from the new set of projections. In mathematical terms, this
method is based on taking the Radon transform of a 2D function
and processing it in Radon space. It is therefore a method
of reducing the dimensionality of a problem for non-separable
images and is completely general. In terms of operators, the
method relies on the application of a 1D processor $\hat{P}_{1D}$ which is
related to an equivalent 2D processor $\hat{P}_{2D}$ by $\hat{P}_{2D} = \hat{R}^{-1}\hat{P}_{1D}\hat{R}$
where $\hat{R}$ and $\hat{R}^{-1}$ are the forward and inverse Radon operators
respectively and $\hat{R}^{-1}\hat{R} = 1$.

1. INTRODUCTION

Super resolution by spectral extrapolation attempts to
generate the spatial frequency components of an image that
lie beyond a certain support defined by the resolution limits
imposed by a particular experiment. A variety of
different spectral extrapolation techniques are available but
attention has recently focussed on the incorporation of prior
knowledge. However, there is one important point that a large
proportion of these processes have in common, which is that
inherent in their method of reconstruction is the inversion of
a matrix. While on a theoretical basis, this point is often
taken for granted, when one is required to design a practical
and stable numerical scheme to implement an algorithm, this
can become a focal issue and puts severe limits on the size and
therefore the type of image that can be processed. In 1D
(i.e. signal processing), inversion of a square matrix is
required whereas in 2D (i.e., image processing) one is required
to invert a block matrix. A block matrix corresponding to
a NxN array can be represented by a $N^2 x N^2$ square matrix. Thus,
the size of the matrix that must be inverted is proportional
to the square of the image size and for a typical working space
of 64x64, one is required to invert a 4096x4096 matrix! This
problem is common to many other image methods. Take for
example, deconvolution, which is one of the most widespread
inverse processes, occurring in all imaging science. The most
accurate method of deconvolving a discrete image is to invert
the relevant block matrix equation. For even standard size
images, this is not generally possible because of the extensive
computations that are required to invert the relevant matrix
equation. Deconvolution is therefore usually carried out
in Fourier space by Weiner filtering. All that is
required for Weiner filtering is a 2D FFT which on a numerical
basis, presents much less of a problem than matrix inversion.

Matrix inversion limits super resolution to either small
images or separable images which is unsatisfactory. The
whole problem can be overcome by processing the image in Radon
space. In this paper, attention is focussed on a particular
super resolution algorithm by processing in Radon space but it
should be stressed that in principle, any suitable 2D process
can be implemented using this method and is most useful
when the 2D process in question is too computationally demanding
to be of practical use.

2. IMAGE PROCESSING IN RADON SPACE

Let us define a 2D processing operator P_{2D} and 1D processing operator P_{1D} which are operationally equivalent (i.e. both operators do the same thing but in different dimensions). If an image $I(x,y)$ is separable, then we can write

$$I(x,y) = I_x(x) I_y(y)$$

and processing can be accomplished using

$$\hat{P}_{2D} I(x,y) = \hat{P}_{1D} I_x(x) \hat{P}_{1D} I_y(y)$$

This is the most common way of overcoming the implementation of a $\hat{P}_{2D}$ that presents computational problems but is strictly limited to images that are separable. This condition is often relaxed and parallel strip processing on non separable images is common but strictly speaking, incorrect. What is required, is a method of reducing the dimensionality of a function without assuming separability and this is where the Radon transform becomes useful. The operator $\hat{R}$ for generating the Radon transform is given by

$$\hat{R} = \int \int dxdy \delta(z - (x^2 + y^2)^{\frac{1}{2}} \cos\theta)$$

where δ is the 1D Dirac delta function and converts a 2D function $I(x,y)$ into a sequence of 1D functions $P_\theta(Z)$ which are projections of $I(x,y)$ taken at different angles of rotation θ about a given co-ordinate. A projection is therefore obtained by summing the components of a 2D function along a given set of parallel lines. The operator $\hat{R}^{-1}$ for generating the inverse Radon transform (i.e. reconstructing a 2D function from a sequence of projections) is given by

$$\hat{R}^{-1} = \frac{1}{2\pi} \int_0^\pi d\theta \hat{H} d_z; \quad Z = (x^2 + y^2)^{1/2} \cos\theta$$

where $\hat{H}$ is the Hilbert transform operator. In practice, the Radon transform is simply a method of converting a digital image into a sequence of 1D arrays from which the image can

be reconstructed. Any suitable 1D process can therefore be
applied in Radon space (i.e. to each projection) and the
resulting set of projections used to reconstruct a new image.
The point to appreciate is that applying $\hat{P}_{1d}$ to the
image in Radon space is equivalent to applying $\hat{P}_{2D}$ to the image
itself. Thus, in terms of operators we have the relationship

$$\hat{P}_{2D} = \hat{R}^{-1}\hat{P}_{1D}\hat{R}$$

This result has been proved rigorously by Blackledge (1986)
for any $\hat{P}_{2D}$ for which the equation

$$\hat{P}_{2D}I(x,y) = A(x,y) \otimes \otimes I(x,y) + B(x,y)$$

can be written where $A(x,y)$ is the point spread
function that is generated by the process and $B(x,y)$ is any
other additional function. The importance of this relationship
becomes clear when we consider a digital process $\hat{P}_{2D}$ which
requires extensive computation such as block matrix inversion
in contrast to the 1D version of the process $\hat{P}_{1D}$ involving
relatively minimal computation such as square matrix inversion.
This relationship is therefore fundamental to processes where
computing $\hat{R}^{-1}\hat{P}_{1D}\hat{R}$ in preference to $\hat{P}_{2D}$ makes a particular
algorithm feasible and cost effective.

3.   APPLICATION TO SUPER RESOLUTION USING PRIOR KNOWLEDGE

Recent solutions to the problem of reconstructing a function
from incomplete Fourier data have been concerned with the
incorporation of prior knowledge. In this case, Fourier data
is extrapolated using a solution that is constrained by a
weighting function which is constructed from prior knowledge.
In 1D, we are faced with the problem of constructing an
estimate $\hat{f}(x)$ for a desired function $f(x)$ that is related to
the spectral samples $\hat{f}(k_n)$; $1 \le n \le N$ via the Fourier transform

$$\tilde{f}(k_n) = \int_{-X/2}^{X/2} dx f(x) \exp(-ik_n x)$$

If we construct a linear weighted estimate for $f(x)$ of the
form

$$\hat{f}(x) = w(x) \sum_{n=1}^{N} a_n \exp(ik_n x) \qquad (3.1)$$

where $w(x)$ is the weighting function, then by minimizing the cost function (Byrne and Fitzgerald 1983)

$$e = \int_{-X/2}^{X/2} dx \mid f(x) - \hat{f}(x) \mid^2 / w(x) \qquad (3.2)$$

the Fourier coefficients a_n, are obtained by solving the square matrix equation

$$\sum_{n=1}^{N} a_n \tilde{w}(k_n - k_m) = \tilde{f}(k_m) \qquad (3.3)$$

where $\tilde{f}(k_m)$ and $\tilde{w}(k_m)$ are the Fourier transforms of $f(x)$ and $w(x)$ respectively. This cost function is only one of many that can be used and is limited to intensity maps because of the condition $w(x) \geq 0 \forall x$ that must be applied to equation (3.2). Equation (3.3) can be inverted as a Toeplitz matrix (i.e. a symmetric matrix whose leading diagonal is a constant —the DC level of the spectrum $\tilde{w}(k_m)$).

In 2D, the estimate $\hat{I}(x,y)$ of some desired image function $I(x,y)$ will be of the form

$$\tilde{I}(x,y) = W(x,y) \sum_{n=1}^{N} \sum_{m=1}^{N} a_{nm} \exp(ik_n x) \exp(ik_m y) \qquad (3.4)$$

Here, $a_{nm} \equiv a(k_n k_m)$ are given by the solution to the block matrix equation

$$\sum_{n=1}^{N} \sum_{m=1}^{M} a_{nm} \tilde{W}(k_n - k_p, k_m - k_q) = \tilde{I}(k_p, k_q) \qquad (3.5)$$

where $\tilde{I}(k_p, k_q)$; $1 \leq p \leq N$; $1 \leq q \leq M$ are the spectral samples which are related to $I(x,y)$ via the 2D Fourier transform

$$\tilde{I}(k_p, k_q) = \int_{-X/2}^{X/2} \int_{-Y/2}^{Y/2} I(x,y) \exp(ik_p x) \exp(ik_q y) dxdy$$

and $\tilde{W}(k_p, k_q)$ is the Fourier transform of $W(x,y)$. Equation (3.5) can be written as a block Toeplitz matrix and in principle, inverted. However, the computational demands are significantly greater than those associated with the inversion of equation (3.3) and as the size of the image increases, these demands can grow to impractical limits. However, in Radon space, equation (3.5) can be written as

$$\sum_{n=1}^{N} a_{n\theta} \tilde{w}_\theta (k_n - k_m) = \tilde{f}_\theta (k_m) \; \forall \; 0 \le \theta \le \pi \qquad (3.6)$$

which for any θ, is identical to equation (3.3). Similarly, equation (3.4) can be written in Radon space as

$$f_\theta (x) = w_\theta (x) \sum_{n=1}^{N} a_{n\theta} \exp(ik_n x) \; \forall \; 0 \le \theta \le \pi \qquad (3.7)$$

which for any θ is identical to equation (3.1). Thus to find $\hat{I}(x,y)$ one can generate all the estimates $\hat{f}_\theta \; \forall \; 0 \le \theta \le \pi$ in Radon space by solving equation (3.6) for all the projections $\tilde{w}_\theta (k_n)$ and $\tilde{f}_\theta (k_m)$ and then apply the inverse Radon transform to obtain $\hat{f}_\theta (x)$. The functions $\tilde{w}_\theta (k_n)$ and $\tilde{f}_\theta (k_n)$ are obtained by taking the Fourier transforms of the projections generated from $W(x,y)$ (the 2D weighting function) and $I_B (x,y)$ respectively where $I_B (x,y)$ is the band limited image whose resolution is governed by the support of $\tilde{I}(k_q, k_p)$. Thus, we arrive at the following algorithm for super resolution by processing in Radon space:

(i) Given a band limited image $I_B(x,y)$ construct a suitable weighting function $W(x,y)$ from prior knowledge on the structure of the image.

(ii) Take the Radon transform of $I_B(x,y)$ and $W(x,y)$ and then take the Fourier transform of each projection to obtain the data $\tilde{w}_\theta (k_m)$ and $\tilde{f}_\theta (k_m)$.

(iii) Generate all the estimates $\hat{f}_\theta (x)$ via equation (3.7) by solving equation (3.6) for each set of projections to find $a_{n\theta}$.

(iv) Take the inverse Radon transform of $\hat{f}_\theta (x)$ to give $\hat{I}(x,y)$.

Using this scheme one is able to super resolve an image without assuming separability and yet avoid the computational demands of block matrix inversion. This method is demonstrated using the synthetic example shown in figure 1. A circular low

pass filter is applied to the image (a) giving the highly
degraded and band limited image (b). A weighting function is
constructed as shown in (c) which is based on part of the known
structure of (a). Application of the above algorithm to image
(b) using (c) yields the super resolved image (d). Notice
that relative to (b), the internal structure of the object
is now clearly defined.

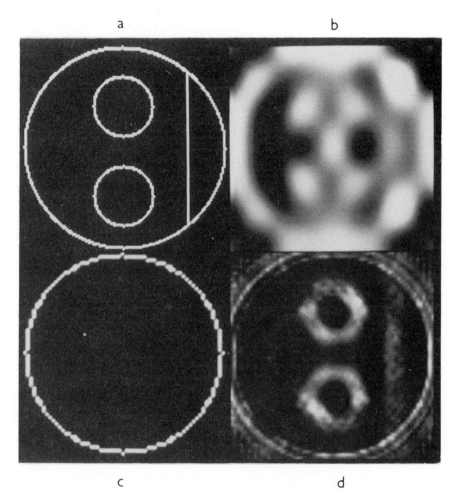

Fig. 1 Demonstration of super resolution by processing in
        Radon space. The image (a) is the original 256x256
        image and (b) is a low pass filtered version of (a).
        The weighting function is shown in (c) and the
        reconstruction in (d).

ACKNOWLEDGEMENTS

   This work is supported by the Ministry of Defence, Royal
Signals and Radar Establishment.

REFERENCES

Blackledge, J.M. (1986)   OPTIK, 73, No. 2, 74-82.

Byrne, C.L. and Fitzgerald, R.M., (1982)   *Siam. Appl. Math.* **42**,
   933-940.

ENHANCEMENT OF MULTI-LOOK SYNTHETIC
APERTURE RADAR IMAGES OF MOVING TARGETS

M.S. Scivier and D.G. Corr
(Systems Designers Scientific, Pembroke House,
Pembroke Broadway, Camberley, Surrey)

We show that speckle reduction by look summation is less
effective for moving scenes than for stationary ones. We
propose a statistical method for multi-looking SAR images of
moving scenes. The proposed algorithm aims to enhance one of
the looks using the local statistics of pixel values in all the
looks. The method represents an extension of adaptive speckle
reduction techniques to take into account the temporal variability of the radar reflectivity.

INTRODUCTION

   Synthetic aperture radar (SAR) generates high resolution
images of the Earth's surface by range and Doppler processing
of radar echoes. SAR images contain speckle noise that ob-
scures the underlying changes in reflectivity. Since the
maximum resolution of SAR in range and azimuth is often better
than is actually needed, it is possible to improve the signal-
to-speckle noise ratio (SSNR) at the expense of resolution.

   In this paper, we consider speckle reduction by look summation in which the available Doppler bandwidth is divided into
N segments and the N images generated from these segments are
summed incoherently. We derive expressions for the effect of
moving targets on the point spread function (PSF) of the SAR
system, and show that if the scene contains moving targets the
incoherent addition of looks may not result in improvements to
image quality comparable to those for stationary targets.
Scenes undergoing random motion (such as the sea surface) will
give rise to variations between looks which are not taken into
account by conventional multi-look techniques.

   We propose a new method for multi-looking SAR images of
moving scenes which is based on the statistical variation of

pixel values from one look to another as well as within each
look. The aim of the algorithm is to enhance one of the looks
by modifying its pixels according to decisions based on the
local statistics within the look and the statistical variation
of corresponding pixels in other looks.

The first part of the paper is a brief review of certain
aspects of speckle and multi-look image processing. The rest of
the paper contains a detailed explanation of the proposed method.

2   SAR SPECKLE

2.1 *Origin of Speckle*

Speckle is characteristic of coherent systems and results
from the coherent addition of many randomly phased contributions.
Typically the radar wavelength is of the order of centimetres
whereas the resolution cell size is several tens of metres
(for spaceborne radars). Consequently, there are usually many
scatterers per resolution cell and, under these circumstances,
the SAR image will have a speckled appearance, even if the
surface is smooth.

Speckle is generally regarded as undesirable (i.e. as noise)
and we are interested in ways of reducing it.

2.2 *Multiplicative Speckle Model*

The probability density function for the speckled image
intensity is a gamma distribution as follows:

$$p(I) = \frac{<I>^N \, I^{N-1} \, e^{-NI/<I>}}{N^N (N-1)!}$$

where I   = image intensity
      N   = number of independent images
    <I> = expected value of intensity.

We can model the image in terms of multiplicative speckle
noise (Frost 1982), i.e.:

$$I = <I> . \, n/2N$$

where n = a random variable obeying a standard chi-squared
           distribution with 2N degrees of freedom.

The random variable n describes the speckle noise. The
quantity <I> is constant over homogeneous areas of the scene

but, more generally, is a slowly varying function of position describing the scene reflectivity.

2.3 Speckle Reduction

One way of achieving speckle reduction is by look summation. This involves the incoherent addition of several independent images of the same scene i.e.:

$$I = \sum_{i=1}^{N} I_i$$

where N is the number of independent images.

As we mentioned in the previous section, the resulting image obeys a gamma distribution. The variance of the distribution is inversely proportional to the number of looks. Therefore, the more looks that are summed incoherently, the greater the degree of speckle reduction.

2.4 Multi-look Processing

In the SAR processor developed by Systems Designers at RAE Farnborough, multi-looking is achieved as follows:

a) We divide the radar beam in the along track direction into a number of segments. This is equivalent to forming a number of subapertures.

b) Each segment of data is processed independently to produce an image.

c) The independent images are summed incoherently to produce the multi-look image.

For SEASAT, each subaperture is separated in time by approximately half a second. If the scene is stationary, the underlying reflectivity in each of the looks is the same. For a dynamic scene, such as the sea surface, this is not necessarily true.

3. THE EFFECT OF TARGET MOTION ON MULTI-LOOKING

3.1 Stationary Target Response

The two-dimensional PSF of a SAR is proportional to the following (Barber 1983):

$$e^{4\pi i f_o \rho / c} \; sinc \; (\Delta F_R t_R) \; sinc \; (\Delta F_A t_A)$$

where ρ $\qquad = at_A + bt_A^2$

$\quad sinc(x) \qquad = sin(\pi x)/\pi x$

$\quad t_R \qquad\qquad = 2y/c, \; y = slant \; range$

$\quad t_A \qquad\qquad = x/v$

$\quad x \qquad\qquad = azimuth$

$\quad v \qquad\qquad = velocity \; of \; radar \; platform$

$\quad f_o \qquad\qquad = centre \; frequency \; of \; transmitted \; chirp$

$\quad a \qquad\qquad = slant \; range \; velocity \; at \; t = 0$

$\quad 2b \qquad\qquad = slant \; range \; acceleration \; at \; t = 0$

$\quad \Delta F_R \qquad\quad = 2\alpha T$

$\quad \alpha \qquad\qquad = rate \; of \; linear \; FM$

$\quad T \qquad\qquad = pulse \; duration$

$\quad \Delta F_A \qquad\quad = Doppler \; bandwidth$

As explained in the previous section, we generate individual looks by processing only part of the available Doppler bandwidth. The full Doppler bandwidth is given by:

$$\Delta F_A = (2f_o/c) v^2 T_A / R_o$$

where $\quad T_A \qquad = integration \; time$

$\quad\quad R_o \qquad = slant \; range \; at \; t = 0.$

The Doppler bandwidth corresponding to a single segment of the radar beam (i.e. a subaperture) is:

$$\Delta F_A = (2f_o/c) v^2 T_A / NR_o$$

For N looks, the azimuth resolution, which is proportional to
ΔF_A, is reduced by a factor of N.

3.2 Moving Target Response

Now we consider the effect of variations in scene reflectivity which occur during the time taken to acquire the image.
The SAR imaging process relies on the relative motion of the
radar and a stationary target. Therefore, if a target is
moving the received signal contains additional terms which are
not taken into account by the processor.

Considering only the range and azimuth velocities, and the
range acceleration, we can write the following expression for
the range to a moving target (Raney 1971):

$$R(t) = R_o - v_R t + \frac{\{(v-v_A)^2 - R_o a_R\}t^2}{2R_o}$$

where v_R and v_A are the range and azimuth velocities respectively, and a_R is the range acceleration.

The effects of target motion can be divided into range and
azimuth dependent terms. If the target has a component of
velocity in the range direction then the image will suffer from
range smear and attenuation of the peak intensity. These two
range effects are look-dependent. In other words, the range
position and peak intensity of the moving point target will
vary from one look to another. The displacement in range
between looks is given by $v_R t_n$ where t_n is the centre time of
the nth look which is equal to $T_A(2n-N-1)/2N$. This effect is
essentially the same as time-lapse imaging (Ouchi 1985).

There are also motion-induced phase terms which lead to
changes in the azimuth response between looks. These should
not be confused with the azimuthal image shift of range
travelling targets, e.g. the displacement of a ship from its
wake, which is unchanged between looks. The look-dependent
azimuthal displacement terms are as follows:

$$\Delta x = \{(a_R R_o/v) + 2v_A\}t_n$$

For a scene undergoing random motion, the reflectivity cannot necessarily be expected to remain unchanged during the time
taken to acquire all of the looks.

4. MULTI-LOOK IMAGE ENHANCEMENT

4.1 *Description of the Algorithm*

When the scene is in motion during the integration time and the reflectivity is intrinsically different between looks, any one of the looks is as good as any other. This is because the ideal image would be a snapshot of the scene at one particular instant in time. However, speckle reduction by look summation is still desirable in those regions of the image where the statistics suggest that the variation from one look to another is due to speckle noise.

Returning to the multiplicative noise for the image, we write

$$z = x.n$$

$$\text{where}\quad z = \text{actual image}$$

$$x = \text{non-speckled image}$$

$$n = \text{speckle noise.}$$

We deal with the non-linearity of the model by using a linear approximation as follows:

$$z = \langle z \rangle + \langle n \rangle \; (x - \langle x \rangle) + \langle x \rangle \; (n - \langle n \rangle)$$

This is equivalent to a first order Taylor expansion about the mean value of z. The form of the equation is the same as an additive noise model. We can now derive an estimate of the ideal image from the measured data by minimising the mean square error. The estimated ideal image is given by:

$$x' = \langle x \rangle - k \; (z - \langle z \rangle)$$

$$\text{where}\quad k = \frac{\langle n \rangle V_x}{\langle x \rangle^2 V_n + \langle n \rangle^2 V_x}$$

The quantities V_x and V_n are the variance of the scene reflectivity and speckle noise respectively.

The method was used by Lee (1980) and Frost (1982) in spatial filtering algorithms. However, in this case, we are concerned with the temporal (rather than spatial) variability of the scene. As shown below, the multi-look SAR data can be considered as a set of stacked data, and the statistical parameters

used in the estimation procedure can be made to relate to the
variation of pixel values across the stack. This approach is
similar to one used for stacked seismic traces which was based
on "maximum a posteriori" estimation, and Gaussian models for
the image and noise statistics (Waltham and Boyce 1985).

The estimated value of the scene reflectivity is the weighted
sum of the mean of x and the value of z. The weighting factor
depends on $<n>$, $<x>$, V_n and V_x. The noise statistics, $<n>$ and
V_n, are deduced from our model for the speckle. For a single
look image, $<n> = 1$ and $V_n = 1$. The scene statistics, $<x>$ and
V_x, are obtained from the local statistics, $<z>$ and V_z.

The collection of the local statistics is illustrated in
Figure 1. A common window is identified in all looks, e.g.
Figure 1 shows 4 looks with a 3x3 window in each. The statistics of all the pixels lying within the windows are measured
to give $<z>$ and V_z. The reflectivity statistics are found by
using the following:

$$<x> = <z>/<n>$$

$$V_x = \frac{V_z + <z>^2 - <x>^2}{V_n + <n>^2}$$

Substituting for $<n>$, $<x>$, V_n and V_x gives:

$$x' = x_M + \frac{V_z - <z>^2}{V_z + <z>^2} (z - x_M)$$

where x_M is the multi-look image. This is the key equation.

When $V_z = <z>^2$, speckle noise accounts for the variation
between looks, and the estimated image tends to the
conventional multi-look image. When $V_z >> <z>^2$, the statistics
suggest that the scenes are intrinsically different, and the
value of the pixel is left unaltered.

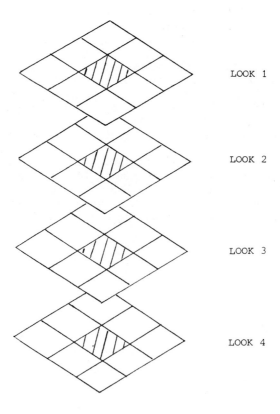

LOOK 1

LOOK 2

LOOK 3

LOOK 4

Fig. 1   To show the pixels used in the calculation of the
         statistics for a multi-look stack of images.

4.2   Concluding Remarks

We have applied the algorithm to SEASAT imagery of the
North Atlantic to obtain some initial results. However, the
improvement is less noticeable than expected and suggests that,
in this particular example, the scene is largely unchanged
between looks. This will not always be the case and we
anticipate that, for other examples of dynamic scenes, the new
method will help to preserve features which would otherwise
be lost after look summation. We are currently investigating
a variety of multi-look data.

We conclude that our work represents a logical extension of adaptive speckle reduction techniques to take into account the temporal variability of the scene when performing look summation.

5. ACKNOWLEDGEMENTS

The authors wish to acknowledge the support of Systems Designers Scientific and Brian Barber (RAE Farnborough).

6. REFERENCES

Barber, B.C., (1983) Some properties of SAR speckle, *Satellite Microwave Remote Sensing,* Ellis Horwood, Chapter 8.

Frost, V.S. et al., (1982) A model for radar images and its application to adaptive digital filtering of multiplicative noise, *IEEE Trans. Pattern Anal. Machine Intell.* Vol. PAMI-4, 157-165.

Lee, J.S., (1980) Digital image enhancement and noise filtering by use of local statistics, *IEEE Trans. Pattern Anal. Machine Intell.* Vol. PAMI-2, 165-168.

Ouchi, K., (1985) On the multi-look images of moving targets by synthetic aperture radars, *IEEE Trans. Antennas Propagat.* Vol. AP-33, 823-827.

Raney, R.K., (1971) Synthetic aperture imaging radar and moving targets, *IEEE Trans. Aerosp. Electron. Syst.* Vol. AES-7, 499-505.

Waltham, D.A. and Boyce, J.F., (1985) Signal-to-noise ratio enhancement in seismic multifold data using Bayesian statistics, *Geophysical Prospecting* Vol. 33.

GLOBAL PERFORMANCE CRITERIA AND CORRESPONDING ALGORITHMS FOR CONTEXTUAL CLASSIFICATION OF MULTISPECTRAL IMAGES

J. Kittler

(Department of Electronic and Electrical Engineering, University of Surrey)

D. Pairman*

(Atmospheric Physics Department, Oxford University)

ABSTRACT

The role of contextual information in segmenting and classifying remotely sensed images has been studied by a number of authors and shown to be very useful. The emphasis of research thus far has been on the mechanics of combining contextual information contained in the image. In contrast, this paper investigates the effect of global performance criteria on the final result of pixel classification. Two criterion functions are considered, namely the maximum a *posteriori* probability of the joint occurrence of assigned pixel labels and an average maximum *a posteriori* probability of assigned class label per pixel. Classification algorithms corresponding to these criterion functions are derived and experimentally compared.

1. INTRODUCTION

Context is an important source of information in automatic image understanding. Its role at higher levels of interpretation has been investigated for some time now but the last decade has also witnessed a considerable proliferation of its use at lower levels of processing often involving pixel classification.

The methods which make use of context range from probabilistic relaxation [1], through augmented vector classification [2], to contextual decision rules. In this

---

* Supported by NRAC (NZ) fellowship while on study leave from the Department of Scientific and Industrial Research, New Zealand.

paper we shall concentrate on the latter approach and in
particular on techniques that found application in
classification of remote sensing imagery.

   Among the first attempts to improve pixel classification
performance by incorporating contextual information into a
decision rule was the work of Swain et al [3] who took into
account the effect of two immediately adjacent pixels, one in
the horizontal and one in the vertical directions, under the
assumption of memoryless noise. Fu and Yu [4] extended the
work of Welch and Salter [5] to consider the context drawn
from the four-neighbourhood and also from the eight-
neighbourhood under some simplifying assumptions. Kittler
and Foglein [6] developed a general contextual decision rule
and investigated a number of special cases that may arise in
practice. An alternative formulation of their approach
(Kittler and Foglein [7] and Kittler and Pairman [8]) led to
the development of an iterative decision making scheme with
close similarities to the methods maximising the aposteriori
probability of joint pixel labelling studied by a number of
authors [9,10].

   As an extension of the work in [8], this paper
investigates two global criteria of image labelling and
compares experimentally the iterative contextual decision
schemes which optimise these indices of performance. In
Section 2 the necessary introductory material is presented
and notation introduced. In Section 3, a heuristic criterion
of pixel labelling is developed and the corresponding
decision rule derived. A similar derivation is carried out
in Section 4 for the aposteriori probability of joint
labelling. The two algorithms are discussed in Section 5 and
applied in Section 6 to the problem of automatically
detecting and identifying clouds in satellite imagery.
Finally Section 7 offers some conclusions that can be drawn
from the results.

2. PRELIMINARIES

   Let us consider a digital satellite image containing N
pixels. With each pixel, j, we associate a multivariate
observation, x_j, the components of which represent the
intensities of radiation detected by a bank of spectral
channels. The nature of the imaging and sampling processes
is such that the pixels are arranged in a square lattice. We
shall assume that the detected radiation is a function of the
spectral properties of an object (land cover) in the part of
a scene the pixel is imaging and that there are m distinct
classes of objects, ω_i, i=1,2,...m. It is convenient to adopt

a probabilistic model for the image generation process and view the multidimensional observations as realisations of a random variable drawn from a mixture distribution composed of m class conditional components representing the objects.

The primary problem in satellite image interpretation is to assign each image pixel into its appropriate object category. Let us introduce at pixel j a random variable Θ_j which denotes the class label for that pixel. Then our task is to decide which of the possible m outcomes ω_i i=1,...m will be assumed by Θ_j. Now let us suppose for the sake of notational simplicity that the measurement made at pixel j is one-dimensional. Conventional techniques base the decision about Θ_j solely on the observation x_j. However, such an approach makes an inefficient use of the data. In contrast, contextual decision schemes attempt to exploit all the evidence that may be available to improve the classification performance. By virtue of the fact that no object in the world exists on its own or independent of its environment, the information conveyed by its neighbours and its setting can be usefully applied to reduce the ambiguity of its interpretation. Thus as far as a single pixel, say j-th pixel, is concerned, there are in principle three sources of information that can be utilised. First of all it is the observation x_j itself. Second, the measurements $x_\ell, \forall \ell \neq j$ on other pixels in the image (neighbours of pixel j), and lastly, probabilistic relation over the labels of neighbouring pixels which convey the prior world knowledge.

From the point of view of modelling these disparate sources of information it is more convenient not to differentiate between the measurements made on pixel j and its neighbours. Instead we shall distinguish between the two main components of each measurement: the ideal spectral signature μ_r representing the class of object imaged by the j-th pixel and the error or noise component, i.e. deviation from μ_r. Taking out any class-to-class differences in the variance of this noise term by dividing by its class conditional standard deviation, σ_r, we obtain a zero mean stochastic variable ξ_j of unit variance, i.e.

$$\xi_j = (x_j - \mu_r) \, \sigma_r^{-1} \tag{1}$$

Now for both the spatial noise process ξ_j and the class label process Θ_j we can reasonably assume that they are Markovian, i.e. the conditional probabilities $P(\xi_j|\xi_\ell, \forall\ell\neq j)$ and $P(\Theta_j|\Theta_\ell, \forall\ell\neq j)$ satisfy

$$P(\xi_j|\xi_\ell, \forall\ell\neq j) = P(\xi_j|\xi_\ell, \forall\ell\in I_j) \qquad (2)$$

$$P(\Theta_j|\Theta_\ell, \forall\ell\neq j) = P(\Theta_j|\Theta_\ell, \forall\ell\in I_j) \qquad (3)$$

In (2) and (3) I_j is the index set of the neighbours to pixel j which convey all the information about that pixel contained in the observed variables.

Markov random fields have recently received considerable attention in image modelling as they permit the generation of valid joint probability distributions on lattice structures. A key specification in defining a Markov process is that of the neighbourhood system I_j, $\forall j$. In the present paper we shall confine our discussion to the four-neighbourhood of figure 1 and consider it equally applicable to the noise and label process throughout the image.

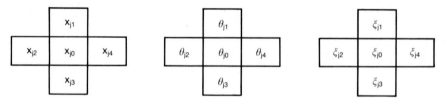

Fig. 1

The physical significance of the noise component ξ_j suggests that it might be appropriate to model it by an autonormal process defined as

$$P(\xi_{j0}|\xi_{j1}, \cdots \xi_{j4}) = [2\pi\sigma^2]^{-\frac{1}{2}}\exp\left\{-\frac{1}{2\sigma^2}[\xi_{j0} - \sum_{\ell=1}^{4}\beta_\ell\xi_{j\ell}]^2\right\}, \forall j \qquad (4)$$

where σ^2 is the conditional variance of ξ_{j0} given its neighbours. For a stationary process the parameters β_ℓ must satisfy $\beta_1 = \beta_3$ and $\beta_2 = \beta_4$.

As far as the label process is concerned we shall not assume any parametric form for it. Suffice it to say that the joint probability of labelling $P(\Theta_1, \ldots \Theta_N)$ must be a Gibbs distribution, i.e. it is defined in terms of interactions between all the possible cliques of pixels that can be formed for the adopted neighbourhood system. Since each site gives rise to four such cliques the structure of the conditional probability function $P(\Theta_{jo}|\Theta_{j1}\ldots\Theta_{j4})$ will be

$$P(\Theta_{jo}|\Theta_{j1}, \ldots \Theta_{j4}) \sim \exp \left\{ G_0(\Theta_{jo}) + \sum_{\ell=1}^{4} G_\ell(\Theta_{jo}, \Theta_{j\ell}) \right\} \qquad (5)$$

where the Markov field consistency conditions [10] require that for two neighbouring pixels Θ_j and Θ_h the G functions satisfy $G_k(\Theta_j, \Theta_h) = G_\ell(\Theta_h, \Theta_j)$ for all k and ℓ such that $\Theta_{jk} = \Theta_h$ and $\Theta_{h\ell} = \Theta_j$.

We shall further assume for the structure of $P(\Theta_{jo}|\Theta_{j1}, \ldots \Theta_{j4})$ that

$$P(\Theta_{jo}|\Theta_{j1}, \ldots \Theta_{j4}) = \frac{P(\Theta_{j1}, \ldots \Theta_{j4}|\Theta_{jo})P(\Theta_{jo})}{P(\Theta_{j1}, \ldots \Theta_{j4})} \sim P(\Theta_{jo}) \cdot \prod_{\ell=1}^{4} P(\Theta_{j\ell}|\Theta_{jo})$$

$$(6)$$

3. HEURISTIC LABELLING CRITERION

In this section we shall develop a family of labelling criterion functions and the corresponding algorithms which aim at maximising the geometric mean of a "per pixel" performance index related to error probability. Our starting point will be the contextual decision rule: assign Θ_j to ω_i if

$$P(\Theta_j = \omega_i | x_j, x_\ell, \Theta_\ell, \forall \ell \neq j) = \max_r P(\Theta_j = \omega_r | x_j, x_\ell, \Theta_\ell, \forall \ell \neq j)$$

$$(7)$$

In (7) and in the ensuing discussion we do not distinguish between a random variable and its realisation unless it is absolutely essential for the sake of clarity.

Using the Bayes formula we can write

$$P(\Theta_j = \omega_r | x_j, x_\ell, \Theta_\ell, \forall \ell \neq j) = \frac{P(x_\ell, \forall \ell | \Theta_\ell, \forall \ell) . P(\Theta_\ell, \forall \ell \neq j, \Theta_j = \omega_r)}{P(x_\ell, \Theta_\ell, \forall \ell \neq j, x_j)}$$

(8)

Now using (1), (2) and (3) we obtain

$$P(x_\ell, \forall \ell | \Theta_\ell, \forall \ell) = p(\xi_\ell, \forall \ell) = p(\xi_j | \xi_\ell, \forall \ell \neq j) . p(\xi_\ell, \forall \ell \neq j) \quad (9)$$

and

$$P(\Theta_\ell, \forall \ell \neq j, \Theta_j = \omega_r) = P(\Theta_j = \omega_r | \Theta_\ell, \forall \ell \epsilon I_j) . P(\Theta_\ell, \forall \ell \neq j) =$$

$$= P(\Theta_\ell, \forall \ell \epsilon I_j | \Theta_j = \omega_r) P(\Theta_j = \omega_r) . \frac{P(\Theta_\ell, \forall \ell \neq j)}{P(\Theta_\ell, \forall \ell \epsilon I_j)}$$

(10)

Substituting (9) and (10) into (8) and ignoring all the terms that do not involve label Θ_j we find

$$P(\Theta_j = \omega_r | x_j, x_\ell, \Theta_\ell, \forall \ell \neq j) \backsim p(\xi_j | \xi_\ell, \forall \ell \epsilon I_j) . P(\Theta_\ell, \forall \ell \epsilon I_j | \Theta_j = \omega_r) . P(\Theta_j = \omega_r)$$

(11)

It follows that the decision rule (7) can be expressed in a more convenient form as

assign Θ_j to ω_i if

$$p(\xi_j^i | \xi_\ell, \forall \ell \epsilon I_j) . P(\Theta_\ell, \forall \ell \epsilon I_j | \Theta_j = \omega_i) P(\Theta_j = \omega_i) =$$

$$= \max_r p(\xi_j^r | \xi_\ell, \forall \ell \epsilon I_j) P(\Theta_\ell, \forall \ell \epsilon I_j | \Theta_j = \omega_r) P(\Theta_j = \omega_r)$$

(12)

where the superscript in ξ_j^r indicates the measurement x_j is normalised under the hypothesis that $\Theta_j = \omega_r$.

While in principle we may wish to assign the j-th pixel to class ω_i according to rule (11), the situation is not so simple. First of all, labels Θ_ℓ, $\forall \ell \epsilon I_j$ are not known but only estimated. Second, any decision regarding Θ_j will also affect all the pixels in its neighbourhood, I_j. We thus need some global criterion which will allow us to compare different labellings Ω assigned to the image and which will lead to a convergent iterative process of readjusting pixel labels at successive sites to find an optimal label assignment.

Since we wish to minimise the error probability at each pixel, it may be readily justifiable to define as a global criterion of labelling the arithmetic or geometric mean of the contextual aposteriori class label probability on the left hand side of (8). However, such an objective function would be extremely difficult to optimise. For computational reasons we shall use a heuristic substitute defined in terms of the right hand side of (12) as

$$J(\Omega) = [\prod_{j=1}^{N} p(\xi_j|\xi_\ell,\forall\ell\epsilon I_j) \ P(\Theta_\ell,\forall\ell\epsilon I_j|\Theta_j) \ P(\Theta_j)]^{1/N} \quad (13)$$

The preference for the geometric mean in specifying $J(\Omega)$ owes to its amenability to further simplification in view of the exponential form of $p(\xi_j|\xi_\ell,\ell\epsilon I_j)$. Taking the logarithm of (13) and negating the resulting sum we obtain an equivalent criterion function

$$J(\Omega) = - \frac{1}{N} \sum_{j=1}^{N} \log [p(\xi_j|\xi_\ell,\forall\ell\epsilon I_j) \ P(\Theta_\ell,\forall\ell\epsilon I_j|\Theta_j) \ P(\Theta_j) \quad (14)$$

which now should be minimised to find the desired labelling.

A suitable algorithm for minimising $J(\Omega)$ in (14) can be found by considering the effect on $J(\Omega)$ of any change in the class label of the j-th pixel. Substituting (4) into (14) it is easy to verify that the analysis leads to the following reassignment rule

assign Θ_j to ω_i if

$$F(\Theta_j = \omega_i) = \min_r F(\Theta_j = \omega_r) \quad (15)$$

where

$$F(\Theta_j = \omega_r) = \frac{1}{2\sigma^2} \xi_{jo}^r [\xi_{jo}^r (1+2\beta_1^2 + 2\beta_2^2) - 4\beta_1 (\xi_{j1} + \xi_{j3}) -$$

$$- 4\beta_2 (\xi_{j2} + \xi_{j4}) + 4\beta_1\beta_2 (\xi_{j5} + \xi_{j6} + \xi_{j7} + \xi_{j8}) +$$

$$+ 2\beta_1^2 (\xi_{j9} + \xi_{j11}) + 2\beta_2^2 (\xi_{j10} + \xi_{j12}) - \qquad (16)$$

$$- \log P(\Theta_{jo} = \omega_r, \Theta_{jk}, \forall k\epsilon I_{jo}) -$$

$$- \sum_{\ell=1}^{4} P(\Theta_{j\ell}, \Theta_{jk}, k\epsilon I_{j\ell} | \Theta_{jk} = \omega_r, k=0)$$

For clarity the local indexation of variables ξ and Θ in the neighbourhood of pixel j adopted in (14) is that of figure 2, i.e. jO replaces j.

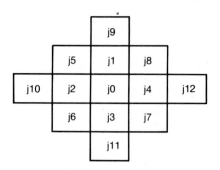

Fig. 2

Denoting by $F_n(\Theta_j = \omega_r)$ the component of (14) due to the noise process, we can write down the equivalent expression for $F(\Theta_j = \omega_r)$ for the probabilities of joint neighbourhood labelling factorised as in (6), i.e.

$$F(\Theta_j=\omega_r) = F_n(\Theta_j=\omega_r) + 3\log P(\Theta_{jo}=\omega_r) -2 \sum_{\ell=1}^{4} \log P(\Theta_{jo}=\omega_r, \Theta_{j\ell})$$

$$(17)$$

4. A POSTERIORI PROBABILITY OF JOINT LABELLING

Instead of optimising a performance index which is related to the probability of each pixel being correctly classified the pixel labelling problem can be formulated as one of maximising the aposteriori probability of particular pixel labels occurring jointly in the image. Formally we wish to find image labelling $\Theta_1, \ldots \Theta_N$ which maximises

$$P(\Theta_1, \ldots \Theta_N | x_1, \ldots x_N) = \frac{P(\xi_1, \ldots \xi_N) \, P(\Theta_1, \ldots \Theta_N)}{p(x_1, \ldots x_N)} \qquad (18)$$

Since the denominator is fixed for a given image, the aposteriori probability can be maximised by optimising the numerator of (18).

It is obviously impractical to optimise the aposteriori probability directly. However a suboptimal algorithm which guarantees to find at least a local optimum can be devised. The approach is based on the work of Besag [10] which shows how the probability ratio of two labelling realisations can be factorised in terms of the conditional probability functions given in (2) and (3). Using Besag's result we can write for the probability ratio of two realisations differing only at one pixel, say pixel j,

$$\frac{p(\xi_1, \ldots \xi_j^r, \ldots \xi_N) \, P(\Theta_1, \ldots \Theta_j = \omega_r, \ldots \Theta_N)}{p(\xi_1, \ldots \xi_j^i, \ldots \xi_N) \, P(\Theta_1, \ldots \Theta_j = \omega_i, \ldots \Theta_N)} =$$

$$\frac{p(\xi_j^r | \xi_\ell, \forall \ell \varepsilon I_j) \, P(\Theta_j = \omega_r, \Theta_\ell, \forall \ell \varepsilon I_j)}{p(\xi_j^i | \xi_\ell, \forall \ell \varepsilon I_j) \, P(\Theta_j = \omega_i, \Theta_\ell, \forall \ell \varepsilon I_j)} \qquad (19)$$

where as before ξ_j^r denotes the noise variable obtained by normalising x_j under the hypothesis that $\Theta_j = \omega_r$.

From (19) it is apparent that maximising the product of the conditional probabilities for the noise and label processes over all possible labels of the pixel currently visited will result in an increase in the aposteriori joint probability of labelling. It can again be easily shown that this leads to the following decision rule assign Θ_j to ω_i if $\qquad H(\Theta_j = \omega_i) = \max_r H(\Theta_j = \omega_r)$ (20)

where

$$H(\Theta_0 = \omega_r) = \frac{1}{2\sigma^2} \xi_{j0}^r (\xi_{j0}^r - 2 \sum_{\ell=1}^{4} \beta_\ell \xi_{j\ell}) +$$

(21)

$$+ 3 \log P(\Theta_j = \omega_r) - \sum_{\ell=1}^{4} \log P(\Theta_{j\ell}, \Theta_{j0} = \omega_r)$$

5.   COMMENTS

Special cases of the criterion function (14) and (18) can be obtained under further assumption relating to the noise and label processes. For instance for memoryless noise, i.e. when

$$p(\xi_j | \xi_\ell, \forall \ell \neq j) = P(\xi_j)$$ (22)

and unconditional pixel label independence, i.e.

$$P(\Theta_j | \Theta_\ell, \forall \ell \neq j) = P(\Theta_j)$$ (23)

The corresponding algorithms can be obtained directly from (16) and (21).

From (16) it follows that the optimisation algorithm corresponding to criterion (14) spans a larger neighbourhood than algorithm (21). However, since parameters β_ℓ are much smaller than unity, any term involving their product can be neglected. Then $F_n(\Theta_j = \omega_r)$ in (17) can be approximated as

$$F_n(\Theta_j = \omega_r) = \frac{1}{2\sigma^2} \xi_0^r (\xi_0^r - 4 \sum_{\ell=1}^{4} \beta_\ell \xi_{j\ell})$$ (24)

and the effective neighbourhoods involved in updating label $\Theta_j, \forall j$ at each step of the algorithm will shrink to the four-neighbourhood of figure 1.

It should be noted that none of the algorithms developed in Section 3 and Section 4 guarantees to find a global optimum. Although stochastic optimisation techniques are available, their convergence is extremely slow and for this reason their application has not been considered practical in this study.

6. EXPERIMENTAL PROCEDURE AND RESULTS

In this section we will demonstrate algorithms developed on some real multispectral satellite data. The data used is a 1024 x 1024 pixel portion from an Advanced Very High Resolution Radiometer (AVHRR) image. The AVHRR instrument is flown on the NOAA-7 satellite and gathers data in five spectral channels at a ground resolution of $\sim$ 1km (sub-satellite). For this study only three spectral bands (1, 3 and 5) are used, which detect radiation in three "atmospheric windows" around 0.65μm, 3.7μm and 12μm respectively. The particular scene used was taken off the coast of Portugal on 26 August 1982 at 15:13 GMT. Calibration data obtained in flight and also measured pre-launch was used to convert the data to physical units. That is % Albedo for the visible (band 1) data and mW/Sr x M^2 cm^{-1} for the two infrared bands. The image contains a mixture of cloud types, a small amount of land, and sea which is considered as two separate classes, with and without sun glint. This image was chosen as one of our interests is the classification of cloud types.

In order to parameterize the various models discussed in earlier sections, and in particular the spatial context models, it is necessary to have a training data set containing large numbers of adjacent correctly classified pixels of all classes. Obviously training fields are unsuitable for this and it is really impractical to obtain such a data set by manual identification of many pixels. In order to overcome the problem we have first classified the data using a non-contextual clustering algorithm [12]. This initial non-contextual classification was felt to be substantially correct but it does have several undesirable features common to non-contextual algorithms which will be discussed presently. Nine classes were identified namely; water, sunglint, land, cumulus and partially filled cumulus pixels, thick low cloud, medium cloud, thin cirrus, thick cirrus and deep convective clouds.

The result of the non-contextual classification algorithm was used as a data set to estimate parameters of the contextual models. Class conditional means and variances, used within the contextual routines to normalise observations into a noise field, are obtained by simply estimating from the data using the clustering result as the conditioning mask.

The parameters of the auto-normal process defined in equation (4) were found using maximum likelihood estimation from the normalised noise field. As maximum likelihood estimation requires independence of the observations, sites were coded (following Besag [11]) so that only sites which were not neighbours of each other were chosen for the estimation. Given the (assumed) Markov property these sites were then mutually independent, given all the other sites.

The spatial context was learnt by forming a list of label combinations that occurred in the training set. Once again sites were chosen to be independent. Rotation and mirror image transformations were performed to merge combinations that were otherwise equivalent as direction was not considered important in the current study and combining such local labellings improves the reliability of the statistics estimated. Spatial context is also represented in equations (17) and (21) by pairs of neighbouring labels. The probability of these pairs was also estimated by choosing independent pairs from the training set. An equal mix of vertical and horizontal pairs was used in this estimation.

The algorithms represented by equations (16), (17) and (21) were tested on the data described using the clustering result as a starting point. In all cases fourteen iterations were performed by which time less than 0.5% of the pixels were still changing class.

Figure 3 shows a portion of the starting image and the results of the three algorithms. Because of the difficulty of showing a detailed classification in Black and White, only a 256 x 256 pixel portion of the results is shown. This small portion of the image only contains four of the classes mentioned above but should still serve to illustrate the salient points.

The small portion of the clustering result, used as a starting point for the contextual routines, is shown in Figure 3a. Among the more obvious undesirable features are:

1.   There are many isolated single pixels or very small groups of pixels which are probably misclassifications. For example there are odd pixels classified as cumulus.

2.   The boundaries between classes are very noisy. The boundary between the (sub-pixel) cumulus and thick low cloud classes illustrates this point. As the transition from one class to another is sometimes continuous this is not unexpected and the"truth" will be somewhat subjective. However clearer boundaries would be desirable if further analysis of the results is contemplated.

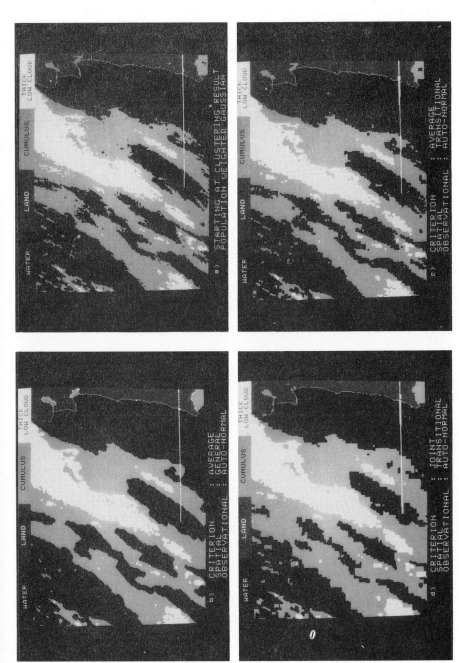

Fig. 3 Contextual Classification Results

3.   There is often a misclassified boundary between two
     dissimilar classes. It is usually several pixels wide
     and is classified differently from either of the adjacent
     classes. This point is best illustrated by the
     boundary between the land and sea classes.

Figure 3b shows the application of the contextual rule in
equation (16) which was derived from the heuristic "average"
probability criterion. Some of the undesirable features of
the previous image have been improved. Isolated pixels and
noisy boundaries have been cleared up considerably. However
the incorrect boundary between the land and sea has not been
removed.

Substituting the "transitional context" model to
represent spatial context gives the result in Figure 3c.
There seems to be a marked tendency to force horizontal and
vertical boundaries. This makes the image subjectively less
appealing compared with Figure 3b, even when the actual
connectivity is similar. The spatial part of the decision
rule involves a smaller neighbourhood so this result may
provide evidence that models based on larger neighbourhoods
would perform better on this data. However such an increase
in the size of the neighbourhood is computationally
prohibitive.

There may also be evidence that the spatial context
labels in a neighbourhood are not altogether independent of
each other, given the central one, as assumed in Section 2.

Figure 3d shows the result using the "joint criterion".
Several of the boundaries seem to have been substantially shifted
where little evidence is found by inspecting the original
data. This would seem to indicate that the joint globally
consistent criterion is rather restrictive for this type of
image data. None of the results seemed to be capable of
suppressing the misclassified boundary between land and
water. We feel that this is primarily due to the way in which
the model parameters were estimated rather than the algorithms
themselves. The original training data set had very few
water and land pixels next to each other due to the
misclassified boundary. Therefore the contextual
algorithms tended to reinforce the boundary rather than
suppress it. The problem of obtaining a suitable training set
requires more work to be done.

7. CONCLUSIONS

The results show that contextual algorithms do have the
potential to improve classification results. In particular

noisy boundaries and isolated misclassifications can be easily corrected. Unlike post processing algorithms, the contextual routines can retain small features where there is good observational evidence of them. The estimation of the contextual model parameters is difficult and can influence the type of undesirable feature which the algorithms can be expected to correct.

The global joint criterion seems to be rather restrictive for the type of data discussed and better results were obtained using the heuristic average criterion. However this may not be the case for other types of data.

REFERENCES

[1] Rosenfeld, A., Hummel, R.A. and Zucker, S.W., (1976), "Scene Labelling by Relaxation Operations", *IEEE Trans. on Systems, Man and Cybernetics*, Vol. SMC-6, No. 6, June.

[2] Weska, J.S., Dyer, C.R. and Rosenfeld, A., (1976), "A Comparative Study of Texture Measures for Image Classification", *IEEE Trans. on Systems, Man and Cybernetics*, pp 269-85.

[3] Swain, P.H., Vardeman, S.B. and Tilton, J.C., (1981), "Contextual Classification of Multispectral Data", Pattern Recognition, Vol. 13, pp 429-441.

[4] Fu, K.S. and Yu, T.S., (1980), "Statistical pattern classification Using Contextual Information", Research Studies Press, John Wiley and Sons Limited.

[5] Welch, J.R. and Salter, K.G. (1971), "A Contextual Algorithm for Pattern Recognition and Image Interpretation" *IEEE Trans. on Systems, Man and Cybernetics*, Vol. SMC-1, pp 24-30, Jan.

[6] Kittler, J. and Foglein, J., (1984), "Contexual Classification of Multispectral Pixel Data", Image and Vision Computing, Vol. 2, No. 1, pp 13-29, Feb.

[7] Kittler, J. and Foglein, J., (1984), "Contextual Decision Rules for Objects in Lattice Configurations", Proc. 7th International Conference on Pattern Recognition, Montreal, pp 270-2.

[8] Kittler, J. and Pairman, D., (1985), "Contextual Pattern Recognition Applied to Cloud Detection and Identification, *IEEE Trans. on Geoscience and Remote Sensing*, Vol. GE-23, No. 6, pp 855-63.

[9] Geman, S. and Geman, D., (1984), "Stochastic Relaxation, Gibbs Distributions, and the Bayesian Restoration of Images", *IEEE Trans. on Pattern Analysis and Machine Intelligence,* Vol. PAMI-6, No. 6, pp 721-41, Nov.

[10] Besag, J.E., (1974), "Spatial Interaction and the Statistical Analysis of Lattice Systems (with discussion)", Journal of the Royal Statistical Society, Series B, Vol. 36, No. 2, pp 192-236.

[11] Besag, J.E., (1986), "On the Statistical Analysis of Dirty Pictures", *Journal of the Royal Statistical Society,* Series B.

[12] Kittler, J. and Pairman, D., (1985), "Segmentation of Multispectral Imagery Using Iterative Clustering", Image Analysis, Proc. 4th Scandinavian Conference, Trondheim, Norway, June 17-20, pp 39-49.

THE EXTRACTION OF GEOPHYSICAL PARAMETERS FROM RADAR ALTIMETER RETURN FROM A NONLINEAR SEA SURFACE

P.G. Challenor and M.A. Srokosz
(Institute of Oceanographic Sciences, Surrey)

1. INTRODUCTION

Radar altimeters have been flown on a number of satellites (GEOS-3, SEASAT, GEOSAT) and been found to give useful geophysical information about the ocean. The basic altimetric measurement is the distance between the satellite and the sea surface, which allows the determination of mean sea surface slopes and hence surface currents, under an assumption of geostrophy. In addition the shape of the return pulse has been found to give information on significant waveheight, while the backscattered power has been used to estimate surface wind speeed.

In order to extract this information, models of the radar return, based on specular reflection from the sea surface, have been proposed by Barrick (1972) and Brown (1977). These assume that waves on the sea surface are linear and therefore the corresponding statistics of surface elevations and slopes are Gaussian. Such models have, for example, allowed good estimates of significant waveheight to be obtained (see Webb, 1981).

In practice, however, it is known that surface waves are nonlinear and this will affect the radar return. Jackson (1979) analysed the effects of nonlinearity on the radar return for the case of undirectional waves. Here we generalise his analysis to allow for directional spreading in the wavefield. Allowing for nonlinear wave effects in the model introduces two more parameters that can be estimated from the radar return, the skewness of the sea surface, which is a measure of the nonlinearity of the wavefield, and a "cross-skewness" parameter, which enables corrections to be made for "sea-state bias" in the altimetric height measurement.

Having derived a model for the radar return we describe how the geophysical parameters (altimetric height, significant waveheight, skewness, cross-skewness and backscattered power, the latter being related to wind speed) can be estimated from an actual return. In order to obtain unbiased estimates of the parameters we employ maximum likelihood estimation, which is known to give asymptotically unbiased estimators with minimum variance. This approach also allows us to calculate the variance-covariance matrix for the estimators and thus determine the accuracy of the estimates.

2. A MODEL OF THE RADAR RETURN

Following Brown (1977) we note that the radar altimeter return from the sea surface $P_r(t)$ may be written as the following three-fold convolution

$$P_r(t) = P_{FS}(t) * q(\zeta) * P_{PT}(t) \tag{1}$$

where P_{FS} is the flat surface response

P_{PT} is the point target response

and q is the probability density function of the elevations of specular points on the sea surface

Here t is the time, measured such that t=0 corresponds to the mean level of the specular points. The height ζ is converted to time via

$$\zeta = \frac{-ct}{2} \tag{2}$$

where c is the speed of light.

Brown (1977) shows that (1) may be reduced to the following expression

$$P_r(t) = \int_0^\infty \int_{-\infty}^\infty \left(\frac{c}{2}\right) P_{PT}(t-\tau) \; q \left\{ \frac{c}{2} (\tau - \hat{\tau}) \right\} \; d\tau \; d\hat{\tau}$$

$$x \begin{cases} P_{FS}(0) & \text{for } t < 0 \\ P_{FS}(t) & \text{for } t \geqslant 0 \end{cases} \tag{3}$$

where

$$P_{FS}(t) = \frac{\alpha \; \sigma^0}{h^3} \exp \left\{ -\frac{4}{\gamma} \sin^2 \xi - \frac{4c}{\gamma h} t \cos 2\xi \right\}$$

$$\cdot I_0 \left(\frac{4}{\gamma} \sqrt{\frac{ct}{h}} \sin 2\xi \right) . \tag{4}$$

Here α and γ are constants depending on the radar parameters (see Brown, 1977), σ^o is the backscattered power and h is the height of the satellite. ξ is the pointing angle of the antenna, ideally equal to zero. In practice the antenna does not always point at nadir so we allow for this effect in our analysis and note that generally $\xi < 0.5^o$. The point target response is given by (Brown, 1977)

$$P_{PT}(t) = \eta P_T \exp\left\{-\frac{t^2}{2\sigma_p^2}\right\}$$ (5)

where η is the pulse compression ratio

   P_T is the peak transmitted power

and σ_p is a measure of the pulse width.

Finally from Srokosz (1986) we have for the distribution of specular points

$$q(\zeta) = \frac{1}{\sqrt{2\pi}\,\sigma_s} \exp\left\{-\frac{\zeta^2}{2\sigma_s^2}\right\} \cdot \left\{1 + \frac{1}{6}\lambda\, H_3(\zeta/\sigma_s)\right.$$

$$\left. - \frac{1}{2}\delta H_1(\zeta/\sigma_s)\right\}$$ (6)

where σ_s is the standard deviation of sea surface elevation

   λ is the skewness of the sea surface

and δ is a "cross-skewness" parameter related to the normalised expectation of the elevation and slope squared (see Srokosz, 1986, for details).

   The H_n are Hermite polynomials with

$$H_3(x) = x^3 - 3x$$
$$H_2(x) = x^2 - 1$$
$$H_1(x) = x \quad .$$ (7)

   The probability density function of specular points (6) is obtained from a weakly nonlinear dynamical model of the surface waves due to Longuet-Higgins (1963). He obtained the probability density function for the surface elevation

$$p(\zeta) = \frac{1}{\sqrt{2\pi}\,\sigma_s} \exp\left\{-\frac{\zeta^2}{2\sigma_s^2}\right\} \cdot \left\{1 + \frac{1}{6}\lambda H_3\left(\frac{\zeta}{\sigma_s}\right)\right\} \qquad (8)$$

as a Gram-Charlier series (modified Gaussian). Longuet-Higgins (1963) also considered the joint distribution of slopes $p(\zeta_x, \zeta_y)$, while Jackson (1979) considered the joint distribution of elevation and slope $p(\zeta, \zeta_x)$. Srokosz (1986) has extended these results to obtain the probability density function of elevation and slopes $p(\zeta, \zeta_x, \zeta_y)$, from which (6) is obtained.

To obtain the form of the return it is necessary to evaluate the double integral in (3) (denoted by $I(t)$). The inner integral may be considered as the convolution of two probability density functions and so may be easily evaluated by adding the cumulants of the two functions (or directly by integration) to yield

$$I(t) = \int_0^\infty \frac{\eta P_T \sigma_p}{\sigma}\left[1 + \frac{1}{6}\lambda_3 H_3\left(\frac{t-\hat{\tau}}{\sigma}\right) + \lambda_1 H_1\left(\frac{t-\hat{\tau}}{\sigma}\right)\right]$$

$$\cdot \exp\left\{-\frac{(t-\hat{\tau})^2}{2\sigma^2}\right\} d\hat{\tau} \qquad (10)$$

where

$$\sigma^2 = \sigma_p^2 + 4\sigma_s^2/c^2$$

$$\lambda_1 = \delta\left(\frac{\sigma_s}{c\sigma}\right) \qquad (11)$$

$$\lambda_3 = -\lambda\, 8\left(\frac{\sigma_s}{c\sigma}\right)^3 .$$

The remaining integral (10) can now be evaluated by a change of variable $v = (t-\tau)\sqrt{2}\,\sigma$ and use of standard results for Hermite polynomials (Abramowitz & Stegun, 1965, chapter 22) to yield

$$I(t) = \frac{\eta P_T \sqrt{2\pi}\,\sigma_p}{2}\left[1 + \mathrm{erf}\left(\frac{t}{\sqrt{2}\,\sigma}\right) - \frac{1}{\sqrt{2\pi}}e^{-t^2/2\sigma^2}\left\{2\lambda_1\right.\right.$$

$$\left.\left. - \frac{1}{3}\lambda_3 H_2\left(\frac{t}{\sigma}\right)\right\}\right] . \qquad (12)$$

Together with (3) this specifies the form of the radar altimeter return from the sea surface.

From equations (3),(4),(11) and (12) it can be seen that the return depends on three wave parameters σ_s (the significant waveheight $H_s = 4\sigma_s$), λ and δ together with σ^o, ξ and h. We note that for $\lambda = \delta = 0$, that is the linear case, the result reduces to that given by Brown (1977).

3. EFFECTS OF WAVE NONLINEARITY ON THE RETURN

To obtain accurate measurements of the height of the mean sea level, to be used in calculating geostrophic currents, it is necessary to allow for "sea-state bias". This bias in the altimetric height measurement is due to nonlinear wave effects. For example, the mean level of the sea surface is given from (8) by

$$\int_{-\infty}^{\infty} \zeta p(\zeta)\, d\zeta = 0 \tag{13}$$

while the mean level of the specular reflectors is given from (6) by

$$\int_{-\infty}^{\infty} \zeta q(\zeta)\, d\zeta = -\frac{\delta}{8} H_s \tag{14}$$

In the linear case $\delta = 0$, so the two results are identical. Fig. 1 illustrates this result diagrammatically.

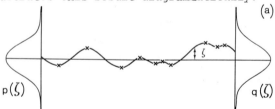

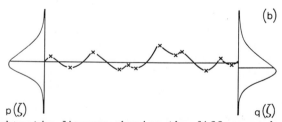

Fig. 1   Schematic diagram showing the difference between the pdfs of surface elevation $p(\zeta)$ and specular points (x) $q(\zeta)$ for (a) linear waves and (b) nonlinear waves. Note that for the nonlinear case the mean level of the specular points differs from that of the surface elevation.

The need for a "sea-state bias" correction arises because the "on-board" trackers used to determine height from satellite altimeter returns make no allowance for nonlinear wave effects. Therefore the trackers incorrectly determine the mean level of the sea surface. Most trackers (for example, those on GEOS-3, SEASAT and GEOSAT) determine the mean level from the position in time of the half power point of the return. This corresponds to determining the median of the distribution of specular points. In the Gaussian case ($\lambda = \delta = 0$) this is the same as the mean and hence the same as the mean level of the sea surface (see (13) and (14)). In general, however, the median and the mean of the specular reflectors differ and they both differ from the mean level of the sea surface. Thus tracking the half-power point of the return does not give the mean sea surface position. From the results given in Srokosz (1986) it can be seen that this leads to errors in the height measurement of order 10 cm, which is the same as the accuracy necessary to estimate geostrophic surface currents.

The method described in the following sections, for extracting geophysical parameters from the altimeter return, avoids these problems as it fits the theoretical return model to the actual return rather than using the half-power point to estimate the mean level of the sea surface.

4.   EXTRACTION OF GEOPHYSICAL PARAMETERS

In order to obtain geophysical information from the return we must estimate the six parameters σ_s, λ, δ, t_o, $\overset{o}{\sigma}$ and ξ. The time origin t_o must be included as it is unknown with respect to the time of transmission of the pulse and it gives the height of the satellite. In developing the model of the return we referred all measurements in time to the mean level, which is one of the parameters we are attempting to estimate. This may be done by replacing t by ($t - t_o$) in the model return and estimating.

Typically the altimeter tracker will produce estimates of the return power at a number of gates M (for SEASAT and for ERS-1 M = 63) situated in time such that the middle gate falls approximately on the mid-point of the leading edge of the return. The model of the return developed in section 2 assumes that the reflections from the sea surface are all in phase, in practice this is not so and it can be shown (Ulaby, Moore and Fung, 1982) that the return power $\hat{g}_i$, from a single pulse, measured at the ith gate has a negative exponential distribution with mean equal to the theoretical return power g_i, thus

$$f(\hat{g}_i) = \frac{1}{g_i} \; e^{-\hat{g}_i/g_i} \; .$$

A reasonable assumption to make if the altimeter moves at least the diameter on the antenna between transmitting pulses and if adjacent gates in the receiver do not overlap is statistical independence of the return power in each gate.

In practice N pulses are averaged together before data is transmitted from the satellite to ground (for SEASAT N = 100 and for ERS-1 N = 50, with a pulse repetition frequency of 1000 Hz). From (15) the average of N pulses will have a gamma, or chi-squared distribution

$$f(\hat{g}_i) = \frac{N^{N-1} \hat{g}_i^{N-1}}{N! \; g_i^{N}} \; \exp \; (-N\hat{g}_i/g_i) \tag{16}$$

where g_i is the return form given in section 2 (that is, $P_r(t)$) and depends on six parameters σ_s, λ, δ, t_0, σ^0, ξ, which will be denoted by θ_j (j=1,..,6) for convenience.

Maximum likelihood estimation gives, asymptotically, minimum variance unbiased estimators of the parameters and so we will use this method to estimate the θ_j. From (16) and the assumption of independence the likelihood of an averaged pulse is given by

$$L = \prod_{i=1}^{M} \frac{N^{N-1} \hat{g}^{N-1}}{N! \; g_i^{N}} \; \exp \; (\; -N\hat{g}_i/g_i) \tag{17}$$

and the log likelihood by

$$LL = \sum_{i=1}^{M} \left\{ (N-1) \ln N + (N-1) \ln \hat{g}_i - N\hat{g}_i/g_i \right.$$

$$\left. -N\ln g_i - \ln N! \right\}. \tag{18}$$

Now g_i is a function of n parameters θ_j (j=1,..n) which we wish to estimate (here n=6, but for generality we will develop the theory for arbitrary n). To derive the maximum likelihood estimators we take derivatives of (18) with respect to θ_j and set the resulting expressions to zero. Thus

$$\frac{\partial LL}{\partial \theta_j} = N \sum_{i=1}^{M} \left(\frac{\hat{g}_i}{g_i^2}\right) \frac{\partial g_i}{\partial \theta_j} - N \sum_{i=1}^{M} \frac{1}{g_i} \frac{\partial g_i}{\partial \theta_j} \qquad j=1,\ldots,n \qquad (19)$$

and so

$$\sum_{i=1}^{M} \left(\frac{\hat{g}_i - g_i}{g_i^2}\right) \frac{\partial g_i}{\partial \theta_j} = 0 \qquad j=1,\ldots,n \qquad (20)$$

are the n simultaneous equations that need to be solved for the maximum likelihood estimators of θ_j.

It is also possible to derive the variance-covariance matrix $\underline{\underline{V}}$ of the estimators. This is the inverse of the Fisher Information matrix $\underline{\underline{F}}$ and so

$$\underline{\underline{V}} = \underline{\underline{F}}^{-1} \qquad (21)$$

where

$$\underline{\underline{F}} = \left\{ E \left(\frac{\partial LL}{\partial \theta_j} \cdot \frac{\partial LL}{\partial \theta_k}\right) \right\} \quad j,k=1,\ldots,n. \qquad (22)$$

Here E denotes the expectation operator (see Cox and Hinckley, 1974, for details both of maximum likelihood estimation and Fisher information).

From (19) we obtain

$$\frac{\partial LL}{\partial \theta_j} \cdot \frac{\partial LL}{\partial \theta_k} = N^2 \sum_{i=1}^{M} \sum_{l=1}^{M} \left\{ \frac{(\hat{g}_i - g_i)}{g_i^2} \frac{(\hat{g}_l - g_l)}{g_l^2} \frac{\partial g_i}{\partial \theta_j} \frac{\partial g_l}{\partial \theta_k} \right\}$$

while from the properties of the gamma distribution and our assumption of independence between gates and pulses we have

$$E[(\hat{g}_i - g_i) (\hat{g}_l - g_l)] = \frac{g_i^2 \delta_{il}}{N}$$

where δ_{il} is the Kronecker delta. Together with (22) these give the following result for $\underline{\underline{F}}$

$$\underline{\underline{F}} = \left\{ N \sum_{i=1}^{M} \frac{1}{g_i^2} \frac{\partial g_i}{\partial \theta_j} \frac{\partial g_i}{\partial \theta_k} \right\} \quad j,k=1,\ldots,n . \qquad (23)$$

From this the inverse may be calculated numerically to obtain the variance-covariance matrix $\underline{\underline{V}}$.

In the above we have described the theoretical basis of maximum likelihood estimation of the geophysical parameters from the return. We have also shown that it is possible to obtain the variance-covariance matrix of the estimators, which gives a measure of the possible error in our estimates. Although the mathematical theory is elegant its implementation is rather more complicated due to complex form of the theoretical return derived in section 2. It is therefore necessary to implement the estimation procedure numerically. This has been done and some results will be presented in the following section.

5. RESULTS AND DISCUSSION

In order to test the method of estimation return pulses have been simulated from the model return given in section 2 together with their statistics described in section 4. Fig. 2 gives an example of both the mean return P_r and the average of 50 individual returns for a specific set of parameters (those of the ERS-1 altimeter; for comparison with an actual SEASAT return see Webb, 1981). The noise on the return can be reduced by averaging a larger number of pulses. In practice it is necessary to balance the number of pulses averaged to reduce noise, against the distance travelled by the satellite in acquiring the data, as the geophysical parameters may vary along the satellite ground track (typically the satellites travel at 7 km per second over the ground).

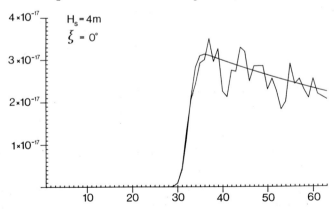

Fig. 2   The mean return P_r (normalised by the transmitted power) plotted against gate number of the ERS-1 altimeter, together with the average of 50 returns to illustrate the departure from the mean of an actual return. (The gate spacing is 3 nanoseconds.)

Fig. 3 illustrates the effect of antenna mispointing on the return pulse. Ideally the antenna should point at nadir but in practice drag on the satellite and differential heating due to solar radiation leads to a small amount of mispointing (for SEASAT about 0.2°) which varies slowly on the time scale of a satellite orbit (approximately 100 minutes). The main effect of the mispointing of the antenna is to reduce the backscattered power and a secondary effect is to change the slope of the trailing edge of the return. As a direct consequence of this it is clear that it is not possible to simultaneously estimate the backscattered power σ° and the pointing angle ξ from the return. This becomes even clearer when we calculate the variance-covariance matrix for a number of cases and hence find that the correlation between σ° and ξ is greater than 0.9999. Furthermore the variance of the estimators of σ° and ξ is large and tends to infinity as ξ tends to zero. This occurs because it is not possible to determine ξ from the amplitude of the return so the only remaining information in the return on ξ is the slope of the trailing edge, but as ξ tends to zero the slope of the trailing edge is almost constant, so it cannot be used to determine ξ. In fact the Fisher Information matrix becomes singular at $\xi=0$ showing that there is no information on ξ in the return.

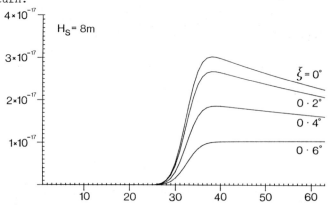

Fig. 3   The mean return P_r (normalised by the transmitted power) plotted against gate number, for various values of the pointing angle ξ. (The gate spacing is 3 nanoseconds.)

This result has implications for the extraction of wind speed information from σ°. If variations in backscattered power due to variations in wind speed and those due to pointing angle cannot be separated then clearly it will not be possible to determine the wind speed accurately from the return. Various authors (Brown, 1977; Barrick and Lipa, 1985) have suggested that the pointing angle can be estimated from the slope of the trailing edge of the return but have failed to note the strong correlation between σ^o and ξ. It would seem that to use the backscattered power to obtain accurate wind speed estimate necessitates the independent determination of the pointing angle ξ. Alternatively a biased estimation procedure or use of information from successive returns (rather than the single return considered here) may allow ξ to be estimated, albeit in a non-optimal fashion. Whether these approaches would give adequate wind speed information from the return remains to be investigated.

The results described above, although preliminary, show the power of mathematical and statistical models to give insight into practical problems. A further area that is currently under investigation, using the techniques described above, is that of the estimation of the nonlinear wave parameters λ and δ and the correction of the sea-state bias in the height measurement described in section 3. If geostrophic surface currents are to be obtained from altimetric height measurements a good understanding of this problem is necessary. It is hoped to report results on this topic in a future paper.

6. REFERENCES

Abramowitz, M. and Stegun, I.A., (1965) Handbook of mathematical functions. Dover Publications, Inc., New York.

Barrick, D.E., (1972) Remote sensing of sea-state by radar. in Remote sensing of the troposphere. Ed. V.E. Derr, Chap. 12, U.S. Government Printing Office, Washington D.C.

Barrick, D.E and Lipa, B.J., (1985) Analysis and interpretation of altimeter sea echo. *Adv. Geophs.*, **27**, 60-99.

Brown, G.S., (1977) The average impulse response of a rough surface and its applications. *IEEE Trans. Ant. Propag.*, **AP-25**, 67-74.

Cox, D.R. and Hinkley, D.V., (1974) Theoretical statistics. Chapman and Hall, London.

Jackson, F.C., (1979) The reflection of impulses from a nonlinear random sea. *J. Geophys. Res.*, **84**, 4939-4943.

Longuet-Higgins, M.S., (1963) The effect of nonlinearities on statistical distributions in the theory of sea waves. *J. Fluid Mech.*, **17**, 459-480.

Srokosz, M.A., (1986)   On the joint distribution of
surface elevation and slopes for a nonlinear random sea,
with an application to radar altimetry.   *J. Geophys. Res.,*
91, 995-1006.

Ulaby, F.T., Moore, R.K. and Fung, A.K., (1982)   Microwave
remote sensing: Active and Passive, Vol. II, Addison-Wesley
Publishing Co., Reading, Massachusetts.

Webb, D.J., (1981)   A comparison of SEASAT 1 altimeter
measurements of wave height with measurements made by a
pitch-roll buoy.   *J. Geophys. Res.,* **86**, 6394-6398.

THE ASSIMILATION OF SATELLITE ALTIMETER DATA INTO OCEAN MODELS

D.J. Webb

(Institute of Oceanographic Sciences, Surrey)

1. INTRODUCTION

Despite the problems that still exist in interpreting the data, satellite measurements of the sea surface are quietly changing the way in which we study the ocean. Two important examples of the way in which GEOS and Seasat altimeter data has been used is the work of Woodworth and Cartwright (1986) in mapping the world wide distribution of tides and that of Cheney, Marsh and Beckley (1983) in mapping the distribution of the synoptic-scale eddy field. During the next five years further ocean surface surveillance satellites are due to be launched which should give more accurate measurements of the surface topography and the wind stress. The latter is important because it drives most of the near surface circulation. Improved radiometer measurements will also give information on surface temperatures and on the radiative heat fluxes which drive the deep thermo-haline circulation of the ocean.

In the U.K., in addition to the work of Woodworth, there has also been an interest in using the altimeter data to study the surface current field. At Imperial College, Marshall (1985) has investigated the problem of distinguishing the effects of ocean currents from those due to errors in the geoid. At Oxford University interest has centred on the oceanography of equatorial regions, where the high speed of planetary waves in the ocean helps to simplify the analysis of the altimeter data (Anderson and Moore 1986).

At I.O.S. we have a special interest in mid-latitude oceans where the Coriolis effects are large. In such regions the surface slope of the ocean is, to a first approximation, balanced by the Coriolis force acting on the surface currents of the ocean. This means that the radar altimeter measurements can be used to determine the surface current field directly.

    To investigate further the potential of such data, we have
been carrying out a program of research to study methods for
assimilating altimeter data into numerical models of the ocean
and to see whether they could be used to determine the deep current
structure of the ocean. If the velocities involved are small,
so that the planetary waves behave linearly, it is known that
the speed of planetary waves depends only on their vertical
structure. Thus by following the changes with time of surface
features produced by the planetary waves it should be possible
to deduce the vertical structure of the currents. However in
a typical mid-latitude ocean, the advective velocities
associated with the planetary waves are larger than their
phase speed. The resulting waves are non-linear and although
the speed at which features propagate should still depend on
their vertical structure, the effect of non-linearities may
degrade the assimilation scheme.

    In an previous paper (Webb and Moore 1986) the properties
and efficiency of such a scheme, designed for use with non-
linear systems, was investigated analytically by making the
assumption that it was being used with a field of linear
planetary waves. The results of this study showed that the
performance of the scheme depended on the difference in phase that
developed between waves of the same horizontal wavenumber
over each assimilation cycle. In the extreme case of two
vertical modes of the ocean with the same phase velocity, it
is impossible from a sequence of surface observations to
distinguish between them. If the differences in their
velocities are small, as usually occurs with the higher
baroclinic modes of the ocean, then it was found that the
assimilation scheme performs better if the assimilation interval
is long than if a larger number of shorter intervals are used.
This arises because the longer interval enables larger phase
differences to develop between the waves.

    In the present paper a more realistic test of the assimilation
scheme is made using a numerical model of the ocean and a
fully developed synoptic-scale eddy field. The initial field
is generated using the full non-linear equations of motion and
the tests then carried out using linear dynamics. The results
confirm the results of the earlier analytic study and also
shows that when the ocean dynamics are almost linear, the
only important currents that cannot be deduced from the
altimeter data are the steady zonal (east-west) currents.
Further studies taking into account friction, forcing and the
full effect of non-linearities are continuing.

2. THE NUMERICAL MODEL

The numerical model used for the present study is the spectral quasi-geostrophic model of C. Rogers and K. Richards (Rogers 1985). The model uses a stream function which is related to the horizontal components of velocity by,

$$u = -\partial\Psi/\partial y \quad , \quad v = \partial\Psi/\partial x. \tag{2.1}$$

The stream function is expanded as a Fourier series in the horizontal direction and in terms of P_n, the normal modes of the ocean , in the vertical,

$$\Psi(x,y,z,t) = \Sigma \ \Sigma \ \Sigma \ P_n(k,l,t)e^{ikx \ + \ ily}. \tag{2.2}$$
$$\qquad\qquad\quad k \ \ l \ \ n$$

The equation satisfied by each component of the expansion is,

$$\frac{\partial}{\partial t} \ (-k^2-l^2-\mu_n) \ P_n(k,l,t) \ + \ i\beta k P_n(k,l,t)$$

$$+ \ J_n(k,l,t) \ = \ 0 \ . \tag{2.3}$$

μ_n is the vertical eigenvalue corresponding to the mode n, β is the northwards gradient of the Coriolis parameter and J is the non-linear advection term. Further details of the scheme are given by Rogers (1984).

The model is designed to represent a rectangular region of ocean and uses periodic boundary conditions on each of the side boundaries. In the runs reported here, these have a length of approximately 1571 kms. The ocean stratification used to initialise the model is taken from Discovery station D10344. The first baroclinic wave has a speed of 2.6 m/s giving a Rossby radius of 27.8 km. The scale of the synoptic eddies is thus about 100 kms (~ π x 28km). The speed of the fastest baroclinic Rossby wave is about 0.7 cm/s so it takes about 170 days for the eddies to move a distance equal to their own diameter.

The model is spun up initially using a scheme proposed by Haidvogel et al. (1981). This feeds in energy at scales of the order of the Rossby radius in much the same way that instabilities of the real ocean feed the eddy field. The two-dimensional field of turbulence then cascades energy to longer and enstrophy to shorter wavelengths (Rhines 1979). The model includes bottom friction and horizontal viscosity and uses bi-harmonic friction to damp out the very short waves.

The model was spun up for 1200 days, by which time the
energy had reached a quasi-steady final state. The surface
height field at this time (figure 1) shows a well developed
synoptic-scale eddy field which is thought to be typical of
mid-latitude oceans. The corresponding wave number spectra
of the barotropic mode (the lowest of the three vertical
modes) is shown in figure 2. The semi-circle line marks the
scale of the Rossby radius. At high wavenumbers the spectrum
drops off rapidly.

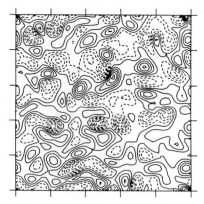

Fig. 1   Ocean surface height at timestep 6000 (day 1200),
         immediately before data assimilation is started.
         Contour separation 1.7 cm.

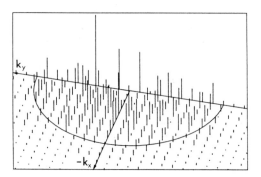

Fig. 2   Spectra in wavenumber space of the barotropic (lowest
         vertical mode) stream function. The semi-circle
         corresponds to the Rossby radius of deformation of the
         first baroclinic mode.

3. THE TWIN EXPERIMENT

The method used to test the assimilation scheme is to treat the first model as a control (i.e. the 'real' ocean) and to use surface height data from it to update a second assimilation model. This second model is started from rest at day 1200 and both models are run in parallel for a further 1200 days with a complete field of height information from the control model being passed to the other at regular intervals. In the tests reported here, new data was assimilated at intervals of either twenty or one hundred days. As we are interested in comparing the numerical results with the analytic theory, the forcing, friction and the non-linear terms are switched off during the second 1200 day period in both models.

The assimilation of new data is achieved by projecting the vector describing the state of the assimilation model onto the surface defined by the new height information. As the ocean is periodic, each horizontal wavenumber is treated separately. If $\underline{m}^-$ and $\underline{m}^+$ are the vectors describing the state of the assimilation model before and after assimilating new data then (Webb and Moore 1986),

$$\underline{m}^+ = \underline{m}^- - \underline{p}(\underline{p}.\underline{h})^{-1} (\underline{h}.\underline{m}^- - \underline{h}.\underline{r}) . \qquad (3.1)$$

The observed height H is given by,

$$H = \underline{h}.\underline{r} \qquad (3.2)$$

$\underline{r}$ is the state vector of the control model and $\underline{h}$ a vector representing the surface displacement produced by each mode. The change in phase of each component of the model vectors between assimilating new data can be represented by a diagonal matrix $\underline{\underline{C}}$. The convergence rate of the assimilation scheme then depends on the eigenvalues of the matrix operator,

$$(\underline{\underline{I}} - \underline{p}(\underline{p}.\underline{h})^{-1} \underline{h}.)\underline{\underline{C}} . \qquad (3.3)$$

If the eigenvectors are given by $\underline{x}_n$ and the corresponding eigenvalues by λ_n, then expanding the error vector $\underline{E}$ at time t,

$$\underline{E}(t) = \sum_n \alpha_n \underline{x}_n . \qquad (3.4)$$

The error vector one assimilation cycle later is,

$$\underline{E}(t + \delta t) = \sum_n \alpha_n \lambda_n \underline{x}_n . \qquad (3.5)$$

The efficiency of the assimilation scheme is determined by the moduli of the eigenvalues, the most efficient scheme in the long run being the one with the smallest maximum eigenvalue.

Even if the eigenvalues λ_n have moduli less than one, the error $\underline{E}$ may increase in the short term, because the eigenvectors are not all orthogonal. However one can show that the error E', defined as,

$$E' = \sum_n (p_n/h_n)(m_n^+ - r_n)^2 , \qquad (3.6)$$

is always reduced.

4.   RESULTS

The reduction in the magnitude of the error vector, for the twenty and one hundred day assimilation cycle times is shown in figure 3. Because three modes are being used there is an approximately one third reduction in the error when the initial projection is made. Immediately following this the twenty day assimilation cycle is seen to be the most efficient, essentially because it is very effective at resolving the long wavelength barotropic modes. However after only a few cycles its inefficiency at other wavelengths and its inefficiency at resolving the higher vertical modes (Webb and Moore 1986) results in it being overtaken by the scheme with the longer assimilation interval. This is eventually better because it allows a greater phase difference to develop between the different modes before new data is introduced.

The errors remaining after assimilating for 1200 days with the one hundred day cycle time are illustrated in figure 4. This shows the error variance of the barotropic mode. The errors remaining are clustered in two main regions, in both of which the assimilation scheme is known to be inefficient.

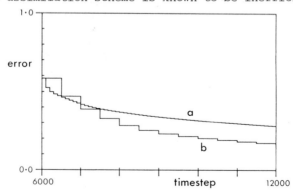

Fig. 3   The relative variance between the control and assimilation
         models (eqn. 3.6) plotted as a function of time. A value
         of one corresponds to the variance of the control model.
         (a) Assimilating new data every 20 days (b) every 100 days.

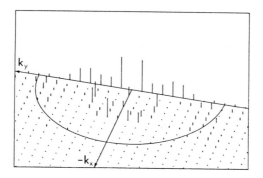

Fig. 4 Spectra in wavenumber space of the difference between
        the barotropic modes of the control and assimilation
        models at timestep 12000 (day 2400).

   The first of these is a small circular region lying on the
k_x axis. As shown in Webb and Moore (1986), at wavenumbers
within this region the barotropic wave changes phase by almost
exactly 2π during each assimilation period. Thus its apparent
phase at the end of the period is very close to that of the
two baroclinic waves. It should be possible to reduce the
errors within this region by occasionally using an assimilation
interval of only fifty days. Calculations which alternated
the assimilation interval between 50 and 100 days showed this
to be the case.

   The second region of large errors lies near the origin along
the k_y axis and corresponds to long wavelength zonal and near
zonal currents. The dispersion relation for Rossby waves
(LeBlond and Mysak 1978) shows that within this region, the
period of the vertical modes is very long becoming infinite
on the k_y axis itself. The errors therefore remain large
because there is insufficient phase separation of the modes
during each assimilation cycle. On the k_y axis the
use of longer assimilation cycle times cannot improve
the errors and information on the vertical structure of the
steady currents can only be obtained from observations
made within the ocean. A single north-south hydrographic
section through the ocean would be sufficient.

   A final feature of the error spectrum is that it shows
that at high wavenumbers, where the assimilation scheme is
again expected to be inefficient because of the low speed of
the modes, the actual errors are small. This arises because
the spectrum drops off rapidly before reaching wavenumbers
where the assimilation scheme is inefficient.

The results of the present study thus show that when
the linear approximation is valid, the altimeter data can be
used to determine the vertical structure of the long period
energetic currents of the ocean. The only energetic currents
which are not resolved are steady or slowly varying currents
of long wavelengths and these may be determined from hydrographic
sections.

5. DISCUSSION AND CONCLUSIONS

The results of the earlier analytic theory have been
confirmed by the use of a twin experiment carried out using a
spectral quasi-geostrophic model with three modes in the
vertical. The advantages of using a long assimilation
period has been confirmed and it has been shown that
when non-linearities can be neglected, the only features that
cannot be resolved by the assimilation scheme are steady or
slowly varying currents of long wavelength. When a constant
assimilation interval is used, some of the higher frequency
modes are not resolved because the barotropic modes have the
same phase increment (modulo 2π) as the slower baroclinic modes.
However by introducing a few intervals of half the normal
length these modes can be resolved. At high wavenumbers the
inefficiency of the assimilation scheme is masked by the rapid
drop off in the amplitude of the spectra.

Further studies are presently being carried out with the
twin models, first with the introduction of forcing and
friction during the assimilation run and secondly with the
introduction of non-linearities. Initial results show that
friction helps to improve the performance of the scheme by
damping out the initial error field. As long as the assimilation
model is given the same forcing as the 'real' ocean, the error
vector should reduce in magnitude even if no altimeter
information is available. This underlines the importance of
accurate measurements of the ocean surface wind stress and heat
fluxes.

The introduction of non-linearities rapidly reduces
the efficiency of the assimilation scheme, primarily because
it reduces the correlation time for the velocity field to
about thirty days. In order to keep track of the real ocean, a
relatively inefficient thirty day assimilation interval has
to be used. Tests are presently being made using a strategy
which assimilates the data going repeatedly forwards and
backwards in time to try to improve the efficiency of the
scheme and so give a method which guarantees convergence.

6. REFERENCES

Anderson, D.L.T. and Moore, A., (1986) 'Data Assimilation' pp. 437-464 in: Advanced Physical Oceanographic Numerical Modelling. (Editor J.J. O'Brien). Dordrecht, D. Reidel, 608pp.

Cheney, R.E., Marsh, J.G. and Beckley, B.D., (1983) Global mesoscale variability from colinear tracks of SEASAT altimeter data. Journal of Geophysical Research 88(C7), 4343-4354.

Haidvogel, D.B., Keffer, T. and Quinn, B.J., (1981) Dispersal of a passive scalar in two-dimensional turbulence: effective diffusivity. Ocean Modelling, No. 41 (unpublished manuscript).

LeBlond, P.H. and Mysak, L.A., (1978) Waves in the Ocean. Elsevier; New York, 602pp.

Marshall, J.C., (1985) Determining the ocean circulation and improving the geoid from satellite altimetry. Journal of Physical Oceanography, 15(3), 330-349.

Rhines, P.B., (1979) Geostrophic Turbulence. p401-441 in: Annual Reviews of Fluid Dynamics II. Palo Alto, Calif : Annual Reviews Inc.

Rogers, C.F., (1985) Quasi-Geostrophic Ocean Models Employing Spectral Methods Part 1 - Theoretical Background. IOS Report No. 191 : Wormley, U.K.

Webb, D.J. and Moore, A., (1986) On the assimilation of altimeter data into ocean models. Journal of Physical Oceanography (accepted).

Woodworth, P.L. and Cartwright, D.E., (1986) Extraction of the M2 Ocean Tide from Seasat altimeter data. Geophysical Journal of the Royal Astronomical Society, 34, 227-255.

MEASURING THE OCEAN WAVE DIRECTIONAL SPECTRUM WITH HF RADAR: THE INVERSION PROBLEM

L.R. Wyatt
*(Department of Applied and Computational Mathematics,
University of Sheffield)*

1. INTRODUCTION

The types of equations that are encountered in many remote
sensing problems are such that analytical solutions are out of
the question. Nor can they be approached with routine numerical
techniques because in general the equations are ill-conditioned.
The temptation, when working in an engineering environment
where mathematical rigour is rather a luxury, is to develop
fitting procedures with as many parameters as possible and
carry on using those techniques until they fail. At which
point it is hoped that sufficient experience will have been·
gained to broaden the parameter range or fitting procedures
to accomodate the problem. This approach to the inversion
problem is described somewhat scathingly by Sabatier (1983) as
'a primitive physical approach' but is exactly the technique
that is described here to interpret HF radar backscatter Doppler
spectra in terms of ocean wave directional spectra.

This paper will attempt to explain why the techniques fail
in particular circumstances by examining the direct problem,
i.e. constructing HF radar Doppler spectra for given ocean
wave spectra. With a better understanding of the structure
of the integral equation that is involved in this problem,
development of more rigorous inversion techiques should be
possible in the future.

The theoretical formulation of the problem of HF backscatter
from the ocean surface has been developed by Barrick (1971, 72),
Barrick and Weber (1977) and Weber and Barrick (1977). They
derive a non-linear, two-dimensional, Fredholm-type integral
equation to relate the Doppler spectrum of the backscattered
echo to the ocean wave directional spectrum. Wyatt (1986)
discusses attempts that have been made to solve this problem

using both mathematical inversion techniques and fitting
procedures. The mathematical techniques developed by Barrick
and Lipa (1979) and Lipa and Barrick (1980) can only be used in
very restricted circumstances at the low HF frequencies required
for long range wave measurement, since they involve a
linearisation of the equation that severely limits the range
of ocean wave frequencies for which a solution can be found.
For this reason a parameter-fitting procedure has been
developed (Wyatt (1986)) and has been extensively tested with
both real and simulated radar data (Wyatt et al (1986)).

Before solving the equation though, some confidence that it
does actually describe the scattering process is required.
Some work in this direction is described in Wyatt et al. (1985)
using radar data obtained with the Birmingham University
FMICW system and comparing it with simulated Doppler spectra
using ocean wave spectra constructed from wave and wind data
measured at points up to 200km away from the radar measuring
point. During NURWEC (see Wyatt et al 1986) a large amount of
co-temporal and co-located radar and buoy data was collected.
Using this buoy data to generate simulated Doppler spectra
provides a much clearer picture of the validity of the equation.
These simulations will also be used here to explore the
integral equation.

2. THE INTEGRAL EQUATION

2.1 The Direct Problem

The Doppler spectrum of the backscatter cross section, σ,
can be written, using the normalisation described in Lipa and
Barrick (1986), as follows:

$$\sigma(\eta) = \sigma_1(\eta) + \sigma_2(\eta) \qquad (2.1.1)$$

where

$$\sigma_1(\eta) = 4\pi \sum_{m=\pm1} S(-2m\,\underline{k}_o)\,\delta(\eta-m) \qquad (2.1.2)$$

$$\sigma_2(\eta) = 8\pi \sum_{m,m'=\pm1} \int_0^\infty \int_{-\pi}^\pi |\gamma_L|^2 \delta(\eta-m\sqrt{k}-m'\,\sqrt{k'})$$

$$x\ S(m\underline{k})S(m'\underline{k})\,kdkd\theta \qquad (2.1.3)$$

where η is the normalised Doppler frequency; $\underline{k},\ \underline{k}'$ are the
interacting wavevectors; $\underline{k}_o$ is the radio wave vector; $S(\underline{k})$ is the
ocean wave directional spectrum

$$\gamma_L = \gamma_H + \gamma_E \qquad (2.1.4)$$

$$\gamma_H = -\frac{i}{2}\left[k+k' - \frac{(kk' - \underline{k}.\underline{k}')(\eta^2+1)}{mm' \sqrt{kk'}(\eta^2-1)}\right] \qquad (2.1.5)$$

$$\gamma_E = \frac{1}{2}\left[\frac{(\underline{k}.\underline{k}_o)(\underline{k}'.\underline{k}_o) - 2\underline{k}.\underline{k}'}{\sqrt{\underline{k}.\underline{k}'} - \Delta/2}\right] \qquad (2.1.6)$$

Δ is the surface impedance

Equation 2.1.3 is non-linear in the ocean wave directional spectrum for which the solution is sought.

Figure 1 shows four comparisons between radar Doppler spectra and simulations with NURWEC data. Three, measured in consecutive 30 minutes, are chosen to show the effect of varying radar frequency and one, some 15 hours earlier, to show the effect of different wave propagation direction relative to the radar beam as indicated by the relative amplitudes of the two first order peaks. The simulations are particularly sensitive to the data used at around the first order Bragg matched frequency. This is not surprising since the first order returns are described by a delta function (eq. 2.1.2) which selects out two components (one propagating directly away from the radar and the other towards) from the entire two-dimensional spectrum. The radar is seeing these components averaged over an area of roughly 7.5 square km and in time, whereas the buoy measurement is a point measurement averaged only in time. Small differences in the directional properties at the Bragg-matched wavelength can lead to large differences between measured and simulated Doppler spectra. Much better agreement is found, and is seen in this figure, if the short wave part of the measured spectrum (where short wave is defined as the Bragg-matched wavelength and shorter) is replaced with a Phillips equilibrium spectrum with a directional distribution determined from the radar data. This is in itself an interesting result but will not be discussed further here. The agreement between real and simulated spectra is repeated in all the cases for which comparison has been possible.

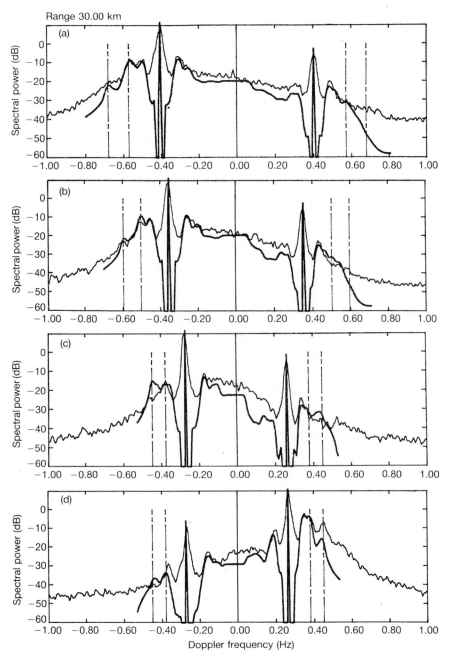

Fig. 1   Radar Doppler spectra measured during NURWEC are
          compared with simulated spectra (heavy line) using
          wave-buoy data.   Radar frequencies of measurement and
          simulation: (a)15.66MHz (b)12.058MHz (c)6.815MHz
          (d)6.903MHz measured 15 hours earlier than (c).

2.2   The structure of the integral

The delta function in equation 2.1.3 describes contours in ocean wavenumber space which select those wavenumber pairs which contribute to a line integral for each Doppler frequency. Close to the first order peaks (normalised Doppler displacements less than 0.4), these comprise one short wave and one long wave. This is the basis for the two-scale model that is used in the Lipa (1977) and Wyatt (1986) techniques and for the linearisation adopted by Barrick and Lipa (1979). Using the two-scale model and making use of the delta function to reduce the equation to a one-dimensional integral, equation 2.1.3 can be written as follows:

$$\sigma_2(\eta) \;=\; \int_{-\pi}^{\pi} K(\eta,\theta)\,S(\eta,\theta)\,d\theta \qquad (2.1.7)$$

$S(\eta,\theta)$ is the long-wave directional spectrum at those vector wavenumbers that satisfy the delta function constraint. The kernel, $K(\eta,\theta)$, in this equation contains the original coupling coefficient modified by the modelled short-wave directional spectrum. The coupling coefficient is symmetrical about zero Doppler and it is the short-wave model which imposes the difference in second order amplitude between positive and negative Doppler frequency just as it sets the two first order amplitudes. The way in which the long-wave directional spectrum interacts with this kernel is summarised in Wyatt (1986) and will now be described in more detail.

The examples to be described below are presented as amplitude distributions in normalised ocean wavenumber space, either as contour plots or as relative amplitudes at each integration point i.e. amplitude expressed as a percentage of the total integral along a particular Doppler contour. This description is confined here to the largest sideband of the Doppler spectrum. The other three sidebands have a similar structure. The positive X-axis is the radar beam direction towards the scattering patch. A normalised wavenumber of one corresponds to a wave with the first-order, Bragg-matched wavelength.

First of all consider the case of a long wave spectrum where the energy is predominantly in the quadrant described by the beam direction plus and minus about 45 degree. Figure 2a shows such a spectrum normalised with respect to a radio frequency of 15.66MHz. Figure 2b shows relative amplitudes of the kernel around the Doppler contours used for the simulations and finally figure 2c shows the relative amplitudes of the contributions to

the Doppler spectrum. Comparing these figures it can be seen
that the relative amplitudes of contributions to the Doppler
spectrum are well correlated with the amplitude distribution
in the long-wave spectrum. The wavenumbers that contribute
significantly to the integral at each Doppler frequency are
clustered around the maximum contribution which is correlated
with the maximum in the long-wave spectrum. This result is
repeated in Fig. 3a,b,c which, by comparison with figure 2,
shows the effect of reducing radar frequency. The limited
ocean-wave frequency range that can be measured at low HF is
illustrated in this figure. A normalised Doppler displacement
of 0.4 corresponds to a smaller absolute Doppler frequency
range and a smaller corresponding ocean wave-frequency range
at 6MHz than at 15MHz. The linearisation used by Barrick and
Lipa confines the search for a solution to a normalised
Doppler displacement of only 0.2 with a correspondingly greater
reduction in ocean wave frequency range at low HF. In fact in
the case illustrated here less than 0.3% of the energy at the
peak of the spectrum is contributing to the Doppler spectrum
at normalised displacements less than 0.2.

Figure 4a,b,c shows the same three diagrams but now for a
case where the long-waves are propagating in a quadrant centred
around the normal to the beam direction. In the integration,
the maximum long-wave amplitudes are multiplied by very small
contributions from the kernel, leading to the compromise in
figure 4c. The correlation between the amplitude distribution
and the relative amplitudes in the integral has gone. A
further difference is that the range of wavenumbers contributing
to the integral at a particular wavenumber is no longer continuous.
It is unlikely then, that methods of solution which implicitly
assume a similar behaviour of the integral in all circumstances,
will be successful in all cases. The model-fitting technique
is a particular example of this problem and was shown not to be
accurate when energetic waves anywhere in the spectrum are
propagating perpendicular to the beam. Another technique
developed by Wyatt (see Wyatt et al, 1986) to provide
measurements of significant waveheight and mean period suffers
from the same problem and can only be overcome by allowing
two possible solutions corresponding to the perpendicular
and non-perpendicular cases respectively, the correct one
being chosen by using additional information. One source of
additional information could be a second radar monitoring the
same patch of sea from a different direction.

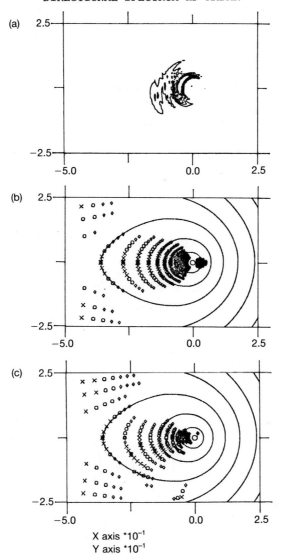

X axis *10^{-1}
Y axis *10^{-1}

Fig. 2   The structure of the integral plotted in normalised
(with respect to 15.66MHz) wavenumber space showing:
(a)   The long wavenumber spectrum. Contours are plotted
at increments of 3% of maximum energy with the largest
contour at 30% of maximum. The remaining energy is
confined within the upper contour.
(b)   The kernel. Maximum contributions along each
Doppler frequency contour are shown by the symbol *;
when added to wavenumber pairs indicated by the symbol
x, 50% of the magnitude of the integral is found; with
the symbol O, 75% and with ◊, 90%. Normalised Doppler
frequency contours are in increments of 0.1 starting
from 0.1 which is the circle around zero wavenumber.
(c)   The integral. Notation as in (b).

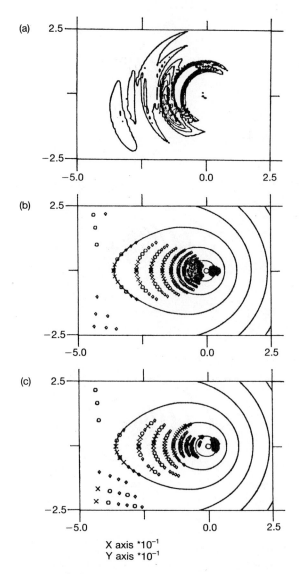

(a)

(b)

(c)

X axis *10^{-1}
Y axis *10^{-1}

Fig. 3   As Fig. 2 but normalised with respect to 6.903MHz.

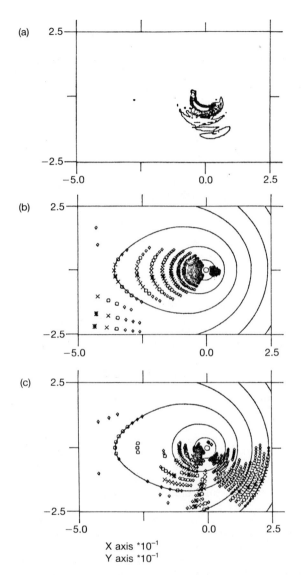

Fig. 4   As Fig. 2 for a different orientation of the long
         wavenumber spectrum with respect to the radar beam.

3.   THE SOLUTIONS

3.1   *The single radar case*

    The results of NURWEC have clearly demonstrated differences
in accuracy achievable with the model-fitting technique for

cases with perpendicular or non-perpendicular wave components.
These confirmed the results of simulation experiments which
show that amplitude is more accurate in the non-perpendicular
cases and that in the other case, direction is more accurate.
In neither case was the directional spreading parameter.estimated
accurately. The reason for the improved direction accuracy
for the perpendicular case is not immediately clear from the
discussion above or from the assumptions that were made in
developing the model-fitting technique. The technique involves
matching measured Doppler spectral amplitudes to simulated
spectra for the appropriate range of long wave parameters.
The perpendicular case is sufficiently anomalous that it is
always identified by this matching. It is the amplitude that
cannot be determined successfully. Another problem with the
technique is a left-right direction ambiguity with respect
to the radar beam.

3.2   The dual-radar case

An extension of the model-fitting technique to handle
dual-radar data is described in Wyatt (1987). The results
of the simulation experiments suggest that improvements in
accuracy are achievable for such a configuration as long as
the two radar beams intersect with an angle greater than about
30 degrees. Of particular importance is the improvement in
accuracy when the technique uses only the upper two sidebands
of each spectrum. This will allow a reduced signal-to-noise
requirement. A further advantage of the method is the ability
to resolve the directional ambiguities.

Figure 5 shows both the single and dual modelfitting techniques
applied to the data shown in figures 2 and 4. The directional
ambiguities in the single-radar solutions have been resolved
using the input data. This particular wave spectrum has a
large narrow peak which neither technique is able to recover
because of the smoothing inherent in the integral equation.
This indicates an upper limit on the frequency resolution
achievable which must be quantified. When averaged over
frequency bands, the amplitudes found using both the single
radar technique for a non-perpendicular beam direction and the
dual radar technique agree well with the input spectrum.
The direction accuracies are improved for the two-radar
configuration. The directional spreading parameter is not
estimated accurately in the single radar case and is only
shown for the dual-radar solution.

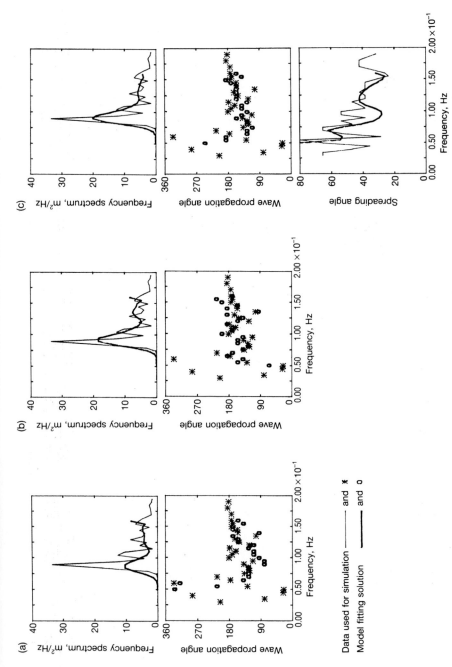

Fig. 5   Model-fitting solutions compared with data input to the simulations. (a) Single radar solution with the simulated radar beam direction perpendicular to the propagation direction of the long waves. (b) Single radar solution for a radar beam along the mean propagation direction of the long waves. (c) Dual radar solution demonstrating the potential for accuracy in all three parameters of the long wave directional spectrum.

4. CONCLUDING REMARKS

HF ground-wave radar has the potential to provide continuous measurement of oceanographic parameters over a large area with good spatial resolution. The radar system developed at Birmingham University has already demonstrated significant waveheight measurement beyond 100km. The work described here demonstrates the need for dual-radar systems for accurate wave measurement. The techniques referred to here will be tested with such a system in the near future.

The techniques for extracting wave data from the Doppler spectra of the backscattered sea echo are showing promise. However the model-fitting technique is limited to the measurement of the long-wave directional spectrum with a high frequency cutoff which depends on radar frequency. The description of the long-wave spectrum is itself not sufficiently general to cope with, for example, two swell components with the same frequency propagating in different directions. Methods are being developed that will provide measurements of higher frequency wave components and more general descriptions of the directional spectrum.

5. ACKNOWLEDGEMENTS

I am grateful for the continuing support and encouragement for this work from my colleagues at Birmingham University, the Rijkswaterstaat and the Institute of Oceanographic Sciences. This work is funded by the Science and Engineering Research Council.

6. REFERENCES

Barrick, D.E., (1971) Theory of HF/VHF propagation across the the rough sea Parts I and II, Radio Science, 6, 517-533.

Barrick, D.E., (1972) Remote sensing of sea state by radar, in Remote Sensing of the Troposphere, ed. V.E. Derr, U.S. Government Printing Office, Washington D.C., 12-1 to 12-46.

Barrick, D.E. and Lipa, B.J., (1979) A compact transportable HF radar system for directional coastal wavefield measurements, in Ocean Wave Climate, ed. M.D. Earle and A. Malahoff, Mar. Sci. 8, 153-201.

Barrick, D.E. and Weber, B.L., (1977) On the nonlinear theory of gravity waves on the ocean's surface. Part II: Interpretation and applications, J. Phys. Oceanogr., 7, 11-21.

Lipa, B.J., (1977) Derivation of directional ocean-wave spectra by inversion of second order radar echoes, Radio Sci., 12, 425-434.

Lipa, B.J. and Barrick, D.E., (1980) Methods for the extraction of long period ocean wave parameters from narrow beam HF radar sea echo, Radio Sci., 15, 843-853.

Lipa, B.J. and Barrick, D.E., (1986) Extraction of sea state from HF radar sea echo: Mathematical theory and modelling, Radio Sci. 21, 81-100.

Sabatier, P.C., (1983) Theoretical considerations for inverse scattering, Radio Sci, 18, 1-18.

Weber, B.L. and Barrick, D.E., (1977) On the nonlinear theory of gravity waves on the ocean surface, Part I: derivations, J. Phys. Oceanogr., 7, 3-10.

Wyatt, L.R., (1986) The measurement of the ocean wave directional spectrum from HF radar Doppler spectra, Radio Science, 21, 473-485.

Wyatt, L.R., (1987) Ocean wave parameter measurement using a dual-radar system: a simulation study, International Journal of Remote Sensing, 8, 881-891.

Wyatt, L.R., Burrows, G.D. and Moorhead, M.D., (1985) An assessment of a FMICW ground-wave radar system for ocean wave studies. Int. Journal of Remote Sensing, 6, 275-282.

Wyatt, L.R., Venn, J., Burrows, G.D., Ponsford, A.M., Moorhead, M.D. and van Heteren, J., (1986) HF radar measurements of ocean wave parameters during NURWEC, J. Oceanogr. Eng., OE-11(2), 219-234.

COMPUTER VISION IN IMAGE INFORMATION SYSTEMS

J.H. Johnson
(Centre for Configurational Studies, The Open University, Milton Keynes)

ABSTRACT

Computer vision, the automatic abstraction of explicit information from digital images, is investigated in the context of Geographic Information Systems. These have the requirement for rapid information-preserving conversion between raster and vector formats. The arithmetic properties of greyscale space are reviewed, and methods for abstracting low-level features from images are presented. A hierarchical architecture for computer vision is outlined, and it is suggested that Geographic Information Systems will have congruent structures. It is explicitly assumed that computer vision will require the kind of information and knowledge stored in image information systems. Thus computer vision and image information systems are each part of the other.

1. INTRODUCTION

Geographic Information Systems (GIS) are arguably the most general kind of image information system because of the variety and formats of their data sources, and the variety of their users and applications:

"The term 'geographic information' refers to any data which relate to specific locations on the Earth. It includes data on natural resources, pollutants, infrastructure such as utility services, land use and the Earth's inhabitants - their health, wealth and employment. Knowing where things are and how they relate to one another is crucial for management, planning and investment decisions taken within both the public and private sectors. ... computers have revolutionised the use of geographic information in applications as diverse as marketing, mineral expolation, flight simulation and development control."

(H.M.S.O. The Chorley Report, 1987)

Digital remotely sensed data are among the most complex
images that could reside in an image information system.
Atmospheric degradation of signals from sensors in the hostile
environment of space or unstable aircraft platforms, geometric
distortion, and other difficulties make all image processing
techniques relevant.

Although this paper wil concentrate on remotely sensed
images in geographic information systems, the arguments it
puts forward will apply to any kind of digital image
information system.

All database systems have the problem that information
can be retrieved in response to complex multi-source queries
only if the data are represented in compatible formats:

> "An important feature of applying Information Technology
> to the handling of spatial data is the ability to link
> data sets; that is, to merge and compare different data
> for the same location. This makes a Geographic Information
> System an analytical and decision making tool fundamentally
> different from a paper map. It is the ability to
> manipulate readily and quickly large volumes of data for
> the same areas which has the potential for adding great
> value to spatial data."
>
> (HMSO, The Chorley Report, 1987, page 84)

Image information systems have two distinct image formats:
vector format which uses point coordinates to represent points,
lines, polygons, symbols and text; and raster format which
effectively presents the image as an array of coloured dots.

Information represented in vector format is explicit.
Although it can be used to draw maps, the information exists
outside the map. Information in raster format is implicit.
Raster remotely sensed images can be considered to be maps,
but their implicit information has no existence outside the
map. If image information systems are to achieve their full
potential it is necessary for a means to be found to convert
rapidly between raster and vector formats.

Computer vision is the term that will be given to automatic
abstraction of information from raster digital images. The
term automatic requires some qualification since no computer
system is totally automatic without any human supervision.
The ideal of creating computer vision systems that can compete
with human vision is a very complex problem: as we learn more
about biological vision and develop new ideas in machine
vision we accept that contemporary computer vision is very
primitive. Currently the computer vision element of automatic

information abstraction in remote sensing is relatively small,
but it will presumably increase as scientists demand more
sensitive and reliable analytic techniques and the demands for
format conversion in GIS increase (Figure 1).

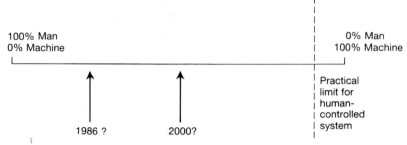

Fig. 1   A possible future for remote sensing Man–Machine
          systems.

A useful distinction can be made between computer vision
and computer-aided human vision. A large class of image
processing algorithms are concerned with the display of digital
image data in a way which enables human scientists to abstract
information. This has proved to be very helpful by highlighting
interesting features but the computer is not involved in any
explicit information abstraction. To clarify the objectives it
will be assumed that human vision plays no part in computer
vision. In principle the explicit information abstracted by
computer vision should be representable in text character and
numerical format, and a colour display should be unnecessary.
The assumption that human vision is totally absent in the
computer vision process is a theoretical constraint which
simplifies and clarifies the objectives and requirement. The
practical application of computer vision will always take place
in an eclectic context of automatic, semi-automatic, and
solely human information abstraction.

Automatic information abstraction begins with evaluating
the information content of the data source. Section 2 sets
up the various mappings which result in a digital image as an
array of numbers on a finite scale. The arithmetic properties
of the scale are investigated in Section 3 and some common
assumptions about n-dimensional greyscale space are questioned.
Section 4 defines pseudohomogeneous polygons as image
primitives which depend on relatively weak arithmetic
assumptions. Textures and non-contiguous (stippled) features
can be assembled from small polygons using subsequent
operators. Raster-to-vector conversion can be achieved by
drawing round the boundaries of polygons, and the application
of low-level vision techniques to abstract coherent lines
from the jaggy contours. This is explained in Section 5 which

outlines a general hierarchical architecture for computer
vision. This is based on algebraically structured hierarchical
languages with features as assemblies of their components at
various level. This algebraic structure is similar to that
in neural networks which suggests that the architecture will
be practical for real-time applications in Geographic
Information Systems. Section 6 discusses raster-to-vector
conversion which can be effected by drawing round polygons
and features, and using vision techniques to correct the raggy
edges. Section 7 concludes that the architecture for computer
vision outlined in this paper is naturally compatible with the
way information is represented and processed in Geographic
Information Systems. Computer vision needs the knowledge
base of GIS, and GIS needs computer vision for format
conversion, so each is part of the other.

2. GREYSCALE MAPPINGS

Let R be a rectangular region, and let P be a set of pixels
which cover this region. Let the centre of pixel p_{ij} be denoted
c_{ij}, and let C be the set of these pixel centres. Then there
is an inclusion $\gamma: C \hookrightarrow R$ which maps the centre of each pixel
to the corresponding point in R. Let s be the (piecewise)
continuous mapping which takes every point in R it is real-valued
intensity, $s: R \longrightarrow \mathbb{R}$. Finally, let τ be the analogue-to-digital
conversion mapping which assigns to each number in $\mathbb{R}$ a quantised
greyscale in a finite set of greyscales, G, written $\tau: \mathbb{R} \longrightarrow G$.
If α is the mapping which takes a pixel to its centre, the
greyscale values on the pixels are given by the sequence of
mappings

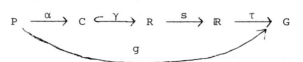

In this context, let the greyscale mapping which assigns a
greyscale to each pixel be denoted g, where $g = \tau s \gamma \alpha$.

With this terminology, the problem of trying to see what is
in the image to sub-pixel accuracy can be stated as that of
reconstructing $s : R \longrightarrow \mathbb{R}$ from $g : P \longrightarrow G$.

A common approach to this problem involves extrapolation
from the sample points at the centre of the pixels, C, to the
whole region R. This may or may not assume that $\tau s = s$ or
that $\tau s \gamma = s \gamma$.

A problem with many images, especially remotely sensed
images, is that the signal may be distorted before it reaches

the sensor. Thus the mapping s may be subject to a degradation
operator, d, and the intensity received by the sensor will be
degraded by d as d(s):R $\longrightarrow$ R. With this notation the
enhancement problem becomes: Given the mapping g for the
pixels P, g = τd(s)γα find an operator d^{-1} with $d^{-1} \circ d = 1$, the
identity mapping on R, such that $d^{-1}(g)=d^{-1}(τd(s)γα) = τd^{-1}d(s)γα =$
τsγα is the 'true' greyscale mapping. The operator d^{-1} is called
an image enhancement operator.

The literature contains many techniques for building image
enhancement operators. This raises the question as to which
operators are particularly suited to which images, since some
operators improve some images but not others. Rao's Paradox
asks "How can an operator which degrades a good image be
expected to enhance a degraded image?" (Johnson and Rao, 1986).
The notation makes the answer clear: in order to enhance a
degraded image by applying an operator d^{-1}, it is first
necessary to know the degradation operator d and to know that
d has an inverse. In practice one rarely knows the precise
form of d.

When the degradation operator is not known, image
enhancement depends on the application of a number of techniques
by a human expert. This approach has been highly successful
in many civil and scientific applications, but depends on
human vision. Drury (1987) gives a comprehensive account of
the degradation problems which are common in remotely sensed
images and the techniques used to correct them. Automating
these techniques will require the automated information
extraction of computer vision.

3.   THE PROPERTIES OF GREYSCALE SPACE

3.1 Greyscale Arithmetic

In the previous section G was defined to be a set of
greyscale values. Usually G is the set of integers {0,1,...,255}.
This set is totally ordered under the relation ≤, i.e. the
less-than-or-equal-to relation is reflexive (g ≤ g for all g ε G),
antisymmetric (g ≤ g' and g' ≤ g imply g = g' for all g,g' ε G),
and transitive (g ≤ g' and g' ≤ g" imply g ≤ g" for all g,g',
g" ε G).

Most image processing algorithms perform arithmetic and
other operations on the set of greyscales, and for many purposes
they give acceptable results. In computer vision special care
is required not to introduce spurious information by the
application of inappropriate operators since we cannot rely on
human vision to reject anomalies:

"Thus we need to know whether the data can be regarded
as 'values' of certain 'variables', and whether these
variables have certain properties: are we at liberty to

embed the data in some 'space', $\mathbb{R}^n$ for example, and to
perform certain operations on them? These are important
questions, and ones that should be asked by the practitioners
of data analysis before, for example, adding or multiplying
ages, names, addresses, professions, etc. - operations that
are often quite unjustifiable. The computer programs that
use them will certainly give results, but this is no
guarantee that the results have any meaning."

(Simon, 1986, page 97)

Lord and Novick (1968) define five different types of measurement
and scale:

Nominal or Classifactory Scale

A nominal scale is a set with no assumed structure. For
example, $\{0,1\}$ can be used to 'measure' male versus female
in a population. The set $\{m,f\}$ could also be used for this
purpose, and might be better since we have no preconception
that $m \leqq f$.

Ordinal or Rank Scale

An ordinal scale is a set with a total order relation, $\leqq$,
defined on it ($\leqq$ is reflexive, antisymmetric, and transitive).

Interval Scale

An interval scale is an ordinal scale, S, which has a
distance function d: S x S $\longrightarrow$ S with $d(s_1, s_2) = d(s_3, s_4)$
iff $d(s_1, s_3) = d(s_2, s_4)$.

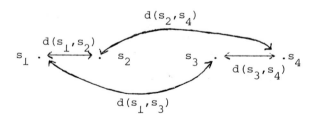

If the scale had an addition operation, an interval scale
would have the property that $d(s,s') = d(s+k, s'+k)$ for all
s, s', k in S. Lord and Novick define interval scales in terms
of them being in one to one correspondence with the real
numbers. The definition given here is deliberately weaker

since it requires the distance function to be defined and
justified on some empirical grounds, and it does not require
the scale to have an addition which would also require
justification.

Ratio Scale

The interval scale is a ratio scale when one of its members is 0.
Then the ratio of pairs of scale members can be taken as
$d(s_1,0)/d(s_2,0)$, and this will be the same whatever the unit
of the scale.

Absolute Scale

An absolute scale is a ratio scale in which the unit of the
scale is fixed.

In the first instance the usual set of greyscale values
$\{0,1,\ldots,255\}$ can be assumed to be at least an ordinal scale.

In digital images the scale is used to measure the number
of photons, or quanta, received by the sensor at a given
wavelength. The greyscale value multiplied by an integer
constant k is the number of quanta recorded. Each quantum
has energy $E = hf = hc/\lambda$ where E is measured in Joules, f is
frequency, λ is wavelength, h is Planck's constant, and c is
the velocity of light (Lillesan and Kiefer, 1979, page 6).

If the information in an image can be assumed to be
completely conveyed by the number of quanta reaching the
sensor it can be assumed that the set of greyscales has
the interval property and the ratio property . Also the
scale has a zero to record no quanta received by the sensor.
With these assumptions the set of greyscale values forms a
ratio scale.

If the constant k, which says how many quanta each
greyscale unit represents, is known the scale is absolute.
However k is not known for the majority of digital images,
and the scale is not absolute for these cases.

Consider a scale which counts the ages of people. The age
difference between two children of two years and seven years
is the same as the age difference between two adults of fifty
two and fifty seven years, namely five years. In simple
chronological terms this scale could be considered absolute.
However, if the scale is measuring something like intellectual
development there is a great deal of difference between five
years at the low end of the scale and five years at the high
end of the scale.

Question

Is it possible that a greyscale difference at one end
of the scale might convey qualitatively different information
from the same greyscale difference at the other end of the scale?

Consider two features in a remotely sensed image, one
associated with relatively low greyscale values and the other
with relatively high greyscale values. Suppose the frequencies
of pixels for each feature were normally distributed, but that
one had difference variance to the other. Then if greyscale
differences are assumed to measure pixel similarity, differences
corresponding to the feature with the smallest variance would be
comparable to relatively larger differences corresponding to
the feature with the greatest variance. In this case we might
conclude that the scale is not interval in terms of information
and interpretation.

3.2 Multi-band Greyscale Space

Remotely sensed images are typically multi-band, with some
bands giving better information on some features than others.
Details can be found in the literature (e.g. Lillisand and
Kiefer, 1979; Drury, 1987; Curran, 1985). Here a multi-band
image will be a set of pixels for a region R, where each pixel
P_{ij} is mapped to an ordered set of greyscale values,
$(g_1(p_{ij}), \ldots, g_n(p_{ij}))$ for an n-band image. Let G^n be the
Cartesian product of the scale G with itself n times, and let
G^n be called greyscale space.

The properties of greyscale space will determine which
algorithms are justifiable. Many algorithms assume that G can
be embedded in the rational numbers (number of the form p/q where
p and q are integers), and can be added, subtracted, multiplied,
and divided. Some algorithms go further and assume that the set
of greyscale values can be embedded in the real numbers so that
$G^n = \mathbb{R}^n$, and the distance between two points a and b is given
by the Pythagorean metric, $d(a,b) = [(a_1-b_1)^2 + (a_2-b_2)^2 + \ldots + (a_n-b_n)^2]^{\frac{1}{2}}$.

Questions

(a)   Under which circumstances is addition of greyscales
      meaningful?
(b)   Under which circumstances is subtraction of greyscales
      meaningful?

(c)   Under which circumstances is multiplication of greyscales meaningful?

(d)   Under which circumstances is division of greyscale meaningful?

(e)   Can the greyscales be meaningfully embedded in the rational numbers?

(f)   Can the greyscales be meaningfully embedded in the real numbers?

(g)   What kind of metrics (distance functions) naturally exist on n-dimensional greyscale space?

As far as addition of greyscales is concerned, this often occurs over a population of pixels when computing statistics such as the mean (when the operation is a combination of addition followed by division by the number of pixels). Subject to the usual caveats of applied statistics, greyscale addition is meaningful in this context.

Consider $g(p) - g(p')$ for a given spectral band. This difference is proportional to the difference in the number of photons recorded by the sensor for pixel p compared with pixel p'. If this difference were relatively small it would indicate that the pixels were similar in this band, and if the number were relatively large it would indicate the pixels were different in this band. In this way the subtraction can be given a coherent interpretation, although the term 'relatively' suggests this interpretation depends upon other pixels in a larger population. As noted previously, it is possible that the same greyscale difference may convey qualitatively different information at different points on the the scale.

When adding or subtracting greyscale values the unit of measurement remains the same (Joules), but multiplication by itself would result in units of Joules^2. Algorithms which assume greyscale space can be embedded in $\mathbb{R}^n$ multiply greyscale differences $(a_i-b_i) \times (a_i-b_i)$ when calculating the distance between points. However, these squared terms are added before taking the square root which again brings the units back to Joules (assuming $G^n = \mathbb{R}^n$).

Consider $g_1(p)$ and $g_2(p)$, the greyscales of pixel p in bands 1 and 2 respectively. The number $g_1(p)/g_2(p)$ is a ratio without units. As a number it belongs to the rationals, but this does not require that greyscale space must be embedded in the rationals. This number establishes a proportional ratio between the two greyscales for this pixel. As with pixel subtraction, the interpretation of the proportion might vary with the place on the scale. For example, let $g_1(p_1) = 4$, $g_2(p_1) = 8$, $g_1(p_2) = 54$, and $g_2(p_2) = 58$. Then $4/8 = .500$ and $54/58 = .931$.

Arguably, all measurement takes place on scales which have at best the properties of the rational numbers. Whether the operations of addition, subtraction, multiplication, and division are defined on greyscale space depends on their interpretation. The assumption that measurement is made on a scale which can be embedded in the real numbers deserves at least some justification.

A number of algorithms in image processing are used to classify pixels according to their greyscale signatures in G^n. It is commonly assumed that G^n can be embedded in $\mathbb{R}^n$, so that the distance between points in the space is defined. Consider the 3-d greyscale points, $p_1 = (5, 20, 110)$, $p_2 = (5, 120, 10)$, and $p_3 = (105, 20, 10)$ where the axes of that space correspond to a blue/green band ($.45-.52\mu m$, good water penetration, strong vegetation absorption), a green band ($.52-.60\mu m$, strong vegetation reflection), and a red band ($.63-.69\mu m$, very strong vegetation absorption). Source: Curran, 1985, Table 5.3). If this 3-d greyscale space is assumed to possess the usual metric, the 'distance' between each pair of points is $100\sqrt{2}$ units. In other words, each pair of points is 'equally similar'. However for some purposes this 'similarity' might be misleading. Points p_1 and p_2 are similar in the first band, but different in the other two. This is a different quality of similarity to p_1 and p_3, which are similar in the second band only.

The similarity measurement of the Pythagorean metric effectively trades off a difference of 100 units on the blue/green axis for a difference of 100 units on the green axis or a difference of 100 units on the red axis. When using algorithms which make the Pythagorean metric assumption, we should be able to justify such trade-offs in order to avoid the possibility of using chalk-and-cheese arithmetic.

A number of highly successful algorithms do make the assumption that greyscale space is $\mathbb{R}^n$. For example, Abrams et al (1988) show how decorrelation stretching, a technique based on principle component analysis, can be used on Landsat Thematic Mapper images to give significant new insights into the geology of the Oman. However it should be realised that these applications depend ultimately on the ability of the human scientist to interpret the transformed image.

It is not clear that image processing techniques which aid image interpretation will be useful in automated computer vision. Often they rotate and stretch greyscale space under one-to-one transformations which, although they make it easier for humans to see what is in the image, do not abstract any explicit information from the image.

The approach adopted here is cautious about the validity of assuming strong arithmetic and metric properties of greyscale space for computer vision. In fact one can abstract some very useful pixel configurations from images using little more than the assumption that greyscale space G^n is partially ordered by the total order on the greyscale set G.

4. POLYGONS, TEXTURED POLYGONS, AND NON-CONTIGUOUS FEATURES

4.1 Pseudohomogeneous Polygons

The simplest kind of feature that can be abstracted from a digital image is a polygon with uniform (homogeneous) greyscale values. Recall that the pixels to the immediate right, left, top, and bottom of a pixel are its 4-neighbours, and these together with the pixels touching its four corners are called its 8-neighbours. For simplicity of presentation a pixel will be said to be adjacent to another if they are 4-neighbours of each other, but the development also applies to 8-neighbours. Let two adjacent pixels be defined to be δ-similar with respect to a set of spectral bands if and only if their greyscale values differ by δ or less units in each band. δ-similarity is reflexive (every pixel is δ-similar to itself) and symmetric (p is δ-similar p' iff p' is δ-similar p). Its transitive closure defined by adding the rule (p δ-similar p' and p' δ-similar p" implies p δ-similar p") is called the pseudohomogeneity relation, and it is an equivalence relation on the pixels in an image. By construction the equivalence classes of pixels under this equivalence relation are simply connected polygons, and they are called pseudohomogeneous polygons (Johnson, 1985). Pseudohomogeneous polygons seem remarkably well suited to segmenting remotely sensed images. This is because the definition requires local homogeneity, but allows the greyscale signature to vary gradually over a polygon. Figure 2 shows common situations in which an unevenly illuminated field will be detected as a pseudohomogeneous polygon.

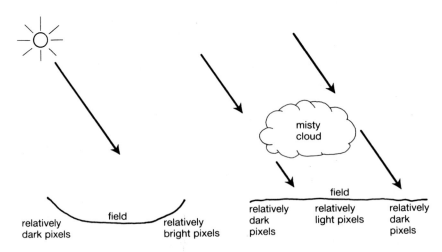

Fig. 2   Irregularly illuminated fields will appear as
         pseudohomogeneous polygons

Pseudohomogeneous polygons have the property that greyscale
signatures may vary considerably in them. In theory it is
possible to have pixels with greyscale values all zero in the
same polygon as pixels with greyscale values all the maximum.
In practice such variation is not encountered.

The local homogeneity of pseudohomogeneous polygons gives
them the highly desirable property of being tolerant to
maverick pixels such a tractor in a field or speckle in an
image: the polygons simply go round the anomaly. This can be
very useful when defining training areas, since maverick
pixels are excluded which would otherwise distort the greyscale
statistics (see for example Johnson, Rao, Denham, 1986).

Selecting appropriate values of δ is a central problem in
the use of pseudohomogeneous polygons within automated
computer vision. Johnson (1988) reports experiments on
automatic image segmentation by pseudohomogeneous polygons
using the greyscale statistics of the image. In general one
seeks polygons with unimodal greyscale distributions and variances
within given limits. Apart from δ varying between images, it
is possible that different values of δ will be appropriate to
abstracting different kinds of features from a single image.
This creates a new problem; if the polygons are found by
region-growing methods what are the effects of selecting the
original seed pixels? When δ is fixed the partition of the
image into pseudohomogeneous polygons is invariant under
change of seed pixels.

Pseudohomogeneous polygons have the property that every
δ-polygon is a sub-polygon of a (δ+1)-polygon, and every
(δ+1)-polygon is the disjoint union of its δ-sub-polygons
(when the polygons are interpreted as sets of pixels). This
means that δ-polygons with relatively small values of δ and
small variances have the potential to be used as building blocks
for constructing larger polygons.

4.2 Textured Features

Some features such as forests or urban areas are highly
textured, i.e. they are made up of many small (possibly single-
pixel) pseudohomogeneous polygons with differing greyscale
signatures. In the literature textures are commonly analysed
using statistics such as mean, variance, skewness and kurtosis
of greyscale values and their differences in an area (Levine,
1985). A problem with methods that compute statistics in a
square neighbourhood of a pixel arises at the edges of textured
areas. For example, Crane and Roberts (1988) developed a method
of population estimation from aircraft images which is reasonably
accurate within evenly textured urban areas, but gives less
reliable results at the edges of areas with different population
densities.

Many textures in remotely sensed images are made up of small
homogeneous regions. For example, urban areas are characterised
by roofs, parking lots, areas of grass, and so on; forests and
even individual trees are made up of small and often very
irregular polygons; and geological features such as the
disturbed ground which characterises limestone are made up of
lighter and darker polygons. Invariably these polygons are
irregular and/or bear no relationship to the pixel grid. Any
analytic technique which uses a mask such as the square which
contains the 8-neighbours will have its accuracy constrained
by the size and shape of the mask. Also the results will
depend on the orientation of the grid with respect to the
features the image contains. Textures are structured sets of
small polygons. The polygon statistics and adjacency structure
form a useful basis for the automatic detection of textured
polygons.

4.3 Non-contiguous and stippled features

Sometimes features occur in images which can be seen as
configurations of pixels (or even polygons) which are spatially
separated. For example an archeological feature such as
Stonehenge would be seen as a set of separate polygons. The
literature contains many examples of features which humans
find very easy to see but more difficult to define. Figure 3
shows detail of the sea off the coast of Wales in a Landsat

image. Although many of the pixels are isolated from each
other, there is clearly a spiral shaped feature in the image.
This will be called a <u>stippled feature</u>. Detecting stippled
features in images will require low-level features be abstracted
based on the spatial relationships between the pixels. In
other words the pixels will have to be assembled to form pixel
configurations such as colinear sets, dense clouds, and
so on. Pattern recognition based on these will continue through
hierarchical aggregation as explained in Section 5.

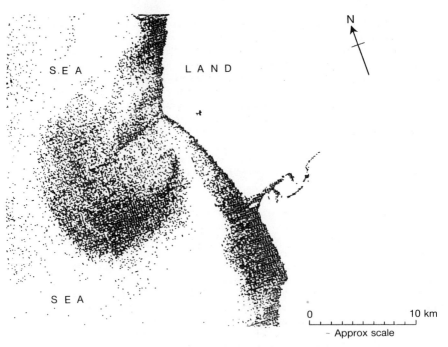

Fig. 3 A stippled spiral feature detected in the sea off the
         coast of Wales, UK. (Landsat Image, pixel dump in band
         ranges 38-50, 20-34, 8-32, 0-25, Dolgellau survey
         areas, Wales).

5. A HIERARCHICAL ARCHITECTURE FOR COMPUTER VISION

"If we wish to understand how a machine or living body
works, we look to its component parts and ask how they
interact with each other. If there is a complex thing
that we do not yet understand, we can come to understand
it in terms of simpler parts that we do already understand."

(Dawkins, 1986, page 11)

This approach to analysis assumes that complex things are
assembled from simpler things, and that in the course of the
analysis we identify component parts. The method is
characterised by the fact that we invariably give names to
the components, and we often give names to the processes of
assembly.

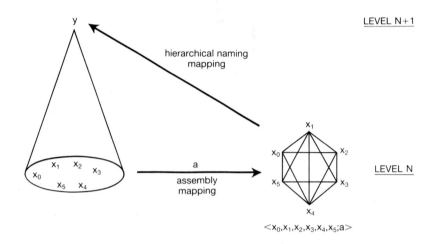

Fig. 4 Hierarchical assembly: the elements $x_0, x_1, \ldots, x_5$ are
assembled by the assembling mapping a to the simplex
$\langle x_0, x_1, \ldots, x_5; a \rangle$ which is given the name y at the next
hierarchical level. Note the hierarchical cone which
has the set of parts of y as its base.

The concept of assembling a structure from the set of its
component parts is illustrated in Figure 4. We write
$a: \{x_0, x_1, \ldots, x_p\} \rightarrow \langle x_0, x_1, \ldots, x_p; a \rangle$, where $\langle x_0, x_1, \ldots, x_p; a \rangle$
will be called an intensionally defined p-simplex. The
intension, or meaning, of this simplex is given by the
assembly mapping a which in general will be a compound predicate
in terms of subsets of $\{x_0, x_1, \ldots, x_p\}$. An extremely important
property of hierarchies is the existence of hierarchical naming
mappings which take a structured, intensionally defined simplex,
and map it to a single named entity at a higher hierarchical
level. At this higher level the structure can be treated as a
single entity and may be assembled with other individuals to
form higher level structures. Figure 4 also shows the
hierarchical cone construction, which identifies an object
with the set of its disassembled parts at some lower level.

(Gould et al, 1984) show how these cones can be used to clarify some common part-whole problems.

In computer vision we start with the names of features of interest and arrays of numbers. The spatial variety and variability of the features usually make immediate feature recognition impossible, so it is necessary to attempt to identify simpler sub-features. Deciding what these features should be is an example of the intermediate word problem, i.e. the problem of finding words which name intermediate features at a level higher than the pixels and lower than the target features. In principle the search for intermediate words and phrases goes much as Dawkins describes. It is remarkable how many intermediate words needed to identify features at all levels already exist in vernacular vocabulary, and how many other necessary terms exist in system-specific technical languages.

Computer vision requires an explicit hierarchical vocabulary in which pixels are aggregated to form relatively simple configurations such as polygons. These are aggregated to form more complex features, these are aggregated to form more complex features, and so on until the highest level features are recognised. At the lowest level the individual pixel greyscales are unlikely to admit any general interpretation. However, as the aggregation proceeds, meaningful features can be recognised.

To illustrate these ideas, consider a limestone region characterised by craters or sinkholes as Drury (1987, page 79) explains:

"Above all else, the distinguishing feature of carbonates in humid climates is the comparative absence of surface drainage. In the case of more-easily dissolved limestones this feature is accompanied by abundant near-circular depressions where rainfall most easily seeps into the bedrock. These sinkholes may assume momumental proportions where underground drainage has carved large caverns which have collapsed. Such solution features sometimes control local centripetal drainage patterns. The net effect is an area of confused drainage on a landscape mottled by sinkholes and related features."

This mottled effect appears as relatively light and dark patches on either side of the sinkholes due to the angle of the sun. Some of these sinkholes have shafts at their centres which show up as dark spots in the image. Figure 5 shows how the pixels aggregate into polygons, the polygons aggregate into craters (possibly sinkholes) and the craters aggregate to identify limestone (and confirm that the craters were sinkholes).

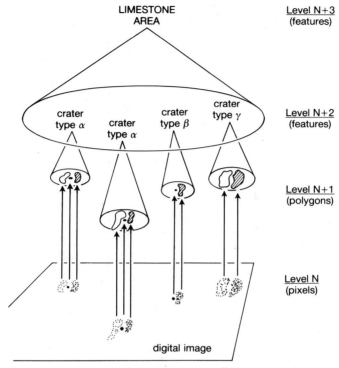

Fig. 5 An aggregation from low level pixel configurations to
         craters, and configurations of craters to identify
         limestone.

In Figure 5 note that none of the polygons making up the
craters have quite the same shape, that some craters have
dark spots while others do not, and that one crater has a
dark area but no light area. In general a feature like a
crater will have no single archetype. Some configurations of
polygons may indicate craters with some uncertainty, and it
may be necessary to consider these configurations in context
before that uncertainty can be resolved.

This proposed architecture for computer vision in remote
sensing begins with bottom-up pattern recognition, each
aggregation into a candidate feature being given a certainty
weighting. At some stage a top-down process begins based on
high-level hypotheses of the features in the image, and system-
specific knowledge relating to those features. This is
illustrated in Figure 6.

In general computer vision has to deal with images which
do not contain perfect information. For example, parts of the

image may be degraded, or features in the image may be
occluded by other features. Johnson (1989) shows how this
hierarchical architecture can be tolerant to missing
information through the design of the hierarchical language.

The simplices in the hierarchical language have a
connectivity structure through their shared faces (Atkin 1974)
which is likely to be useful in controlling the bottom-up
recognition and top-down decision making processes. Simplices
which are faces of many others in large star-hub structures
have a significant strategic place in the structure (Johnson,
1986).

The bottom-up part of the architecture outlined here has
some similarities to neural network structures (Wassermand and
Schwartz, 1987, 1988) which suggests they can be designed to
work in parallel, and hence very fast. The bottom-down
decision making will be some form of rule-based system, possibly
in the context of a blackboard system controlling the bottom-up
and top-down interaction (Shrihari, 1987, Hayes-Roth, 1983).

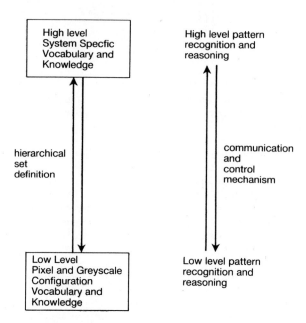

Fig. 6   An architecture for computer vision

6.   RASTER TO VECTOR CONVERSION

Section 4 showed how polygons can be abstracted from images,
and Section 5 showed how polygons can be assembled to form
features. Once these polygons have been abstracted it is a

simple matter to compute their boundaries as a sequence of
horizontal and vertical lines at the edges of their pixels.
Figure 7 illustrates this for the Barton Broad area of Norfolk.
This raster-to-vector conversion shows many features
such as the lake, fields, marshlands, and so on. It also
illustrates how different textured features are made up of
polygons of differing sizes, for example the villages are made
up of many small polygons (not drawn) while the orchard is made
up of larger and more irregular polygons.

Lake

Marshland

Orchards

River

Catfield
Village

Ludham
Airstrip

Ludham
Village

Fig. 7   Raster-to-vector converted map of Barton Broads, UK
         (Daedalus airborne scanner image, with a nominal 10m
         ground resolution).

This raster-to-vector conversion results initially in
vector polygons with irregular ragged edges. Johnson (1989)
shows how more coherent lines can be abstracted from these
edges giving a kind of vertex model (Levine, 1985, page 524).
It is important to be able to abstract the edge features of
polygons because their shape and relative size often convey
important information. For example, the fields in Figure 7
have relatively simple and regular shapes.

Registering images to one another is an important problem
in Geographic Information Systems. A minimum requirement for
this is being able to match up features in the images. However,
it is also necessary to be able to register the features, which
involves matching special points such as polygon corners. Very
often this matching has to take into account that one image
is an affine transformation of the other (or worse in the case
of aircraft image distortions), which makes abstracting point
information from the features essential.

Apart from registering images, abstracting vector features
such as points, lines, and polygons makes reasoning about the
image easier. In fact the kind of hierarchical aggregation
shown in Figure 5 can be applied to lines and polygons. For

example, appropriate configurations of lines can aggregate to
special features such as rectangles. Johnson (1989) shows
how digitised line drawings can be hierarchically assembled
to give error-tolerant pattern recognition. It is intended
that the pattern recognition process illustrated in Figure 6
will start with all available image primitives, including
lines.

7. CONCLUSION: COMPUTER VISION IN GEOGRAPHIC INFORMATION
    SYSTEMS

   Section 5 outlined a general architecture for computer
vision. That architecture assumed the existence of an
explicit hierarchically structured vocabulary, and the
existence of a great deal of pattern recognition knowledge
expressed in terms of that vocabulary. In Section 1 it was
argued that computer vision is necessary for Geographic
Information Systems, and now it is clear that a well organised
database is necessary for computer vision.

   A major problem with hierarchical databases is that data
aggregated for one purpose is very often unsuitable for
another. Indeed the Chorley Report (HMSO, 1987) explicitly
recommends that "data suppliers should both keep and release
their data in as unaggregated a form as possible". This
underlines the fact that we have great freedom in building
hierarchical languages, and once a language is built for a
given purpose it will enable information to be aggregated in
a self-consistent way. However, comparison with information
aggregated in another hierarchy is usually impossible in any
consistent way unless one can go back to the unprocessed data.
The kind of architecture described here for computer vision
is a more general architecture for building languages to record
data on and reason about complex systems. If this is so it is
a natural architecture for the language of all information
systems. In other words, there is a natural congruence between
Geographic Information Systems and the theory of automatic
computer vision.

REFERENCES

   Abrams, M.J., Rothery, D.A. and Pontual, A., (1988)
Mapping in the Oman ophiolite using enhanced Landsat
Thematic Mapper images, Tectnophysics, 151, 387-401.

   Atkin, R.H., (1974) Mathematical structure in human affairs,
Heinemann Educational Books, London.

   Crane, R. and Roberts, C., (1988) Private Communication,
Department of Geography, Pennsylvania State University,
Pa 16802, U.S.A.

Curran, P.J., (1985) Principles of remote sensing,
Longman, London.

Dawkins, R., (1986) The blind watchmaker, Penguin Books,
London.

Drury, S.A., (1987) Image interpretation in geology, Allen
and Unwin, London.

Gould, P., Johnson, J.H. and Chapman, G.P., (1984) The
structure of television, Pion, London.

H.M.S.O., (1987) Handing Geographic Information. Report
of the Committee of Enquiry chaired by Lord Chorley,
H.M.S.O., London.

Hayes-Roth, B., (1983) The blackboard architecture: a
general framework for problem solving?, Report HPP-83-30,
Stanford University, U.S.A.

Johnson, J.H., (1985) On hierarchical picture languages,
Proc. Image Understanding in Remote Sensing, University
College London, J-P. Muller, M. Duff (Eds).

Johnson, J.H., (1986) A theory of stars in complex
systems, in Casti and Karlqvist (Eds). Complexity, language
and life: mathematical approaches, Springer-Verlag, Berlin.

Johnson, J.H. and Rao, S.V.L.N., (1986) Configurational
Logic: abstracting truth from digital images, Proc. IEE
conference on Image Processing and its Applications,
Imperial College, London.

Johnson, J.H., Rao, S.V.L.N. and Denham, C.M., (1986)
Configurational Logic: experiments in image analysis,
NERC Remote Sensing Research Project, Report 3, Centre for
Configurational Studies, Open University, MK7 6AA, England.

Johnson, J.H., (1988) Pseudohomogeneous Polygons, mimeo
in preparation, Open University, Milton Keynes.

Johnson, J.H., (1989) Pixel parts and picture wholes, in
from the pixels to the features, J.-C. Simon (ed), North
Holland, Amsterdam.

Levine, M.D., (1985) Vision in Man and Machine, McGraw
Hill, London.

Lillesand, T.M. and Kiefer, R.W., (1979) Remote Sensing
and image interpretation, John Wiley, New York.

Lo, C.P., (1986) Applied Remote Sensing, Longman
Scientific and Technical, Harlow.

Lord, F.M. and Novick, R.N., (1968) Statistical theories
of mental test scores, Addison-Wesley, London.

Shihari, S.N., Wang, C-H, Palumbo, P.W. and Hull, J.J., (1987)
Recognising address blocks on mail pieces, AI Magazine,
Vol. 8, No. 4, 25-40.

Simon, J-C., (1986) Patterns and Operators, Translated
by J. Howlett, North Oxford Academic, KoganPage, London.

Wasserman, P.D. and Schwartz, T., (1987) Neural Networks,
Part 1, IEEE Expert, Winter 1987, 10-13.

Wasserman, P.D., and Schwartz, T., (1988) Neural Networks,
Part 2, IEEE Expert, Spring 1988, 10-15.

THE HOR QUADTREE
AN OPTIMAL STRUCTURE BASED ON A NON-SQUARE 4-SHAPE

S.B.M. Bell

(NERC Unit for Thematic Information Systems, Reading University),

B.M. Diaz

(Department of Computer Science, Liverpool University)

and

F.C. Holroyd

(Faculty of Mathematics, The Open University, Milton Keynes)

ABSTRACT

This paper presents a tesseral addressing and arithmetic system (Bell et al, 1983) for the Hexagonal or Rhombus (HoR) structure. The system is based on the observation that an identical lattice is formed by amalgamating the centroids of 4 hexagons and amalgamating 4 rhombi. When identical addresses are given to the tiles, the authors note that the structure can be used to avoid the limitation that hexagons cannot be further divided (the limit property of the hexagons) while being able to capture, when required, the unit adjacency of the hexagons. This ability to exploit the advantages of two underlying tilings renders the HoR structure optimal for image processing and automated cartography. Example algorithms using the arithmetic and possible methods of using features of the structure are also discussed.

1. INTRODUCTION

A tiling of the plane (or tessellation) is termed isohedral if all the tiles are equivalent under the symmetry group of the tiling. (Bell et al, 1983) have considered 11 of the 81 isohedral tilings of the Euclidean plane identified by (Grunbaum and Shepherd, 1977, 1981) and the tile hierarchies that may be built including them as the base or "atomic" level tile. They show that some are of special interest in computer graphics and image processing. Representatives of these 11 with straight edges, "Laves nets" have properties among which they identify unit adjacency (the adjacency number of a tiling is the number of different inter-centroid distances between any tile and its immediate neighbours) and tiling hierarchy unlimitedness (the absence of a definable atomic tiling) as being the most valuable. They note that while the $[3^6]$ hexagonal tiling possesses unit adjacency it

has a tiling hierarchy which is limited (Holroyd, 1983). In
contrast the other contender for "best tiling", the [4^4] square
tiling, is unlimited but does not have unit adjacency (Figure
1). Finally, they note that there should be ways of
reconciling the need for these two properties by admixing the
underlying centroid lattices of the two tilings.

(a)

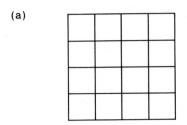

[4^4] (square) tiling

(b)

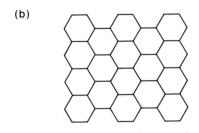

[3^6] (hexagonal) tiling

Fig. 1

    In this paper we present a centroid lattice over which both
the hexagonal (Figure 2a) and rhombic (Figure 2b) tilings may
be placed. The rhombic tiling like the square (Figure 2d) can
be amalgamated unlimitedly by a 4-shape, and has the same
adjacency number. The hexagonal tiling can be amalgamated by a
4-shape which is limited at the hexagonal level. The identical
lattice implies that where necessary use of this Hexagonal or
Rhombic (HoR) structure can yield the advantageous properties
of either base atomic tiling. It is the authors' belief that
such a structure is an optimal one for image processing and
computer graphics.

It is important to note that the hexagonal lattice of the HoR structure retains the same orientation angle at all levels in the hierarchy. In contrast the orientation angle of the hexagonal lattice based on the 7-shape, changes at each level in the hierarchy (Figure 2c).

The HoR Quadtree

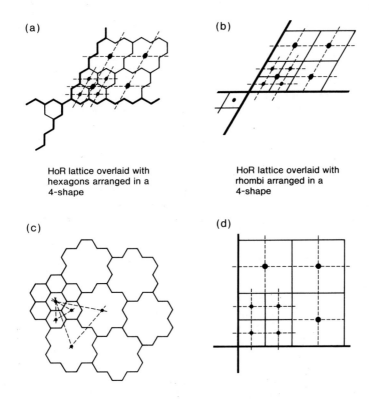

(a)

HoR lattice overlaid with hexagons arranged in a 4-shape

(b)

HoR lattice overlaid with rhombi arranged in a 4-shape

(c)

Hexagonal 7-shape

(d)

Square 4-shape

Fig. 2

2.   PROPERTIES OF THE HOR STRUCTURE

Table 1 contrasts the four hierarchies presented in Figure 2 using the list (repeated here) of tile properties identified by (Bell et al, 1983) as being of importance in image processing and automated cartography, with one addition, which the new HOR example suggests as useful criterion.   However, these properties depend on the tiling chosen, and not on the underlying lattice of tile centres alone, although this also supports a tesseral hierarchical structure   (Holroyd, 1985). For example, we see that the hexagonal (Figure 3a) and rhombic (Figure 3b) tilings have different adjacency structures.   The hexagonal 4-shape touches a neighbouring tile across L and S, the rhombus across E, V and V'.   We decide which adjacency and other tile properties to associate with the lattice by creating courtesy tiles in a way that depends on the lattice alone. These are formed from the Dirichelet region (or Voronoi polygon) ("a polygon whose interior consists of all the points in the plane which are nearer to a particular lattice point, e.g. P, than any other lattice point" (Coxeter, 1980).   They are shown in Figure 4 for the atomic level (level 1).   The lattice points are joined by a solid line, and the tile boundaries by dotted lines.   The lattice points adjacent to P which affect the boundary of the Voronoi polygon are labelled P_c.   It should be noted however, that the lattice cannot be generalised to level "k" > 1 before the Voronoi polygons are created.   Otherwise lattice points at a level less than "k" that are subsumed within the same lattice point at level "k" may not be included within its Voronoi polygon.   We prefer to associate a generalised Voronoi polygon with a lattice point P' at level "k" > 1.   The interior of the polygon consists of all points in the plane which are nearer to any of the lattice points at a level less than "k" from which P' is formed than any other lattice point.   We call this polygon the lattice tile at level "k".   With this definition lattice points on a hexagonal lattice can be amalgamated so that the lattice tiles at all levels are identical to those of the $[3^6]$ 4-shape or 7-shape.   The properties of the lattice hierarchy of interest are therefore those of the 4-shape.

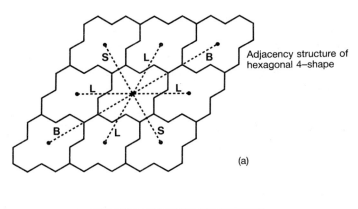

Adjacency structure of
hexagonal 4–shape

(a)

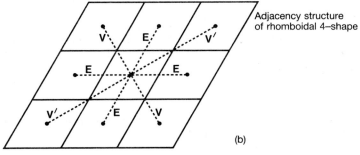

Adjacency structure
of rhomboidal 4–shape

(b)

Fig. 3

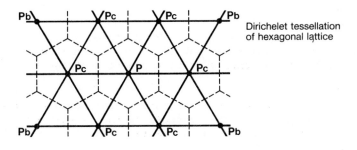

Dirichelet tessellation
of hexagonal lattice

Fig. 4

Adjacency

Two tiles are considered to be neighbours if they are
adjacent either along an edge or at a vertex. A tiling is
uniformly adjacent if the distances between the centroid of

one tile and the centroids of all its neighbours are the same.
More generally the adjacency number of a tiling is the number
of different intercentroid distances between any one tile and
its neighbours. (If we consider tiles to be neighbours only
if they are adjacent along an edge, then an analogous
definition can be made.)

In image processing terms, the lower the adjacency number
the easier it is to identify those tiles which constitute the
same logical object.

Rotation

The rotation number of a tiling is the largest order of a
rotational symmetry of the tiling (where a rotation of order n
is a rotation by 360/n degrees).

It is an advantage to avoid floating point arithmetic and
the larger the rotation number the more likely it is that this
will be possible.

Aperture

The aperture of a level "k" tile is the number of level "k
- 1" tiles needed to construct it.

The higher the aperture the easier it is to generalise the
"pattern types" which may be generated for any level in the
hierarchy. Depending on application this may or may not be an
advantage.

Circularity

The circularity of a tile is the difference in area between
the smallest circumscribed and the largest inscribed circle
which can be drawn, expressed as a fraction of the area of the
tile.

It is an advantage for all points on the boundary of a tile
to be equidistant from the centroid of the tile, thus reducing
the number of different algorithms required to process these
points.

Average Circularity

The average circularity of a tile is the average distance of
a point in the tile from the tile centroid, expressed as a
fraction of the square root of the area of the tile. It
depends only on the shape of the tile. For a circle it is
0.376.

The smaller this measure, the less the average error in treating the whole tile as though it were located at its centroid, as is frequently assumed. Average circularity is proportional to the number of tiles needed to cover a given region within a given error.

Convexity

The convexity of a tile is the difference in area between the tile and its closed convex hull, expressed as a fraction of the area of the tile.

It is an advantage for the points furthest from a tile centroid not to be closer to another tile centroid.

Orientation

Tiles with the same orientation can be mapped onto each other by translations of the plane which involve no rotation or reflection. The tiling is said to have uniform orientation if all tiles have identical orientation. More generally the orientation number of a tiling is the number of distinct tile orientations which are possible.

If two tiles do not have the same orientation, integer addressing of the tiles will not be possible.

Limit

If a tiling has tiles at level "k + 1" which are not "similar" (see below) to those at level "k" then that hierarchy is said to be limited. Unlimited tilings do not have a definable atomic tiling because every atomic tile can be subdivided into smaller tiles which then become the atomic tiles of a new hierarchy. Although courtesy tiles formed from the Voronoi polygons of a lattice can always be redefined for a subdivided lattice, it may be necessary to change the level 2 tile to which a level 1 tile belongs, on subdividing to level -1. This difficulty does not arise for the hexagon lattice and the 4-shape: it does however for the 7-shape.

If a tiling is limited no change of scale lower than the limit atomic tiling can be envisaged without difficulty.

Similarity

The tiles at level "k" and "l" are "similar" if they have identical shape, i.e. the tiles at level "l" are scaled images of those at level "k".

Regularity

A tiling is regular if the atomic tiles are composed of regular polygons.

Isohedrality

An isohedral hierarchy is one where all the molecular tilings (those composite tiles at levels above the atomic) are isohedral. Clearly all unlimited tilings are isohedral. Further, (Holroyd 1983) has shown that if the first three levels of molecular tiling are isohedral then all subsequent levels will also be isohedral.

It is a disadvantage to have separate algorithms to deal with differing tile types as would occur if the hierarchy were not isohedral.

Democracy

Democracy among level "k" tiles exists when in the level "k + 1" tile it is impossible to differentiate between the level "k" tiles.

This is a property which may or may not be an advantage depending on the application. Thus the polygonal coherence afforded to "pattern types" obtained by having a clear central tile is an advantage of undemocratic molecular tiles; however it is an advantage to be able to treat all democratic tiles with the same algorithms.

Table 1 reveals that the advantages of the $[4^4]$ 4-shape over the rhombic 4-shape lie in circularity and rotation (since the lack of democracy is due to a trivial distinction between rhombus diagonals). Section 6 discusses why these advantages are more apparent than real.

Table 1

	$[4^4]$ {4}	$[3^6]$ {7}	$[3^6]$ {4}	rhombus {4}
Adjacency				
Atomic	2	1	1	2
Molecular	2	1	1	2
Rotation				
Atomic	4	6	6	2
Molecular	4	6	2	2
Aperture	4	7	4	4
Circularity				
Atomic	0.785	0.302	0.302	2.041
Molecular	0.785	2.073	1.360	2.041
Average Circularity				
Atomic	0.388	0.377	0.377	0.403
Convexity				
Atomic	0	0	0	0
Molecular	0	0.143	0.167	0
Limited?	No	Yes	Yes	No
Uniform Orientation?	Yes	Yes	Yes	Yes
Similarity?	Yes	No	No	Yes
Regularity?	Yes	Yes	Yes	No
Isohedrality				
Atomic	Yes	Yes	Yes	Yes
Molecular?	Yes	Yes	Yes	Yes
Democracy	Yes	No	No	No

Notes.   All molecular entries are for level 2. All
lattice entries are for the atomic level. {}
indicates the shape, e.g. {4} = 4-shape.

3. THE HOR ADDRESSING SCHEME

All tiling hierarchies described above (Figure 2) can give
rise to hierarchical addressing schemes. Such systems for the
$[4^4]$ tilings have been proposed by (Gargantini 1982), and
others (Morton, 1966; Rosenfeld and Samet, 1979; Woodwark,
1982; Oliver and Wiseman, 1983) and as the quadtree has
received considerable attention (Samet, 1984). An important
advantage of tesserally constructed addressing schemes (those
which apply to the entire Euclidean plane) is that it is
possible to perform simple arithmetic operations with the
addresses and to imbue these operations with geometric
interpretation. In constructing such a scheme for the HoR
structure we must preserve the link between arithmetic
and geometric interpretations without jeopardising our ability
to move between the underlying tilings.

We achieve our aim by giving the molecular tiles of both the
hexagonal (Figure 5) and rhombic 4-shape hierarchies (Figure 6)
the same addresses. A generalisation of 2's complement for
negative numbers uses the repeating digit or "beaked notation"
described by (Diaz, 1984) to ensure that the entire Euclidean
plane is addressed (Bell et al, 1988b). Clearly this implies
that we can exploit the unlimitedness of the rhombic 4-shape
while retaining the ability to consider the HoR structure centroids
as centroids of hexagons. Several advantages can be accrued by
this scheme. We note in particular the advantages over the
7-shape namely that the tiles 1, 10, 100, ...; 2, 20, 200, ...;
3, 30, 300, ... form straight lines rather than digital
spirals. The spiralling the case of 7-shape requires us to
know which level in the hierarchy the digit of an address
occupies when comparing it with Cartesian addresses. In
contrast all the digits of the HoR addressing structure can be
treated identically in any conversion algorithm. For example
the algorithms for inter-conversion from tesseral to cartesian
address in two's complement binary form follow. We may use
integer arithmetic operations (note: not tesseral) for the
first algorithm save at the end.

The HoR Quadtree

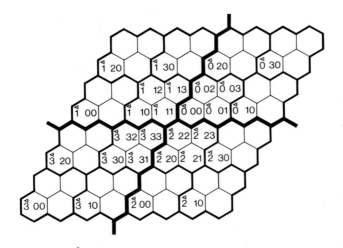

Addressing for HoR structure hexagons

Fig. 5

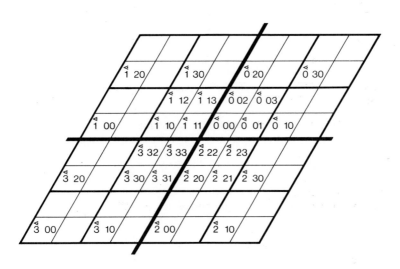

Addressing for the HoR structure rhombi

Fig. 6

HoR-to-Cartesian

Integer:

```
x, y = 0
for i = least significant (=0) to most significant tesseral
        digit (t_d) do:
[    s = integer part of (t_d/2)
     x = x + (t_d - s*2) * 2**i
     y = y + s * 2**i
]
```

Floating point:

```
y = y * 3^½/2
x = x + y / 2
```

For the opposite conversion we need to use floating-point arithmetic only at the beginning:

Cartesian-to-HoR

Floating point:

```
a = 2y / 3^½              and round to nearest integer
x = 2x                    and round to nearest integer
```

Integer:

```
b = integer part of (a/2)
c = integer part of ({x - a}/2)
d = integer part of (c/2)

for i = least significant (=0) to most significant tesseral
        digit (t_d) do:

[ t_x = c - d*2
  t_y = a - b*2
  t_d = tx + ty*2
  {i is column position of tesseral digit}

  a = b
  b = integer part of (b/2)
  c = d
  d = integer part of (d/2)
]
```

These algorithms are considerably simpler than their 7-shape counterparts (Gibson and Lucas, 1982). The difference

between them and their equivalents for the quad-tesseral
structure is in the multiplication (Note, not tesseral
multiplication) by $3^{\frac{1}{2}}$ (see Bell et al, 1986b).

The second advantage lies in the curve defined by addresses
of increasing magnitude, when considered as non-tesseral
numbers to base 4. This "HoR-tesseral raster" (Figure 7a) is
similar to the quad-tesseral raster described by (Bell et al,
1988a) and can be related to the Peano curve (Morton, 1966;
Goodchild and Granfield, 1983) as illustrated in Figure 7b.

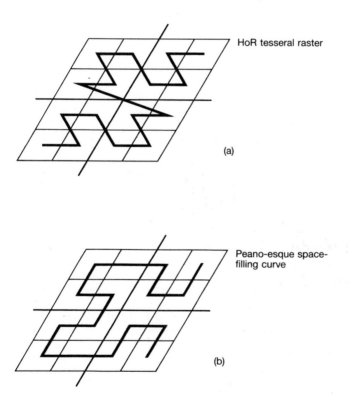

HoR tesseral raster

(a)

Peano-esque space-
filling curve

(b)

Fig. 7

4. HOR STRUCTURE ARITHMETIC

- *addition and subtraction*

The addition of two tesseral addresses results in a translation of the first number by an amount equal to the vector from O to the second number. This process has been described by (Bell et al, 1983) and has been shown to extend equally well to the negative (beaked) quadrants in the case of the quad-tesseral addressing structure (Bell et al, 1988b). By using the addition (Table 2) it is possible to achieve translations of the HoR-tesseral addresses in an identical manner. Furthermore, because only a linear transformation of tile centroids is required to transform the lattice of quad-tesseral centroids into the lattice of the HoR structure centroids, the addition table is identical to that for the quad-tesseral structure.

Table 2

+	O	1	2	3	<1	<2	<3
O	O	1	2	3	<11	<22	<33
1	1	10	3	12	O	<23	<22
2	2	3	20	21	<13	O	<11
3	3	12	21	30	2	1	O
<1	<11	O	<13	2	<10	<33	<32
<2	<22	<23	O	1	<33	<20	<31
<3	<33	<22	<11	O	<32	<31	<30

Note. A beaked or negative digit d is written thus: <d

The HoR / quad-tesseral addition table

In the example addition below, we see that the addition of 31 to <223 involves five table lookup operation, and in Figure 8 the geometric interpretation (as a translation) of the addition takes us from tile<223 to tile 102.

<223 + 31

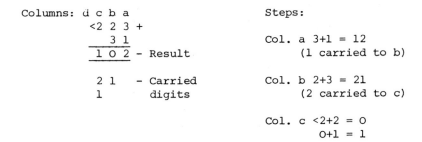

```
Columns: d c b a              Steps:
         <2 2 3 +
            3 1               Col. a 3+1 = 12
         ─────────               (1 carried to b)
         1 0 2  - Result

           2 1   - Carried    Col. b 2+3 = 21
           1       digits        (2 carried to c)

                              Col. c <2+2 = O
                                    O+1 = 1
```

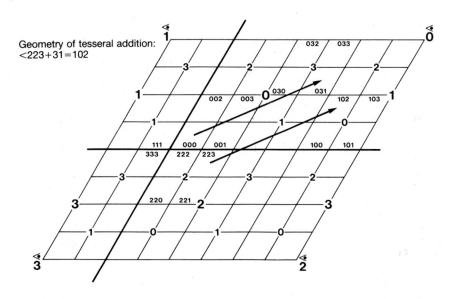

Geometry of tesseral addition: <223+31=102

Fig. 8

Table 3

:	Source digit	: O :	1 :	2 :	3 :	<1 :	<2 :	<3 :
: (a) Negative		: O :	<1 :	<2 :	<3 :	1 :	2 :	3 :
: (b) Complement		: 3 :	2 :	1 :	O :	- :	- :	- :

Then add 3

Subtraction of a tesseral number is a translation equal to
the vector from that number to O. Performing a subtraction
involves finding the negative of the tesseral number to be
subtracted and then adding this number using the HoR addition
table (Table 2). Obtaining the negative of a tesseral number
is the process of locating the point an equal magnitude from O
but in the opposite direction. This may be achieved by
replacing every digit in the number by its negative or
complementing (by use of Table 3(a) or 3(b)). 3 must be added
after complementing. This number may contain internal beaked
numbers which must be removed if the absolute tile address is
to be found (Diaz, 1984). However, "unbeaking" may be left
until after the completion of addition since the addition
table contains the beaked digits as entries. To unbeak,
separate the number into two parts, one containing the positive
digits (P) and one the negative (N), (A = P + N). Remove the
beaks from N thus converting N to its negative as in Table 3(a)
(A = P - (-N)). Complement -N as in Table 3(b), and add 3
(A = P + (-(-N))). Add P and the new, beakless representation
of N (A = P + N).

Figure 9 presents the geometric interpretation (as a
translation) of the HoR subtraction 102 - 31. The calculation
itself is as follows. Table 3(a) is used to complement
31: 102-31 = 102+<302+3, then a tesseral addition is performed.

<102 + <302 + 3

Columns: d c b a   Steps:
              1 0 2 +
              <3 0 2   Col. a 2+2+3 = 20+3 = 23
                  3   (2 carried to b)
          ‾‾‾‾‾‾‾‾‾
          <2 2 3 - Result
                                   Col. b 0+0+2 = 0+2 = 2
              2   - Carried
                  digits   Col. c 1+<3 = <2

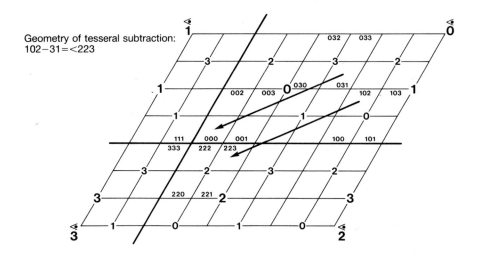

Geometry of tesseral subtraction:
102−31=<223

Fig. 9

5. HOR STRUCTURE ARITHMETIC

− multiplication and division

Tesseral multiplication, n by m, scales and rotates the
multiplicand vector (the line from O to n) by an amount
proportional to that required to transform the vector O1 (the
line from O to 1) into the multiplier vector Om. To perform
such multiplication with the HoR addresses it is necessary to
use Table 4, which applies equally well to both hexagonal and
rhombic arrangements. However, because the quad-tesseral
centroid lattice cannot be obtained by a scaling and rotation
of the centroids of the HoR structure lattice the
multiplication tables for the HoR and quad-tesseral structures

are different. Table 5 contains the quad-tesseral multiplication table for reference purposes. It should be noted that the usual rules of multiplication apply. These may be used to speed alorithms. For example:

$2*3 = 2*1 + 2*2$ $<2*3 = 2*<3 = -(2*3) = -(<131) = <202+3 = <21$

Table 4

*	0	1	2	3	<1	<2	<3
0	0	0	0	0	0	0	0
1	0	1	2	3	<11	<22	<33
2	0	2	<13	<131	<22	<23	<21
3	0	3	<131	22	<33	<21	<202
<1	0	<11	<22	<233	1	2	3
<2	0	<22	<23	<21	2	<13	<131
<3	0	<33	<21	<202	3	<131	22

The HoR-tesseral multiplication table

Table 5

*	O	1	2	3	<1	<2	<3
O	O	O	O	O	O	O	O
1	O	1	2	3	<11	<22	<33
2	O	2	<11	<13	<22	1	<23
3	O	3	<13	20	<33	<23	<20
<1	O	<11	<22	<33	1	2	3
<2	O	<22	1	<23	2	<11	<13
<3	O	<33	<23	<20	3	<13	20

The quad-tesseral multiplication table

The example multiplication shown below shows how the multiplication and addition tables are used in conjunction to obtain the rotation and scaling shown in Figure 10.

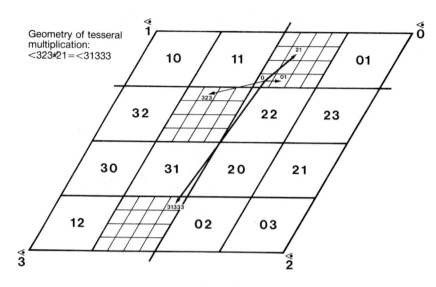

Geometry of tesseral
multiplication:
<323*21=<31333

Fig. 10

```
<323 * 21

Columns: e d c b a                    Steps:
         <3 2 3 *
            2 1                        Multiplication:
        ───────────
         <3 2 3 - Partial
        <1 3 1 O   Results            <323*1 = <323
        <1 3 O O                      Col. b 2*3  = <131
       <2 1 O O O                     Col. c 2*3  = <13
       ───────────                    Col. d 2*<3   <21
       <3 1 3 3 3   Result

          <1        Carry             Addition:
                    from the
                    addition          Col. a   3+O = 3
                                      Col. b   2+1 = 3
                                      Col. c  <3+3 = O
                                               3+O = 3
                                             <1+<1 = O
                                             carry <1
                                               1+O = 1
                                      Col. e <2+<1 = <3
```

A division is shown geometrically in Figure 11. The
arithmetic for a method similar to that described in
(Bell et al, 1986b) follows first, with the HoR-tesseral
multiplication table replacing the quad-tesseral one.
The procedure exactly as described in the reference may not

converge in all cases when applied to the HoR-tesseral.   The
arithmetic for a second, preferred, algorithm follows.   The
multiplicative inverse produced by these means may extend
beyond the radix point, but is generated entirely within the
tesseral number system using only tesseral digits.   For further
details of alternative tesseral number systems and a more
detailed explanation of the derivation of the multiplicative
inverse the reader is directed to (Bell et al, 1986).

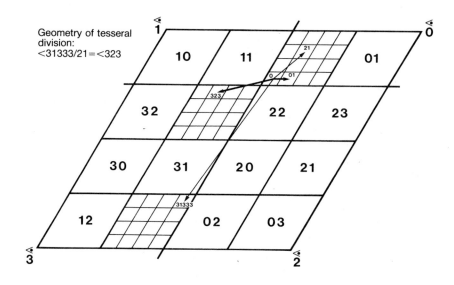

Fig. 11

Example division <31333/21

Step 1.   Generate all the multiples of 21.

$$
\begin{array}{ccc}
21* & 21* & 21* \\
\underline{\phantom{<}1} & \underline{\phantom{<}2} & \underline{\phantom{<}3} \\
21 & <132 & <1313
\end{array}
$$

Using table 3(b), find the negative of these numbers,
i.e. exchange the digits as shown there, and add 3 to
the least significant digit.

21*<1 = 31   (-21*1)
21*<2 = <212   (-21*2)
21*<3 = <2023   (-21*3)

```
To subtract 21*1      add 31
To subtract 21*2      add <212
To subtract 21*3          <2023
To subtract 21*<1         21
To subtract 21*<2         <132
To subtract 21*<3         <1313
```

Step 2.   Bring down the first 3 digits (<313) and subtract all
          multiples:

```
<313      <313      <313      <313      <313      <313
<331      <212      <2023       21      <132      <1313
 132      <2021     <2000     <222      <101      <130
  11       12         3       <11       32        13
```

We select <3.
The remainder is <130.   Diagrammatically we have:

```
        <3
21/<31333
    <1313
    <130
```

Step 3.   Bring down the next digit (3) and repeat step 3.

```
<1303     <1303     <1303     <1303     <1303     <1303
<3331     <2212     <2023       21      <132      <1313
<122      <131      <2110     <1332     <13011    <13120
<311      <2 2        2         1       <1322     <1313
```

We select 2 as the new result digit, thus:

```
       <32
21/<31333
    <1313
    <1303
    <2212
    <131
```

Step 4.   Bring down the next digit (3) and repeat step 3.

```
<1313     <1313     <1313     <1313     <1313     <1313
<3331     <2212     <2023       21      <1132     <1313
<1132      21         0        222      <2101     <13130
 311      <312       333       111      332       <1313
```

We select 3 to complete the division, thus:

```
          <323
    21/<31333
     <1313
     ‾‾‾‾‾‾‾
     <1303
     <2212
     ‾‾‾‾‾‾‾
     <1313
     <2023
     ‾‾‾‾‾‾‾
          0
```

The second algorithm for division is called p-adic tesseral
division. Positive digits only are used. Let the division
equation be:

   A * X = B

where A and B are known, X unknown. The principal of the
algorithm is to find the digits of X in the reverse order, from
right to left. First any zero digits at the right end of A are
stripped. They are reintroduced by adjusting the radix point
in X once X is found. The last digit of X is chosen so that
the last digit of A*X will cancel, additively, with that of B.
The algorithm then iterates. The process ceases when a
sequence of digits at the most significant end of X begins to
repeat.

   Two factors ensure the success of the algorithm. Firstly,
Table 5, the multiplication table, shows that whatever the last
digits of A and B (except when that of A is zero) a positive
digit can be found which, when it multiplies A will convert the
last digit to that of B. Secondly, any repeating string of
digits to the right of the radix point, potentially
representing fractions of magnitude less than a digit, can be
written in a form repeating to the left of the radix point.
Let the repeating strings be:

       <ab .... c: and   :ab .... c>

where the letters represent tesseral digits. Then:

       <ab .... c: .   :ab .... c> = 0

where "." is the radix point, the algorithm is illustrated
below and is described in (Bell et al, 1986).

   Example division 21 * X' = <31333, X' unknown.

   X' is replaced by XO + x where x is the last digit of X'

Step 1.

        Choose x = 3.

        X' = (.... 3)

```
21 * (XO + 3) = <31333
21*XO + 21*3  = <31333
21*XO + <1313 = <31333
21*XO         = <31333 - <1313
21*XO         = <31333 + <2023
21*XO         = <20020
21*X          = <2002
```

Step 2.

        Choose x = 2, redefine X to be (XO + x) and repeat step 1.

        X' = (.... 23)

```
21 * (XO + 2) = <2002
21*XO         = <2002 - <132
21*X          = <2023
```

Step 3.

        Choose x = 3, redefine X to be (XO + x) and repeat step 2.

        X' = (.... 323)

```
21 * (XO + 3) = <2023
21*XO         = <2023 - <1313
21*X          = <2023
```

        The algorithm now loops, hence:

        X' = ... 3323 = <323

6. SUMMARY AND DISCUSSION

The HoR quadtree structure can be seen as a method of reconciling the requirement for an unlimited tiling hierarchy while at the same time providing unit adjacency. This is achieved by having identical hierarchical addressing schemes for the hierarchy based on the hexagonal and rhombic 4-shape and stems from the observation that they share the same lattice. Clearly, we could have a database where the addressing was based on the rhombic 4-shape and therefore was

unlimited; and a display consisting of hexagonal pixels (the
screen) where addressing was based on the hexagonal 4-shape.
Data movement between the two, database and screen, would
remain a simple procedure and providing the operations with the
screen image restricted themselves to the atomic level - there
is no problem for compatible re-storage of the screen image in
the database.

   The HoR structure as a whole compares favourably with the
$[4^4]$ 4-shape (the quad-tesseral) structure. The sole
disadvantage of the rhombus over the square is that its
circularity is double that for the square , because circularity
weights extreme points more heavily. Since the most extreme
part of the square is 1.41, as compared with 1.61 for a rhombus
of equal area, this results in an unfavourable comparison being
made. However, the average circularity (the average error in
locating all points in the rhombus at the centroid) is not
significantly greater (+5%). The greater symmetry possessed by
the square tile also shows itself in the greater rotation
number. However, the rotation number of the square lattice is
equal to that of the square tile. This is not true of the
rhombus. Here the underlying HoR lattice has greater symmetry,
reflected in the unit adjacency and larger rotation number of
the hexagon. These features show great promise in the design
of algorithms to fulfil common goals. They unite geometric
advantage, for example only one type of adjacency need be
considered in determining connected components, with arithmetic
advantage. Tesseral integer calculations use addition and
multiplication to control both scale and angle. The high
rotation number gives finer independent control of angle and
scaling effects for a given size of tile (which cannot be
duplicated using Cartesian coordinates).

   HoR structure tesseral arithmetic is Abelian and has a
multiplicative inverse. It has a geometric interpretation
which is identical to that for the quad-tesseral structure and
indeed shares the same addition table. Furthermore, because
the operations conserve image similarity, provision of fast
interpolation and reflection algorithms within the arithmetic
(and indeed the exploitation of tesseral hardware and parallel
processing) are uniquely provided for. It is highly likely
that many image processing algorithms when expressed in
tesseral arithmetic terms, will be simpler and possibly also
faster than their Cartesian counterparts. Although in this
paper we do not present a rule-based arithmetic operating on
whole registers rather than tesseral digits for the HoR
structure, such an arithmetic is clearly possible and should
allow rapid HoR to Cartesian inter-conversion as well as rapid
tesseral arithmetic operations on the spatial addresses.

An important property of the addressing system is that the points 3, 30, 300, ... are in a straight line. Consequently "point shift" either as multiplication or division is merely a change in scale. In contrast, a "point shift" in the case of the 7-shape over the $[3^6]$ tiling, where the line 3, 30, 300, ... is a digital spiral, is both a scaling and rotation. Two consequences flow from this property. The first is practical and means that existing integer arithmetic hardware can be used for HoR-tesseral, indeed such algorithms already exist. The second consequence is that we can envisage an extension of HoR-tesseral arithmetic operations to encompass a radix or tesseral point. This second consequence clearly has much interesting research potential both in its mathematical as well as computational interpretation.

Although the proposal to adopt the HoR-tesseral structure for image processing is rather radical, it is the authors' belief that it has several advantages and should be seriously considered. Work to determine if there are better hierarchies, say based on a 9-shape are being researched at the Open University by Holroyd in collaboration with Dr. J. Johnson. This work will consider both algorithm speed and conceptual simplicity. Meanwhile, much research is being directed at the database implications of the use of tesseral addresses and hierarchies - especially along the lines of a "concept" rather than "image" store. Finally, and perhaps most importantly the mathematical and computational aspects of such a radical change in approach is being researched by all three authors.

7. REFERENCES

[1]   Bell, S.B.M., Diaz, B.M. and Holroyd, F.C., (1988a) "Capturing image syntax using tesseral addressing and arithmetic", In J.P. Muller, (ed.), *Digital Image Processing in Remote Sensing*, Taylor and Francis Limited, Basingstoke.

[2]   Bell, S.B.M., Diaz, B.M. and Holroyd, F.C., (1988b), "Tesseral addressing and tabular arithmetic of quadtrees", *Communications of the A.C.M.)*.

[3]   Diaz, B.M. and Bell, S.B.M., (Eds.), (1986), *Spatial Data Processing using Tesseral Methods*, Natural Environment Research Council, Swindon, UK, 425pp.

[4]   Bell, S.B.M. and Holroyd, F.C., (1986) "Constructive tesseral algebra", *in* [1]; 233-256.

[5]   Bell, S.B.M., Diaz, B.M. and Holroyd, F.C., (1986) "Constructive tesseral inverses", in [1]; 285-312.

[6]   Bell, S.B.M., Diaz, B.M., Holroyd, F.C. and Jackson, M.J.,
       (1983), "Spatially referenced methods of storing and
       handling raster and vector data", *Image and Vision Computing*
       1(4); 211-220.

[7]   Coxeter, H.S.M., (1980), *Introduction to Geometry,* 2nd
       ed. John Wiley and Sons Limited, New York, 469 pp.

[8]   Diaz, B.M., (1984), "Tabular tesseral method and
       calculators", in [1]; 99-110.

[9]   Gargantini, I., (1982), "An effective way to represent
       quadtrees", *Communications of the A.C.M.,* 25(12); 905-910.

[10] Gibson, L. and Lucas, D., (1982), "Vectorisation of
       raster images using hierachical methods", *Computer Graphics
       and Image Processing,* 20; 82-89.

[11] Goodchild, M.F. and Granfield, A.W., (1983), "Optimising
       raster storage: An examination of four alternatives",
       *Proceedings of the 6th International Symposium on Computer
       Assisted Cartography (Auto-Carto VI);* 108-118.

[12] Grunbaum, B. and Shephard, G.C., (1977), "The eighty-one
       types of isohedral tilings in the plane", *Procs. Cambridge
       Philosophical Society,* 82; 177-196.

[13] Grunbaum, B. and Shephard, G.C., (1981), "The Geometry of
       planar graphs", *Procs. 8th British Combinatorial Conference,*
       124-150.

[14] Holroyd, F.C., (1985), "Addressing systems for digital
       pictures", In Sarah Bell, (ed.), *Tesseral Newsletter No. 2*
       September 1985, NUTIS, University of Reading, Whiteknights,
       P.O. Box 227, Reading, RG6 2AH, UK.

[15] Holroyd, F.C., (1983), "The geometry of tiling
       hierarchies", *Arcs Combinatoria,* 16-B; 211-244.

[16] Lucas, D., Steinfeld, P. and Anderson, J.W., (1978),
       System X User's Manual, Martin Marietta Corp., USA, 65 pp.

[17] Morton, G.M., (1966), "A Computer oriented geodetic
       database and a new technique for file sequencing", *Internal
       report for IBM Canada Limited,* Ottawa, Canada.

[18] Oliver, M.A. and Wiseman, N.E., (1983), "Operations on
       quadtree encoded images", *The Computer Journal,* 26, (1),
       83-92.

[19]   Rosenfeld, A. and Samet, H., (1979), "Tree structures
     for region representation", *Procs. 4th International
     Symposium on Computer Assisted Cartography (Auto-Carto IV);*
     108-118.

[20]   Samet, H., (1984), "The quadtree and related hierarchical
     data structures", *A.C.M. Computing Surveys,* 16(2); 187-260.

[21]   Woodwark, J.R., (1982), "The explicit quadtree as a
     structure for computer graphics", *The Computer Journal,* 25,
     (2); 235-238.

APPENDIX

Average distance of tiles from centroid

   This may be calculated by dividing the tiles into triangles,
and calculating the product of (i) the average distance of the
points in a triangle from the centroid of the tile and (ii) the
area of the triangle. All such products are summed, and then
divided by the total area of the tile. Let one of the
triangles be ABC, and we will assume an arbitrary point P is
the centroid of the tile. We calculate the product of the
average distance from P and the triangle area for triangles
ABP, BPC, PAC. Let these be [ABP], [BPC], and [PAC]. Then the
required answer is:

$\{$[ABP] $\pm$ [BPC] $\pm$ [PAC]$\}$/area of triangle ABC, and with the signs
depending on the position of P.

   To find a typical example, [BPC], of the terms appearing
above we divide the arbitrary triangle into two right-angled
triangles by dropping a perpendicular from P to BC at O. We
find [OPC] and [BPO], then

$$[PBC] = [OPC] + [BPO]$$

Let Q be an arbitrary point in OPC. Let OP be h. Let y be
the distance from Q to PO, and h − x the distance from Q to OC.
Let angle OPC be p. Then [OPC] is the integral of:

$$(x.x + y.y)^{\frac{1}{2}} dy.dx$$

with limits of x.tan(p) and O on y, and h and O on x. We
substitute:

          $y = x.\tan(a)$ and the integral becomes:
          $x.x.\sec(a).\sec(a).\sec(a)$ da.dx

with limits of p and O on a, and h and O on x. We integrate
and obtain the result:

$$[\text{OPC}] = (h.h.h/3) \cdot \{\tan(p)\sec(p)/2$$
$$+ \log \{\sec(p) + \tan(p)\}^{\frac{1}{2}}\}$$

where log is a natural logarithm. This completes our top-down derivation of average distances of points in tiles from the tile centroid.

Should we wish to find the average distance of points in triangle OPC from P, we can divide by the area

$$h.h. \tan(p)/2 \qquad \text{obtaining:}$$

$$(h/3) \cdot \{\sec(p) + 2.\cot(p) \cdot \log\{\sec(p) + \tan(p)\}^{\frac{1}{2}}\}$$

THE APPLICATION OF SYNTACTIC PATTERN RECOGNITION
TO REMOTE SENSING

J.F. Boyce and J. Feng
(Wheatstone Laboratory, King's College,
London, WC2R 2LS)

ABSTRACT

The formalism   of object identification utilising
attributed relational directed graphs or relational data
structures is reviewed. An example of the possible application
of the latter to feature identification in aerial photographs
is given.

1. INTRODUCTION

Scene analysis requires the amalgamation and organisation of
low level information derived directly from pixel intensities
into collections of spatially extended features. Image
understanding, which implies object classification and
recognition, may then be based upon the features. However,
only a limited class of objects may be recognised solely on
the basis of their feature attributes, most require contextual
information for unambiguous identification. Syntactic pattern
recognition is concerned with how this additional information
may be obtained and utilised. It has evolved from the
syntactic grammars which may be employed in language analysis,
(Fu, 1974), and has been extensively applied to problems which
have a natural expression in sequential form (Gonzalez and
Thomason, 1978, Fu, 1982(a)). It has developed further to
encompass nonsequential problems of two and three dimensions,
characteristic of those encountered in scene analysis. It
is these latter developments which we shall review.

The problem of defining features and their contextual
relations is connected both to that of their representation
and to their extraction from an image. Segmentation, for
example, may depend upon feedback from the results of higher
level processing. We shall, nevertheless, assume that, as a
first approximation, segmentation and recognition may be
considered as separate problems, in order to concentrate upon
representation and recognition. Actual systems of image

understanding, of which MSYS (Barrow and Tenenbaum, 1976),
VISIONS, (Hanson and Riseman, 1978), SIGMA, (Huang, Davis,
and Matsuyama, 1985), and that due to Nagao and Matsuyama
(Nagao and Matsuyama, 1980), are examples, combine the
two processes of extraction and classification using feedback.

A natural mathematical structure for representing contextual
information is a graph, $G = (V, A)$, the vertices, V, being
the features and the arcs, A, being relations between features.
Object recognition then becomes a problem of isomorphism
between a graph derived from the observed scene and a member
of a library of graphs representing known objects. Account
must be taken, however, of the following complicating factors.
Features are not all of the same class, the vertices must
therefore be labelled; they possess attributes, possibly with
continuous ranges of values. Relations between features may
have attributes. Features may be constrained by third or
higher order constraints rather than simply in pairs. Due to
noise and to avoid an explosion of library storage any
recognition algorithm should be sufficiently robust to accept
imperfect matching between candidate and library data
structures. Since the graph isomorphism problem is known to
be NP-complete (Salomaa, 1985), i.e. non-polynomially bounded
computation time with increasing order of the graph, it may
be expected that computational complexity will be a major
problem.

As a result, generalisations of the above model have been
developed. One approach (Haralick, Davis, Rosenfeld, and
Milgram, 1978) has extended a body of work arising in
artificial intelligence on the solution of constraint
satisfaction problems. It represents objects by relational
structures, (Barrow and Popplestone, 1971), and expresses
object recognition as the comparison of such structures via
its identification as a consistent labelling problem. An
alternative approach (Fu, 1974, Eshera and Fu, 1984) is to
maintain the underlying graphical structure but to permit
the vertices and arcs to have attributes associated with them,
the resulting structure being known as an attributed
relational graph. In these terms object recognition requires
both graph and attribute matching.

We shall describe these two approaches in turn.

2.   RELATIONAL STRUCTURES

2.1 *Definition of a Relational Structure*

We shall consider a set of features or pattern primitives,

$P = \{p_i \mid i = 1,\dots,I\}$. They may possess attributes which are exclusive, so that possession may be represented by set membership

$$p \in P(A) \iff p \text{ possesses property } A \qquad (2.1)$$

where $P(A) \subseteq P$.
Constraints between pairs of features may be similarly expressed as a property of their ordered pair

$$(p_1,p_2) \in R(1,2) \subseteq P(1) \times P(2) \iff p_1 \text{ and } p_2 \text{ are constrained}$$
by $R(1,2)$ $\qquad (2.2)$

Such a constraint constitutes a binary relation, (Montanari, 1974). In general it is not symmetric in its arguments. From equation (2.1) it follows that possession of an attribute may be regarded as a unary relation.

If we consider a collection of unary relations

$$\hat{P} = \{P(i) \mid i = 1,\dots,M\} \qquad (2.3)$$

and a collection of binary relations

$$\hat{B} = \{R(i,j) \mid i,j = 1,\dots,M\} \qquad (2.4)$$

then there is a natural identification between $(P,\hat{P},\hat{B})$ and a graph with vertices $p_i \in P$, which are self-connected by labelled arcs from $\hat{P}$ and interconnected by labelled directed arcs from $\hat{B}$. This graph provides the basis for the attributed relational graphs which we shall consider in §4.

Higher order constraints may be similarly expressed as subsets of the Cartesian product of the appropriate domain sets, an N-ary relation being a collection of N-tuples,

$$(p_1,\dots,p_N) \in R^N \subseteq P(1) \times P(2) \times \dots \times P(N) \iff p_1,p_2,\dots,p_N \text{ are}$$
constrained by R^N $\qquad (2.5)$

Such a relation is not, in general, directly representable as a graph when $N \geq 3$. However it may be expressed as a set of binary relations (Fu, 1974), when each member of the set may be so represented. If the domain sets $P(i)$, $i = 1,\dots,N$, are identical and the relation is invariant under permutations of the arguments of the N-tuples which are its elements then it is, by definition, symmetric. The usual set operations of union, intersection, complement, difference, product, projection, and quotient may be extended to relations (Delobel and Adiba, 1985). Of these we shall need only product and projection.

Given two relations,

$$R^N = \{(p_1,\ldots,p_N)\} \subseteq P(1) \times \ldots \times P(N) \qquad (2.6)$$

and

$$R^M = \{(q_1,\ldots,q_m)\} \subseteq Q(1) \times \ldots \times Q(M) \qquad (2.7)$$

then their Cartesian product is defined as

$$R^N \otimes R^M = \{(p_1,\ldots,p_n,q_1,\ldots,q_m)\} \subseteq P(1) \times \ldots \times P(N) \times Q(1)$$
$$\times \ldots \times Q(M) \qquad (2.8)$$

Given the N-ary relation, R^N, defined by equation (2.5) on P, the product of the domain sets $P(i)$, $i = 1,\ldots,N$,

$$\tilde{P} = P(1) \times P(2) \times \ldots \times P(N) \qquad (2.9)$$

then, for any sub-product

$$\tilde{P}' = P'(1) \times P'(2) \times \ldots \times P'(M) \quad M \leq N \qquad (2.10)$$

where

$$P'(i) \in \{P(1),\ldots,P(N)\} \qquad (2.11)$$

R^N defines a natural M-ary relation, the projection of R^N on $\tilde{P}'$.

In particular, $N(N-1)/2$ binary relations may be so projected from an N-ary relation, one binary relation, R^2_{ij}, between each pair of domain sets $S(i)$ and $S(j)$. Such a set forms a network of binary relations (Montanari, 1974). A network may be represented by a directed graph $\left[\{P(i)\}, \{R^2_{ij}\}\right]$, in which the vertices are the domain sets and the arcs are the binary relations. Given such a network then the collection of N-tuples $\rho = \{(p_1,\ldots,p_N)\}$ with $p_i \in P(i)$, $i = 1,\ldots,N$ and such that $(p_i,p_j) \in R^2_{ij}$, $i,j = 1,\ldots,N$ is an N-ary relation. However it is not, in general, isomorphic to the N-ary relation from which the network was generated by projection, hence it is not possible to compare general N-ary relations by comparing the networks of binary relations which may be derived from them.

A collection of relations, $\hat{R} = \{R_i\}$, $i = 1,\dots,K$, whose members may be of arbitrary orders, is a relational structure. Objects will be represented by such structures. The problem of identifying an object thus becomes that of comparing a candidate relational structure derived from scene analysis with a library of such structures. A criticism which may be made of such a representation is that attributes which have continuous values must be discretised in order to be represented by relations. In any application, however, when comparing variable values, it is normal to set limits of acceptability to the allowed values, so that such discretisation is not unnatural.

The formalisation of object identification is the subject of the next section. In it we shall restrict consideration to relations defined on products of identical sets, though the relations need not be symmetrical.

2.2 *Comparison of Relational Structures*

We wish to compare a candidate relational structure with a library of such structures in order to recognise or classify the candidate. A preliminary problem is therefore to consider the comparison of a single candidate relation with a library of relations of the same order. (For binary relations this is a graph matching problem). We thus need to define a mapping between relations which will permit their comparison. Due to viewing constraints and noise the amount of information which is available about the candidate may be less than that for an ideal exemplar, hence the matching will not be one to one. Indeed, since the library relations will most likely be representatives of classes of objects, it will be necessary to define a metric between relations in order to select optimal rather than definite recognition or classification.

Consider a candidate object, having feature set P and feature interrelations which are expressed by an N-ary relation, T, defined on P^N. The candidate is to be compared with a library object having feature set Q together with an N-ary relation S defined on Q^N. Suppose that there is a mapping h: P → Q, which compares the features of a candidate object with those of a library object. In order to compare the interrelations between the two sets of features we need to similarly map the relations. Following Shapiro and Haralick (Shapiro and Haralick, 1981), the composition, T o h, of T with h is defined by

$$T \circ h = \{(q_1, \ldots, q_N) \in Q^N \mid \exists (p_1, \ldots, p_N) \in T \text{ with } h(p_i) = q_i,$$

$$i = 1, \ldots, N\} \qquad\qquad (2.12)$$

We may now compare the composed relation of the candidate to that of the library object. $T \to T \circ h$ is defined to be a relational homomorphism from T to S if

$$T \circ h \subseteq S \qquad\qquad (2.13)$$

while, if $h: P \to Q$ is a bijective mapping and $T \to T \circ h$ is a relational homomorphism, then $T \to T \circ h$ is a relational isomorphism.

Homomorphism implies that the candidate object is relationally a component of the library object. Isomorphism implies relational equivalence between the two. The above definitions extend immediately to relational structures, with one structure relationally homomorphic (isomorphic) to another if each component relation is homomorphic (isomorphic) to a component relation of the other in a one to one manner.

The problem of determining whether a candidate object is homomorphic to a library object may be formalised by identifying it as a particular case of a more general problem known as the consistent labelling problem, (Haralick and Shapiro, 1979, 1980), which may be conveniently stated in terms of a compatibility model.

3.   THE CONSISTENT LABELLING PROBLEM

3.1   The Compatibility Model

Let $U = \{u_i\}$, $i = 1, \ldots, I$ be a set of units, $L = \{l_i\}$, $i = 1, \ldots, K$, a set of labels, and $h: U \to L$ a mapping of U into L . Since h assigns at most one label to each object, it defines an unambiguous labelling of the objects. Now it may happen that the objects are constrained so that only certain combinations are permitted, in particular, they may be required to satisfy a constraint defined by the N-ary relation $T = \{(u_1, \ldots, u_N)\} \subseteq U^N$. Moreover, only certain labellings, $(l_1, \ldots, l_N)$, of the N-tuples $(u_1, \ldots, u_N)$, of objects may be permitted, this latter condition being expressed by the 2 N-ary relation

$$R = \{(u_1, l_1, u_2, l_2, \ldots, u_N l_N)\} \subseteq (U \times L)^N \qquad (3.1)$$

The 4 tuple (U,L,T,R) defines a compatibility model. The
labelling h may be composed with the object constraint relation
T to yield a relation $T \circ h \subseteq L^N$. The labelling defined by
h is compatible with the model if the product of two relations
T and T o h, (the product being as defined by equation (2.8)),
is a sub-relation of R.

$$T \otimes (T \circ h) \subseteq R \qquad (3.2)$$

Given a compatibility model, the consistent labelling
problem, (Haralick and Shapiro, 1979, 1980), is to find the
set of unique labellings which are compatible with the model.
We may identify the relational homomorphism problem as a
particular case of a consistent labelling problem. The
candidate feature set, P, may be identified with the set of
units, U, and the library feature set Q, with the set of
labels, L. The combined unit-label constraint relation, R,
factorises into the product of an N-ary unit constraint
relation, T, with an N-ary label constraint relation, S viz.,

$$R = T \otimes S \qquad (3.3)$$

T being the N-ary constraint which describes the interrelations
between the features of the candidate, while S is that of
the library object. Hence, the condition for compatibility
with the model (P,Q,T,T $\otimes$ S) is simply

$$T \circ h \subseteq S \qquad (3.4)$$

the condition for a relational homomorphism.

The benefit of making the above identification is that it
enables the relational homomorphism problem to be placed in
the context of the class of constraint satisfaction problems
and thereby permits the application to it of the general
techniques which have been developed for the solution of such
problems.

3.2 Methods of Solution

The consistent labelling problem, as defined in §3.1, is
that of finding the set of labellings which are consistent
with a compatibility model. Related problems are that of
the existence of solutions, finding the number of solutions,
or finding any one solution. We shall focus, however, on
what is, for us, the main problem, that of finding all
solutions. Most workers in the field have restricted
consideration to binary relational constraints, when the problem

is equivalent to subgraph isomorphism (Ullman, 1976). As
such it is, in turn, equivalent to a graph colouring problem,
Haralick et al., 1978) and hence is NP-complete. Even when
considering relations of order higher than two, the algorithms
have largely been generalisations of those developed for the
binary case.

The same set of consistent labellings may arise from
different relations of the same order. Such relations may
therefore be regarded as being equivalent. Given a relation,
an effective strategy might be to determine the smallest
relation to which it is equivalent and then solve the labelling
problem for this latter relation. Unfortunately, as shown by
Montanari for binary relations, (Montanari, 1974), although
such a relation does exist, the task of finding it is itself
NP-complete.

Tree search techniques are guaranteed to find all of the
solutions of a consistent labelling problem. They suffer,
however, from 'thrashing', when inadmissible object-label
assignments are repeatedly tested and rejected. Such behaviour
may be alleviated either by forward or backward checking,
(Gaschnig, 1977, Freuder, 1978) or by filtering techniques,
which, by analogy with the subgraph isomorphism problem, may be
considered as the application of 'vertex', 'arc', and 'path'
consistency conditions to the labelling (Mackworth, 1977,
McGregor, 1979). Vertex consistency consists of restricting
the set of labels at a given vertex to that set which is
obtained by projecting (as in equations 2.9 - 2.11) onto the
domain set of the vertex. For binary relations, arc
consistency implies that for any labelling of u_i by l_i with

$(u_i, u_j) \in T$, then there exists an l_j such that $(l_i, l_j) \in S$.
Rosenfeld et al (Rosenfeld, Hummel and Zucker, 1976) have shown
that, for binary relations, arc consistency can be achieved via
discrete relaxation, an inherently parallel procedure, which may
be implemented using dedicated hardware, (Dixit and Moldovan, 1984).
Path consistency over paths of length m within a graph is
achieved by imposing simultaneous arc consistency on each
segment of any path of length m (Mohr and Henderson, 1986).

Both arc and path consistency have been generalised to
N-ary relations (Haralick et al, 1978) together with algorithms
for their implementation (Haralick and Elliott, 1980). By
classifying consistent labelling problems into classes of
complexity, heuristics have been derived (Nudel, 1983) for
finding hybrid search strategies which are an optimised
combination of tree-search (depth) and filtering (breadth)
algorithms.

An algorithm based upon the matching of maximal cliques, and having a relaxation implementation has been proposed by Cheng and Huang (Cheng and Huang, 1981, 1984). Algorithms for solving a consistent labelling problem have been derived (Nishihara and Ikeda 1984, 1985) by expressing the problem as an equivalent constraint network. These latter algorithms are parallel and permit the simultaneous consideration of a set of relational constraints rather than dealing with each constraint as a separate problem. As such they are naturally suited to the comparison of relational data structures.

Even when the relational homomorphism problem has been solved it will, in general, be necessary to choose between a number of matches between a candidate object and the data structures of a library. The problem of defining a metric on relational data structures which would permit such a selection has been considered by Shapiro and Haralick (Shapiro and Haralick, 1985), who define a suitable distance which satisfies the axioms required of a metric.

4. ATTRIBUTED RELATIONAL GRAPHS

A collection of features having unary and binary relations may be represented by an attributed relational graph of vertices and directed arcs. Higher order relations may first be expressed as sets of binary relations (Fu, 1974) before being so represented.

Following Tsai and Fu (Tsai and Fu, 1979), the features, P, form a set of vertices, with which is associated an alphabet, A, of attributes, together with a set of functions, $\mu: P \to A$, which generate the attributes. Similarly the binary relations, B, form directed arcs, with an associated alphabet, E, of arc attributes, together with a set of functions, $\varepsilon: B \to E$, which generate them. The elements of the alphabets take the form of a tuple formed by uniting an identifying syntactic symbol with a semantic vector of attributes. The 6-tuple $(P,B,A,E,\mu,\varepsilon)$ is an attributed relational graph. The advantage of this representation is that the attributes may now possess continuous values, rather than being expressions of an exclusive relation.

Since in this formulation of scene analysis objects are represented by attributed relational graphs, the problem of object recognition becomes that of comparing a candidate attributed graph with the members of a library of such graphs. A distance measure between such graphs may be defined (Eshera and Fu, 1984), in terms of the minimum of the cost function which depends on the transformations which must be performed on one graph in order to produce the other.

The procedure for calculating the distance between two
attributed graphs, G, G', requires that each graph be
decomposed into a collection of basic graphs, $\hat{G}$, these being
tree graphs having one root vertex and a single generation of
arcs and leaf vertices. Starting from a pair, $(\emptyset,\emptyset)$, of empty
graphs, the graph pair (G,G') is then reconstituted by
embedding matched pairs of basic graphs from the two
collections $\hat{G}$ and $\hat{G}'$. This procedure may be represented as a
directed lattice in a space of states; the initial state being
$(\emptyset,\emptyset)$, the final state being (G,G'), and the possible
reconstruction sequences being directed paths of the lattice
between them. As a consequence of the above construction,
any set of transformations which transform G into G' may be
represented by a path of directed arcs on the lattice. Hence
the cost of transforming G into G' may be defined along any
set of arcs by adopting suitable definitions for the costs of
individual transformations. The inter-graph distance is
defined to be the minimum such cost. Although this distance
does not satisfy all of the axioms of a metric, (the triangle
inequality, in particular), due to the directed lattice
structure, it may be obtained by dynamic programming, with
consequent efficiency of algorithmic implementation. A
direct comparison of this approach with that based upon
consistent labelling has yet to be made.

In order to permit the variability of attributes, Wong
and You (Wong and You, 1985) regard the vertex and arc
attributes as random variables, with associated probability
distributions, the resulting structures being known as random
attributed graphs. They define the entropy of a random
graph and utilise this quantity to define a distance measure
between random graphs by defining the change of entropy caused by
transforming between them. This permits the techniques of
statistical pattern recognition to be applied, for example,
to the definition of a characteristic representative of a
cluster of graphs, i.e. an exemplar of a class of objects.
A further benefit of the inclusion of randomness is that it
permits the continuous variability of object attributes to be
naturally encompassed by the formalism.

5. AN ILLUSTRATIVE EXAMPLE

Application of syntactic pattern recognition techniques to
remote sensing has been made to structure location, (Nevatia
and Price, 1982), to image classification, (Faugeras and Price,
1981), to scene analysis (Kitchen and Rosenfeld, 1984), and to
drainage network classification, (Haralick, Campbell and Wang,

1985; Haralick, Wang, Shapiro and Campbell, 1985). In order
to demonstrate the concepts introduced above, we consider the
interpretation of an aerial image of a prototype airport.
Real images of this kind have been analysed by McKeown, Harvey
and McDermott, using a rule-based system SPAM, (McKeown, Harvey
and McDermott, 1985). The notional image is shown in Fig. 1,
and comprises a configuration of features; terminal building,
hanger, aircraft parking apron, taxiways, grassy areas, and
runways. We wish to show how syntactic pattern recognition
techniques may be employed to distinguish between them. Let
us begin by supposing that a complete classification has been
achieved, consider how each separate feature may be
distinguished from the configuration, and then examine how
the same information may be employed during the classification
of the entire image.

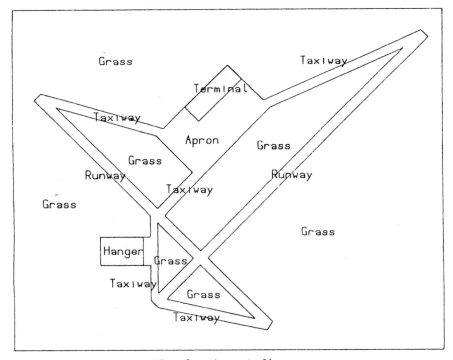

Fig. 1 Airport diagram

The features will be represented by relational data
structures formed from unary, binary, and ternary relations.
A suitable choice for the case under consideration is:

Unary relations; Pixel intensity, Texture value, Region width,
Region length;
Binary relation; Adjacency;
Ternary relation; Connectedness.

The unary relations (attributes) of the features are
shown in Table 1.

Feature	Intensity	Texture	Width	Length
Terminal	$>\Delta_I$	–	–	$<\Delta_L$
Hanger	$>\Delta_I$	–	$>\Delta_W$	$<\Delta_L$
Runway	$<\Delta_I$	$<\Delta_T$	$<\Delta_W$	$>\Delta_L$
Taxiway	$<\Delta_I$	$<\Delta_T$	$<\Delta_W$	$<\Delta_L$
Apron	$<\Delta_I$	$<\Delta_T$	$>\Delta_W$	$<\Delta_L$
Grass	$<\Delta_I$	$>\Delta_T$	–	–

Table 1.   Unary Relations.

Δ_I, Δ_T, Δ_W, and Δ_L are thresholds for intensity, texture,
width, and length respectively.

Based upon the attributes of pixel intensity and texture,
the image may be segmented using region growing algorithms into
disjoint regions labelled as, – grass, – buildings, – tarmac.
The three corresponding images are shown in Figs. 2, 3, and 4,
respectively. The tarmac region may now be skeletonised,
and then segmented into regions, each associated with a
linear segment of the skeleton. These regions, together
with the previously segmented building and grassy regions
constitute the basic set of features for the subsequent
analysis. The average width and length of each region can
now be defined.

The symmetric binary relation of adjacency between
features is shown in Table 2, where 1 implies that the
relation is satisfied, and 0 the converse. Notice, for
example, that no apron feature is adjacent to any runway
feature, but that a grass feature can be adjacent to any
other type.

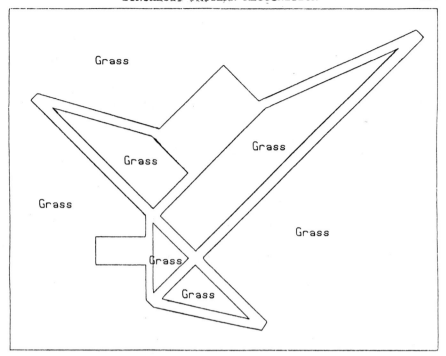

Fig. 2.   Grass Regions

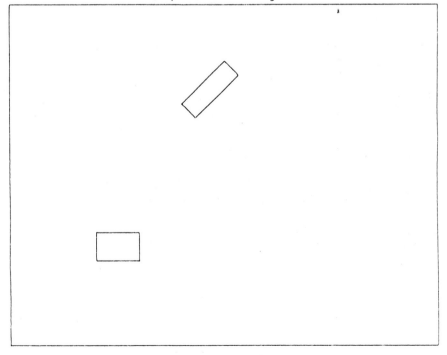

Fig. 3 Buildings

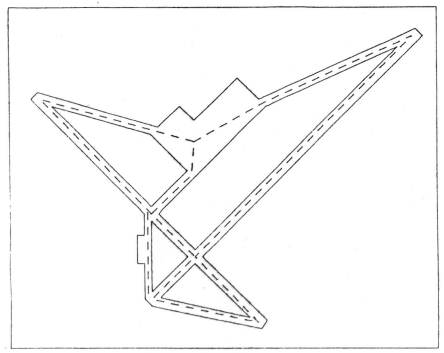

Fig. 4. **Tarmac, with** skeleton

	Terminal	Hanger	Runway	Taxiway	Apron	Grass
Terminal	1	O	O	O	1	1
Hanger	O	1	O	1	O	1
Runway	O	O	1	1	O	1
Taxiway	O	1	1	1	1	1
Apron	1	O	O	1	1	1
Grass	1	1	1	1	1	1

Table 2. Binary Relation of Adjacency

The ternary relation of connectedness, based upon the skeletal image is shown in Table 3. This relation is

symmetrical in the connected features, but is not, in general,
symmetric under permutations which involve the connecting
feature. In particular, a taxiway may connect an apron to a
runway, but an apron may not connect a taxiway to a runway.

Feature	Connected Features	
Apron	Apron	Taxiway
	Taxiway	Taxiway
Taxiway	Apron	Taxiway
	Apron	Runway
	Taxiway	Runway
Runway	Taxiway	Taxiway
	Taxiway	Runway
	Runway	Runway

Table 3. Ternary Relation, Connectivity

   The analysed image will consist of a set of features,
(image regions), each with an associated relational data
structure, i.e. a set of unary, binary and ternary relations.
Each such relation will be a sub-relation of those defined by
Tables 1, 2, and 3. If one region remained unclassified, then
its class could be determined by comparing its data structure
with the library of structures represented by the three tables.
The analysis of the entire image could thus proceed iteratively,
(Price, 1985), with overall consistency being progressively
introduced. Given the initial image, the set of allowed labels
for each feature is restricted by the unary relations, (Kitchen
and Rosenfeld, 1984). By discrete relaxation labelling,
(Rosenfeld, Hummel, and Zucker, 1976), these allowed labels
are iteratively restricted via the binary and ternary relations
of Tables 2 and 3 until stable sets of relations are obtained.
The image will then be consistently labelled relative to the
relational data structures. Any non-uniqueness in this
resulting labelling would indicate the necessity to augment
the relational data structures by additional relations.

   General questions, such as the choice of the particular
relations utilised for a specific problem, together with their
consistency, redundancy, or optimality, are at present
answered heuristically.

CONCLUSIONS

The techniques of syntactic pattern recognition now
available have application to problems of remote sensing,
particularly when spatially organised non-local structures are

of significance. Notable difficulties which remain, however,
are firstly the problem of computational complexity, as
expressed, for example, by the graph homomorphism problem,
and secondly that of the optimum choice of relational structure
for a specific application.

REFERENCES

Barrow, H.G. and Popplestone, R.J., (1971). "Relational
Descriptions in Picture Processing", *Machine Intelligence*,
6, Edinburgh University Press.

Barrow, H.G. and Tenenbaum, J.M., (1976). "MSYS: A system
for reasoning about scenes", Tech. Note 121, SRI International,
Menlo Park, CA.

Cheng, J.K. and Huang, T.S., (1981). "Image Registration
by Matching Relational Structures", Proc. IEEE Conf. on
Pattern Recognition and Image Processing, Dallas, 542-547.

Cheng, J.K. and Huang, T.S., (1984). "Image Registration
by Matching Relational Structures", *Pattern Recognition*,
17, 149-158.

Delobel, C. and Adiba, M. (1985) "Relational Database
Systems", North Holland, Amsterdam.

Dixit, V. and Moldovan, D.I., (1984). "Discrete Relaxation
on SNAP", First Conf. Artificial Intell. Applications,
IEEE Computer Society, 637-644.

Eshera, M.A. and Fu, K.S., (1984). "A Graph Distance
Measure for Image Analysis", *IEEE Trans. Syst. Man. Cybern.*,
14, 398-408.

Faugeras, O. and Price, K., (1981). "Semantic Description
of aerial Images using Stochastic Labelling", *IEEE Trans.
Pattern Anal. Machine Intell.*, PAMI-3, 638-642.

Freuder, E.C., (1985). "A Sufficient Condition for
Backtrack-Bounded Search", *J. Assoc. Comput. Mach.*, 32,
755-761.

Fu, K.S., (1974). "Syntactic Methods in Pattern
Recognition", Academic Press, New York.

Fu, K.S., (1982, a). "Syntactic Pattern Recognition and
Applications", Prentice-Hall, New Jersey.

Fu, K.S., (1982, b). "Syntactic Pattern Recognition",
Applications of Pattern Recognition, Ed. K.S. Fu, CRC Press,
Florida, 37-64.

Fu, K.S., (1982, c). "Application of Pattern Recognition
to Remote Sensing", Applications of Pattern Recognition,
Ed. K.S. Fu, CRC Press, Florida, 65-106.

Gaschnig, J., (1977). "A General Backtrack Algorithm
that Eliminates Most Redundant Tests", Proc. 5th, Intl.
Joint Conf. on Artif. Intell., Cambridge, MA, 457.

Gonzalez, R.C. and Thomason, M.G., (1978). "Syntactic
Pattern Recognition: An Introduction", Addison-Wesley,
Reading, Mass.

Kitchen, L. and Rosenfeld, A., (1984). "Scene Analysis
using Region-Based Constraint Filtering", Pattern Recognition,
17, 189-203.

Haralick, R.M., Campbell, J.B. and Wang, S., (1985).
"Automatic Inference of Elevation and Drainage Models from
a Satellite Image", Proc. IEEE, 73, 1040-1053.

Haralick, R.M., Davis, L.S., Rosenfeld, A., Milgram, D.L.,
(1978). "Reduction Operations for Constraint Satisfaction",
Information Sciences, 14, 199-219.

Haralick, R.M. and Shapiro, L.G., (1979). "The Consistent
Labelling Problem: Part I", IEEE Trans. Pattern Anal.
Machine Intell., PAMI-1, 173-184.

Haralick, R.M. and Shapiro, L.G. (1980). "The Consistent
Labelling Problem: Part II", IEEE Trans. Pattern Anal.
Machine Intell., PAMI-2, 193-203.

Haralick, R.M., Wang, S., Shapiro, L.G., and Campbell,
J.B., (1985). "Extraction of Drainage Networks by using
the Consistent Labelling Technique", Remote Sensing of
Environment, 18, 163-175.

Mackworth, A.M., (1977). "Consistency in Networks of
Relations", Artificial Intelligence, 8, 99-188.

McGregor, J.J., (1979). "Relational Consistency Algorithms
and Their Application in Finding Subgraph and Graph
Isomorphisms", Information Sciences, 19, 229-250.

McKeown, D.M., Harvey, Jr, W.A., and McDermott, J., (1985). "Rule-Based Interpretation of aerial Imagery", *IEEE Trans. Pattern Anal. Machine Intell.*, **7**, 507-585.

Mohr, R. and Henderson, T.C., (1986). "Arc and Path Consistency Revisited", *Artificial Intelligence*, **28**, 225-233.

Montanari, U., (1974). "Networks of Constraints: Fundamental Properties and Applications to Picture Processing, *Information Sciences*, **7**, 95-132.

Nevatia, R. and Price, K., (1982). "Locating Structures in Aerial Images", IEEE Pattern Anal. Machine Intell., PAMI-**4**, 476-484.

Nishihara, S. and Ikeda, K., (1984). "A Constraint Synthesising Algorithm for the Consistent Labelling Problem", Proc. IEEE Seventh Intl. Conf. on Pattern Recog., Montreal, 310-312.

Nishihara, S. and Ikeda, K., (1985), "Consistent Labelling Methods using Constraint Networks", ISE-TR-85-49, Institute of Information Sciences and Electronics, University of Tsukuba, Japan.

Nudel, B., (1983). "Consistent-Labelling Problems and their Algorithms: Expected-Complexities and Theory-Based Heuristics", *Artificial Intelligence*, **21**, 135-178.

Price, K.E., (1985). "Relaxation Matching Techniques - A Comparison", IEEE Trans. Pattern Anal. Machine Intell., PAMI-**7**, 617-623.

Rosenfeld, A., Hummel, R.A. and Zucker, S.W., (1976). "Scene Labelling by Relaxation Operations", *IEEE Trans. Syst. Man. Cybern.*, SMC-**6**, 420-433.

Salomaa, A., (1985). "Computation and Automata", Cambridge University Press, Encyclopedia of Mathematics and its Applications, Vol. 25.

Shapiro, L.G. and Haralick, R.H., (1981). "Structural Descriptions and Inexact Matching", IEEE Trans. Pattern Anal. Machine Intell., PAMI-**3**, 504-519.

Shapiro, L.G. and Haralick, R.H., (1985). "A Metric for Comparing Relational Descriptions", IEEE Trans. Pattern Anal. Machine Intell., PAMI-**7**, 90-94.

Shi, Q.Y. and Fu, K.S., (1983). "Parsing and Translation of (Attributed) Expansive Graph Languages for Scene Analysis", *IEEE Trans. Syst. Man. Cybern,* SMC-**9**.

Tsai, W.H. and Fu, K.S. (1979). "Error-Correcting Isomorphisms of Attributed Relational Graphs for Pattern Analysis", *IEEE Trans. Syst. Man. Cybern,* SMC-**9**, 757-768.

Ullmann, J.R., (1976). "An Algorithm for Sub-Graph Isomorphism", *J. Assoc. Comput. Mach.,* **23**, 31-42.

Wong, A.K.C. and You, K., (1985). "Entropy and Distance of Random Graphs with Application to Structural Pattern Recognition", *IEEE Trans. Pattern Anal. Machine Intell.,* **7**, 599-609.

ANALYSIS OF SYNTHETIC APERTURE RADAR IMAGES OVER LAND

S. Quegan[*], A. Hendry and J. Skingley
(GEC-Marconi Research Centre, Chelmsford)

1. INTRODUCTION

When a human observer looks at an image, he imposes a structure on that image, in which knowledge of the real world and what he is looking for are essential ingredients. In scene analysis we attempt to emulate human performance by automatic means. This is not a well-defined aim, since human performance varies between observers, and may depend on the reasons for examining the image. An essential first step in scene analysis is therefore the definition of a more tightly-defined set of scene elements thought to be relevant to image structure. These may be specific objects, e.g. icebergs, hills, vehicles, towns, etc or general scene-building elements such as edges, thin lines, corners, segments, etc.

In this paper we concentrate on the latter approach, and in particular on the problems of edge-detection, segmentation and thin line detection. The major themes of this paper are the development of

(i)   quantitative methods for comparing segmentation
      schemes, and

(ii) thin line detection schemes.

This is a recognition of the urgency of developing such methods. From the point of view of application of remotely sensed data, there is a clear need to specify 'best' methods of carrying out image interpretation, and to put some meaning to the word 'best'; there is also a need to quantify how often our algorithm will detect what it purports to find, and how often mistakes will be made.

---

* now Department of Applied and Computational Mathematics, University of Sheffield.

The methods discussed will be applied to synthetic aperture radar (SAR) data. A full discussion of the structure of a SAR image is beyond the scope of this paper and we note just two essential properties. The first is that the nature of coherent imaging causes a phenomenon known as speckle, in which interference from sub-resolution size scatters causes individual pixels to have negative exponential sampling statistics, if displayed as a power image (amplitude images exhibit Rayleigh statistics). Such a distribution is very far removed from the much-discussed Gaussian noise model, and special methods are therefore required for dealing with it. The second property of a SAR image is that if sampling occurs at the Nyquist limit, neighbouring pixels are uncorrelated; however, there is often some degree of over-sampling - this is the only system effect we will consider.

The first half of the paper will deal with methods of comparing segmentation schemes; the second with thin line detection. Rigorous universally accepted definitions of a segment or a thin line are not easy to construct, nor is it obvious how to make a clear distinction between the two scene elements. Our working definition will characterise a <u>segment</u> as a uniform connected region surrounded by edges, thin lines, or a combination of the two. An <u>edge</u> will denote a 'sharp' change in radar reflectivity at the abutment of two extended regions. A <u>thin line</u> will refer to a 'narrow' (1-2 pixels) region of enhanced or reduced reflectivity relative to the background. Statistical separation of thin lines from the background is likely to require searching along particular directions, i.e. they are essentially one-dimensional structures, whereas segments leave significant signatures in two-dimensional windows. This conceptual separation, though not rigorous, reflects the fact that different methods are often necessary to find the two sorts of feature; our description provides paradigms on which to construct algorithms.

2. SEGMENTATION

2.1 Segmentation Methods

Numerous methods of image segmentation have been suggested, of which we recognise three basic types

(i)   segment-growing

(ii)   edge-detection

(iii)   hybrids.

Segment-growing methods recognise the segment, rather than its edge, as the object of the search. For SAR images, various

authors have suggested using split-and-merge techniques
(Nooren et al., 1985), constrained optimisation methods (Delves,
private communication), and methods based on local bonding of
pixels, usually preceded by filtering (Cruse et al, 1984).
Direct edge-detection, regarding the segments as secondary
artefacts, can be carried out by a battery of operators (see,
e.g. Frost et al, 1982; Hendry et al, 1985). Hybrid methods
try to use both edge and segment concepts to arrive at the
final segmentation. (White, 1985).

It is desirable to define a general framework within which
to compare all segmentation schemes, but we have been unable
to achieve that. Instead, we have considered comparison of
schemes based on edge-detection. In common with most of the
papers dealing with edge-detection in SAR images, we treat
edge-detection as a two-step process. The first step filters
to remove speckle; the second thresholds a local statistic of
the filtered image to produce an edge map. We assume segments
are formed from the edge map simply by identifying closed
boundaries, though the procedure we describe is applicable to
more sophisticated methods. The definition of the segmentation
procedure therefore requires selection of a filter, and a
criterion for thresholding some quantity associated with the
filter output. We first deal with the possible form of the
filter.

2.2 Image Filters

There are a large number of filters suggested in the
literature for use with speckled images, and we can identify
three basic types

(i)   speckle-specific

(ii)   edge-preserving

(iii)   general.

Speckle-specific filters (Lee, 1981; Frost et al, 1982;
Kuan et al, 1982) use the known statistical properties of
speckle to define a minimum mean square error filter under
different assumptions about the form of the statistics of the
underlying scene. All the above-mentioned filters are
adaptive and non-linear; they use statistics in local windows
to describe the mean and variance of the local area; and each
has a tendency to smooth less near edges. Their outputs
therefore tend to reduce the contrast (s/μ, where s is the
local standard deviation and μ is the local mean) in uniform
areas and to leave large values of contrast near edges.

While the above-mentioned filters are edge preserving in the same sense that they have a tendency to preserve the original data near edges, other filters are specifically edge-preserving in that they attempt to sense edges and smooth selectively (e.g., Lee, 1983 and Oddy and Rye, 1983). This is carried out by averaging over only some of the pixels within a window, the assumption being that those 'too far' in intensity from the central pixel are probably from a different statistical population.

By general filters we denote those filters which use no special knowledge of image structure or statistics, but simply express some form of average behaviour of pixels within a window. The most-widely used examples are the mean and median filters.

Once a filter has been applied, it is necessary to find a criterion by which to detect edges in the output. This should, of course, use the properties of the filter output, but several papers in the literature have paid little attention to such a consideration (e.g. Kirby, 1981). The edge-detector must be matched to the filter. This involves investigating the filter output at edges and in uniform regions, and development of discriminators of the two classes of output pixels. We do not discuss this further in the general case. Instead, we suggest a discriminator matched to the Frost et al (1982) filter, and proceed to show how relatively simple methods lead to an optimal way of applying this discriminator in a segmentation scheme.

2.3 The Frost Filter

As already noted, the filter developed by Frost et al (1982) is a minimum mean square error filter designed for use in speckled imagery. Under a specific assumption about the form of the autocorrelation function of the scene reflectivity, Frost et al derive a filter with impulse response of the form

$$m(t_1.t_2) = C(\alpha) \ e^{-\alpha} \ (|t_1|+|t_2|) \quad (t_1,t_2 \in R).$$

where $C(\alpha)$ is a normalising factor and $\alpha = K\sigma_I/I$; σ_I and I are the standard deviation and mean of the pixel values over some window, and K is a free parameter which we call the Frost parameter (it is intended to reflect the width of the scene reflectivity autocorrelation function).

The form of α implies that less smoothing is done in regions where image contrast (σ_I/I) is large. This will tend to be the case at edges. Hence an obvious detector for Frost-filtered images is a threshold, T, placed on the local value of output image contrast. However, this is a sample statistic, and fluctuations will occur both at edges and in uniform regions, leading to missed edge pixels and false edge pixels in uniform regions. We seek ways of optimising such a scheme, over K and T, using a fixed window size of 5 x 5 pixels.

2.4 Algorithm Optimisation

Optimisation implies the existence of a performance measure. There are obviously many possible measures which can be applied to segmentation schemes (area preservation, edges in the right places, corners preserved, etc), but the two most fundamental questions are

(i) How often are real edges destroyed, and

(ii) How often are false edges generated?

We construct performance criteria based around these two questions.

When seeking to compare filter performance for different values of K, we need to fix some quantity. We choose false alarm rate. To do this, we apply the segmentation scheme to simulated regions of pure speckle, where any segments are the result of statistical fluctuation. For a fixed value of K, we alter the threshold (T) which decides whether a given local value of σ/μ denotes an edge pixel or not. Figure 1 shows a cumulative frequency plot for the output distribution of σ/μ, for K = 5. Figure 2 shows how this translates into false alarms after the bonding operation (which will reject isolated pixels or pixels not enclosing an area). We show only the tail of the distribution since we are interested in low false alarm rates. From this figure we can choose T to fix the false alarm rate at a desired level. False alarm curves like Figure 2 have been generated for several values of K.

Now that a false alarm rate can be defined as a function of K and T, we address the problem of edge detections and edge preservation. In order to give a value to this we generate an image consisting of many edges. Figure 3 shows such an image. The quantities alongside the horizontal bands in Figure 3 correspond to relative mean intensities. The speckled test pattern is subdivided into vertical stripes (for the purpose of this paper they are all 12 pixels wide), and we then apply the segmentation scheme to each stripe. The values of T are chosen to fix the false alarm rates (and hence depend on K), and we vary K. Figure 4 shows the result of such an exercise,

with a fixed false alarm rate. What is clear is that
detections improve as K decreases; this is true for all false
alarm rates. The highest number of detections occurs for K = 0;
under these conditions, the Frost filter reduces simply to the
mean filter. This is, in fact, consistent with the work of
Frost et al (1982), since for uniform regions their filter also
reduces to the mean filter. It is not clear whether a similar
conclusion would apply to textured regions. This conclusion is
important in the context of SAR images gathered over land,
since analysis of airborne SAR data has shown that fields of
crops and grass show no texture; in this case, one should
simply use the mean filter, not the more complex Frost filter.

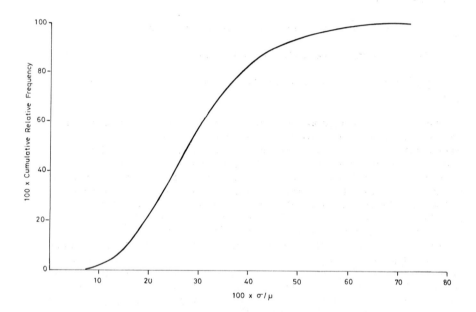

Fig. 1 Distribution of σ/μ after filtering a 256^2 pure
speckle region using Frost parameter $K = 5$.

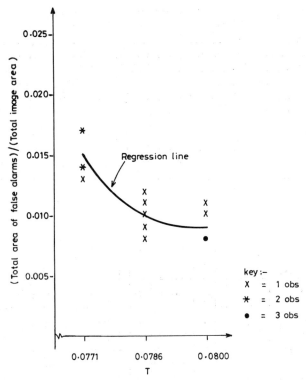

Fig. 2 False Alarm Rates vs Edge Detector Thresholds from the Segmentations of Pure Speckle Regions.

Fig. 3 Speckle Test Pattern (showing relative mean intensities)

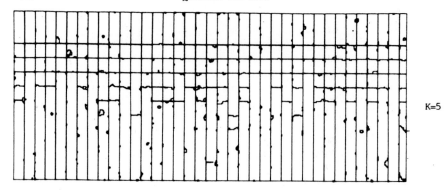

K=5

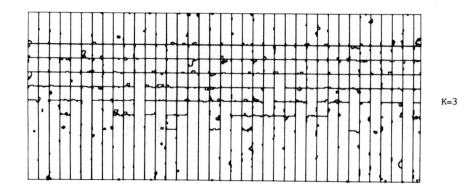

K=3

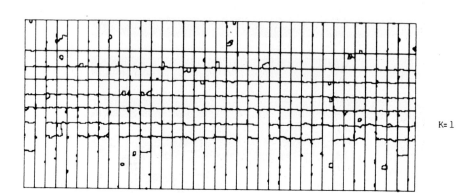

K=1

Fig. 4 Segmentation of striped speckled test pattern for a
fixed false alarm rate.

The above extended example illustrates a method of comparing
filters based on two crucial properties of any useful
segmentation scheme. We have developed it considerably and
applied it to several segmentation schemes and types of image
(in particular, over-sampled images). Space prevents full
description of this work. It is important to note that it is
not, in practice, generally applicable. The great difficulty
is definition of the false alarm rate. For some schemes
examined, the variation in the false alarm rate as defined
above is very large, and would require very many simulations to
arrive at statistically significant results. At the moment
we wish to avoid such large-scale computing. As a final word,
it should be noted that we turned to simulation with some
reluctance after attempting analytic methods. These rapidly
lead to considerable complexity, and we have been unable to
establish many useful results.

3.   THIN LINE DETECTION

3.1   Thin Line Detectors

The potential user of a thin line detection scheme is
faced with a variety of possible schemes (though the choice is
not as bewildering as for segmentation). We can identify
three basic types

(i)   local operators (masks)

(ii)   dynamic programming

(iii)   global transforms.

The use of local operators has been described by, e.g.,
Vanderburg (1976) and Gurney (1980). This method effectively
involves 'convolving' the image with a set of masks representing
idealised line elements at different orientations, so that
each pixel is accorded the value 1 (i.e. it belongs to a line)
if it meets some threshold conditions, otherwise it has the
value O. Post-processing is required to link 'line' pixels
together. Their application to SAR images has been described in
Hendry et al. (1985), and they have been found not to perform
well when the ratio of line to background intensity is small.
This is very often the case in real images.

Dynamic programming methods seek to maximise or minimise
a linear cost function over a set of paths through an array.
Such methods have been applied to SAR images by Wood (1985),
who used as his cost function mean intensity along a path,
weighted by curvature penalties. Little quantitative work on
the performance of this method has yet been carried out. Such

methods need no special extra conditions in order to track
curved lines.

Line detectors using global methods have been based on the
Radon or Hough transforms (Deans, 1981; Duda and Hart, 1972).
The Hough Transform has been shown to be a special case of the
Radon Transform (Deans, 1981). It has been applied to SAR
images (Skingley and Rye, 1985; Hendry et al, 1985) with
useful results. One of its attractive features is that a
description of its performance is analytically tractable. The
following sections deal with the results of such an analysis.

3.2   The Hough Transform

The behaviour of the Hough Transform is conceptually very
simple. For a finite rectangular image, any straight line
joining points on two different sides of the image has a
unique parameterization (see Figure 5) in terms of R (its
distance from the image centre) and θ (its orientation). The
Hough Transform assigns to each (R,θ) point the mean intensity
of the line parameterized by R and θ. In practice, quantization
of R and θ (and, of course, the fact that we are using digital
images) complicates matters, but we will not discuss that
further.

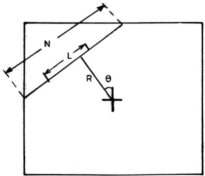

Fig. 5   Parameterization of Image Space

Individual points in the original image domain are mapped to
sinusoids in the transform domain. Lines are mapped to single
points. The detection of bright or dark lines therefore becomes
a problem of peak and trough detection in the transform domain.
Comparatively straightforward arguments based on the Central Limit
Theorem allow an analytic expression for the detection probability
of a line to be derived (Skingley and Rye, 1985). Under the
assumption that the background and line both display Rayleigh
statistics (this is reasonable for the background, but not
enough is known about the statistics of thin lines to fully

justify this), this expression has the form:

$$p = 1 - \Phi \left[\frac{\alpha \, K\sqrt{N} - L(r-1)}{\left\{ LK^2(r^2 - 1 + NK^2) \right\} \, 1/2} \right] \qquad (1)$$

where N is the total possible line length (Figure 5)
      L is the number of line pixels (Figure 5)
      p = detection probability
      K = $\sigma/\mu \sim$ 0.52 (Rayleigh speckle)
      r = ratio of mean amplitudes of line and background (m_1/m_b)

      $\Phi(a) = P(X>a)$ = standardised normal distribution function

and α is a parameter which fixes the detection threshold and hence the false alarm rate. (By false alarms we mean those line detections caused by the statistical fluctuations in the speckle). The threshold, T, is related to α by the relation

$$T = m_b + \frac{\alpha \sigma_b}{\sqrt{N}}$$

where σ_b is the standard deviation of the background. Note that N varies as a function of R and θ, and the expression (1) is not valid for small values of N.

Figure 6 shows the detection probability, p, against r for lines of lengths 25 to 150 with a fixed threshold given by $\alpha = 4$ (chosen to give an expected false alarm rate of 2, i.e., there would be 2 'line' detections due to the fluctuations in the speckle statistics). As expected, detection probabilities decrease with line length and line to background ratio. The Hough Transform can, however, locate low intensity lines with a high degree of probability, as long as there are enough pixels in the line. In this respect it is far superior to local methods.

These theoretical results have been compared with the results of simulation. Detections fitted the theoretical values very well, but there were far too many false alarms. This has been traced to the effects of correlation between pixels in the transform domain and statistical bias in the procedure used to estimate the threshold T (which relies on local estimation of m_b and σ_b). The details are not supplied here.

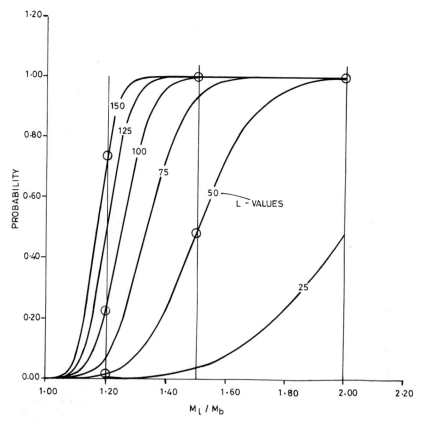

Fig. 6   Probability of detecting a line (N=150 α = 4, k = 0.52)

The simple treatment of the Hough Transform given above has
neglected several important elements in its use.   The Hough
Transform gives no indication of where along a given line of
pixels the detected line lies;   nor does it give any indication
of its length, since short bright lines give the same detection
probability as long faint lines.   As a result, use of the
Hough Transform requires methods of detecting ends of lines.
These are still under development, as also are methods for
detection of false alarms.

4.   SUMMARY

Application and use of remotely sensed data require the
development of image analysis algorithms to detect relevant
information-bearing elements of images.   These algorithms need
to have a known level of performance, both so that optimum
algorithms can be identified (if such exist) and so that some
degree of belief in their outputs can be defined.   Segmentation

and thin line detection are two related scene analysis
techniques of considerable importance. We have shown how
segmentation schemes based on edge-detection can be compared,
and a measure of performance defined. This allowed optimisation
of a scheme based on a filter due to Frost et al (1982). These
methods seem inapplicable to other types of segmentation, in
part because of the difficulty of defining a reliable false
alarm rate without using very large amounts of computer time.
We must stress that the performance measures used are of the
simplest sort (detections vs false alarms) and applications-
orientated algorithm evaluation may need more refined measures.

Mathematical analysis of segmentation schemes proved too
complex to yield useful results and we were obliged to use
simulation. By contrast, line detection using the Hough
Transform is susceptible to analysis, and detection rates can
be set. False alarm rates prove more troublesome, but can be
accounted for in terms of correlation and bias in the sampling
statistic used to set the false alarm theshold.

The incomplete and rudimentary results presented above
represent early attempts at an important and difficult problem.
Remote sensing will only begin to yield its potential when the
nature of the information it contains is understood, and
reliable methods for extracting that information are developed.
Reliability must be defined, quantified and measured. Results
presented here constitute part of our continuing effort to
achieve this aim.

5. ACKNOWLEDGEMENTS

The authors would like to thank the European Space Agency,
the Royal Aircraft Establishment, Farnborough, and the Royal
Signals and Radar Establishment for support of various aspects
of the work presented in this paper.

6. REFERENCES

Cruse, D., Quegan, S. and Wright, A., (1984) 'Detection and
    Recognition of Targets in SAR Images', Marconi Tech. Rpt.
    MTR 84/28.

Deans, S.R., (1981) 'Hough Transform from Radon Transform'.
    IEEE Trans. Pattern Analysis and Machine Intelligence,
    Vol. PAMI-3, No. 2, pp. 185-188.

Duda, R.D. and Hart, P.E., (1972) 'Use of the Hough
    Transformation to Detect Lines and Curves in Pictures'.
    Comm. ACM. Vol. 15, No. 1, pp. 11-15.

Frost, V.S., Stiles, J.A., Shanmugan, K.S. and Holtzman, J.C.,
     (1982) 'A Model for Radar Images and its Application to
     Adaptive Digital Filtering of Multiplicative Noise'.
     IEEE Trans on Pattern Analysis and Machine Intelligence,
     Vol. PAMI-4, No. 2, pp. 157-165.

Frost, V.S., Shanmugan, K.S. and Holtzman, J.C., (1982) 'Edge
     Detection for Synthetic Aperture Radar and Other Noisy
     Images'. Proc. IGARSS '82, Munich.

Gurney, C.M., (1980) 'Threshold Selection for Line Detection
     Algorithms'. IEEE Trans. Geoscience and Remote Sensing,
     Vol. GE-18, No. 2, pp. 204-211.

Hendry, A., Quegan, S., Rye, A.J., Skingley, J. and Churchill, P.,
     (1985) 'Application of Image Processing Techniques to
     SAR 580 Data'. Marconi Tech. Rpt. MTR 84/59.

Kirby, M.E., (1981) 'Applications of Robert's Gradient
     Operator for the Digital Enhancement of Icebergs from
     SAR Imagery. Proc. 7th Canadian Symposium on Remote
     Sensing, Winnipeg.

Kuan, D.T., Sawchuk, A.A., Strand, T.C. and Chavel, P., (1982)
     'Adaptive restoration of Images with Speckle'. SPIE Vol.
     359 Applications of Digital Image Processing IV.

Lee, J., (1981) 'Speckle Analysis and Smoothing of Synthetic
     Aperture Radar Images'. Computer Graphics and Image
     Processing 17, pp. 24-32.

Lee, J., (1983) 'A Simple Speckle Smoothing Algorithm for
     Synthetic Aperture Radar Images'. IEEE Trans. on Systems
     Man, and Cybernetics, Vol. SMC-13, No. 1.

Nooren, G.J.L., Attema, E.P.W., de Loor, G.P., van der Lubbe,
     J.C.A. and Krul, L., (1985). 'Use of SAR in Agriculture and
     Forestry'. Thematic Applications of SAR Data, ESA SP-257.

Oddy, C. and Rye, A.J., (1983) 'Segmentation of SAR Images
     using a local similarity rule'. Pattern Recognition Letters
     1, pp. 443-449.

Skingley, J. and Rye, A.J., (1985) "The Hough Transform applied
     to SAR Images for Thin Line Detection'. Proc. BPRA Conf.
     St. Andrews.

Vanderburg, G.J., (1976) 'Line Detection in Satellite Imagery',
     IEEE Trans. Geoscience Electronics, Vol. GE-14, No. 1 pp.
     37-44.

White, R.G., (1985) 'Low Level Segmentation of Noisy Imagery'.
    RSRE Memo. No. 3900.

Wood, J.W., (1985) 'Line Finding Algorithms for SAR'. RSRE
    Memo. No. 3841.

THE USE OF SPATIAL CONTEXT IN THE CLASSIFICATION
OF REMOTELY SENSED DATA

J.J. Settle
(NERC Unit for Thematic Information Systems,
Department of Geography, University of Reading)

1. INTRODUCTION

The classification of multispectral remotely sensed imagery
is mainly carried out by methods that compare the spectral data
of a pixel to the spectral signatures of a number of
predetermined classes. The class whose signature is in some
sense closest to the pixel data will be assigned to label the pixel.
Usually these methods can be formulated in terms of discriminant
functions, one for each class. These take for their argument
the pixel vector spectral data, X, and the method of
classification is to award the pixel to class i (label z_i) if:

$$g_i(X) > g_j(X) \quad \text{for all } j \neq i \qquad (1.1)$$

g_i being the discriminant function for class i. Perhaps the
most popular of these methods is Maximum Likelihood
classification; closely related to it is Bayes' optimal
classification. The discriminant functions for the latter are
the a posteriori probabilities of the pixel belonging to each
class, given the spectral information:

$$g_i = \Pr(z_i|X) = \Pr(X|z_i)\Pr(z_i)/\Pr(X)$$

or equivalently:

$$g_i = \Pr(X|z_i) \cdot \Pr(z_i) \qquad (1.2)$$

since $\Pr(X)$ is the same for all classes and so does not affect
the decision. $\Pr(X|z_i)$ is the distribution of feature vectors
for the ith class, regarded as a probability density function.
$\Pr(z_i)$ is the prior probability for the ith class measuring,

in the absence of any spectral evidence, how likely we think
the pixel is to belong to that class. Bayes' classification
has the attractive property of being the spectral classifier
with the smallest expectation of error per pixel.

Now one very obvious failing of spectral methods is that
they do not take into account the spatial information contained
in the imaged scene. We could shuffle up all the image pixels
and then classify correctly as many as in the original scene,
despite destroying all the spatial evidence. Clearly, a method
that attempts to incorporate both spectral and spectral
information ought to be preferable to a scheme which relies
only on one sort. This is what we have attempted to do, by
using spatial information to modify the prior probabilities in
(1.2).

2. CONTEXTUAL PRIORS (ICM)

From this point on we shall denote by $N(Q)$ a set of pixels
close to pixel Q, and which will be examined when we come to
form the prior for Q, and by $C(Q)$ the labelling on those
pixels (or the context of Q).

Suppose for a given pixel we knew the true class
assignments (labels) for $N(Q)$, $C_t(Q)$, say. How should this
information be used to influence the decision process? An
obvious approach is to use as discriminant function the
posterior probability $Pr(z_i | X, C_t)$ or, equivalently,

$$g_i = Pr(z_i | C_t) Pr(X | z_i, C_t) \qquad (2.1)$$

where we have suppressed terms which are common to all classes
and which therefore do not affect the classification. If the
probability distributions depend only on the class being
considered, then this expression simplifies further to:

$$g_i = Pr(z_i | C_t) Pr(X | z_i). \qquad (2.2)$$

The discriminant function thus factors into a purely spectral
term and a purely spatial term. Classification using these
discriminant functions is the basis of the method we have
developed. We see that (2.2) is very similar to (1.2), the
difference being that the prior probabilities are conditioned
on the context information. We might choose to call this method
classification by "contextual priors".

Of course, at first we do not have any spatial information and the obvious way to get some is to carry out a non-contextual classification of the image using, say Maximum Likelihood classification. Contextual priors can be extracted from the resulting image and combined with the spectral data to obtain an improved classification, using the discriminant functions (2.2), except that we use the observed context $C(Q)$ instead of the unknown $C_t(Q)$. This classification can in turn be used to provide (we hope) a yet more accurate classification, and so on. This defines a simple iterative procedure which one hopes will converge reasonably quickly, and give significantly better classification results than any non-contextual method. To complete the definition of the method we need to specify a form for $\Pr(. | C(Q))$.

An identical method has been proposed by Besag (1986), who has shown that it is very effective when applied to some simulated 1-dimensional date. His approach is conceptually slightly different from ours, in that he starts by taking g_i to be the conditional probability of z_i given all the signals in the image and the labelling of all other pixels. He then makes simplifying assumptions which reduce the problem to discrimination using (2.2). The resulting iterative algorithm is that outlined above and Besag calls the method classification by Iterated Conditional Modes (ICM).

The method is a special case of the more general method considered by Kittler and Foglein (1984); the model they consider reduces to ICM when conditional independence of the spectral data is assumed.

3. PAIRWISE MARKOV PRIORS

We can approach the problem of obtaining a suitable $\Pr(. | C(Q))$, as Besag has done, by appealing to the theory of Markov Random Fields (MRFs). If the conditional probability for label z, at pixel Q, given the labels for all the other pixels, should depend on just a subset of those other labels, then the lattice of labels forms a Markov Random Field (Besag 1974, Kelly, 1979). The pixels whose labels condition that at Q are known as the neighbours of Q. The conditional probabilities cannot be chosen arbitrarily and must satisfy certain consistency relations, as a result of which it follows that the definition of neighbour is symmetric: if pixel A is a neighbour of pixel B then B must also be a neighbour of A. We define a clique to be a set of pixels which consists of just a single pixel, or is such that each member is a neighbour to all the other pixels in the set. The consistency relations imply that the probability distribution for the complete image labelling, z, must be given by:

$$p(z) \propto \prod_v \phi_v(z_v) \qquad\qquad (3.1)$$

(Kelly, 1979), where v is a clique and z_v the labelling on it; the product is taken over all the cliques that the picture contains. The function ϕ_v is non-negative but otherwise arbitrary.

This expression can be greatly simplified if we assume translation invariance, so that ϕ_v is independent of position, and assume also that only cliques of 1 and 2 pixels make an effective contribution to the above product. The resulting distribution is an example of what Besag calls a pairwise interaction Markov Random Field; we shall call the corresponding conditional probabilities Pairwise Markov Priors. We find that they are given by:

$$Pr(z_i | C(Q)) \propto \exp\left\{\alpha_i + \Sigma_j \beta_{ij} n_j\right\} \qquad\qquad (3.2)$$

for some constants α_i, β_{ij} $(= \beta_{ji})$ where n_j is the number of pixels in $N(Q)$ labelled as z_j. A further simplification occurs if we set $\alpha_i = $const, $\beta_{ij} = 0$ for $i \neq j$; and $\beta_{ii} = \beta$. We then obtain:

$$Pr(z_i | C(Q)) \propto \exp (\beta n_i) \qquad\qquad (3.3)$$

The constant of proportionality is the same for all classes and so can be ignored for the purposes of classification.

(3.3) is the prior used to effect by Besag (1986) on artificial datasets, and which we have used in our own early investigations, reported here. Of course, it is not necessary to adopt any such form for the prior probabilities; in any practical application the size of neighbourhood will be chosen for computational convenience, not because we believe that the pixels in this set convey all the information that is to be extracted from the image. It is only in the last case that the restrictions of MRF theory would need to be considered. However, we will see later that this is probably the most appropriate form to choose for small neighbourhoods.

4. IMPLEMENTATION

Our implementation of this algorithm works in a Model 75 Image Processor, made by International Imaging Systems. This machine can perform basic operations on 512x512 images (weighted addition of image bands, etc) in about 1/30 second. We have already succeeded in implementing in this image processor Maximum Likelihood classification; this runs 100 times faster than the equivalent CPU program (in a VAX 750) and correspondingly large savings can be expected for the contextual classifier. This vast saving in time means we shall be able to run and test the method on a wide number of datasets, trying different parameter settings and extensions to the basic method. For now, however, we have just tried to identify trends with a single test image, using the simple pairwise prior (3.3).

A minor complication with our implementation is that we set a rejection threshold on the probabilities (class distribution functions): any pixel for which $Pr(X|z_i)$ is smaller for each

i than some limit will not be classified, no matter how strong the spatial evidence (these are essentially pixels for which $Pr(X)$ is small). This is an essential feature of building the algorithm in an 8-bit image processor using integer arithmetic; an acceptable range of probabilities has to be scaled to a range of integers. Our value of β is chosen so that the rejection criterion, defined initially for the non-contextual Maximum Likelihood classification, and applied now to the spatial-spectral discriminant functions, will refuse to classify a pixel to a class if all its neighbours were assigned to other classes. We could probably succeed just as well with a somewhat lower value of β than is implied by this, but higher values seem inappropriate, leading only to an increase in the number of pixels rejected. The presence of unclassified pixels in an image does not invalidate the MRF approach, which just requires a notional modification at each iteration so that the field is defined only over non-reject pixels. There is no inconsistency lurking when we use the same classification method on the reject pixels.

Finally we should note that there is no guarantee that the method will properly converge; it is quite possible for the method to end up oscillating between classification maps.

5.   TESTING

We have so far tested the method quantitatively on one dataset. This is part of some airborne multispectral scanner imagery of a part of southern England, obtained during NERC's aircraft campaign for 1984. We took a 5-band, 512x512 pixel subset of the initial data, which were of 5 metre resolution. When this image was obtained a team of researchers was out in the field collecting ground reference data, and the resulting map of field data has now been digitised. It is because of this complete coverage of reference data (often miscalled "ground truth") that this imagery, rather than any satellite data, has been chosen for our early tests. A pixel by pixel comparison of a classification map with this reference map gives us immediately a figure for the overall accuracy of the classification without our having to go through the uncertain business of estimating these accuracies from small testing sets. We were also able to use this reference map to obtain an upper limit to the accuracy that is possible with the method, as we will later describe.

We ran the method using a number of different sized neighbourhoods for 1, 3, 5, 11 and 19 iterations; the resulting classification accuracies are shown in table 1. In each case we started off with the non-contextual Maximum Likelihood classification, which had an accuracy of about 75%.

Table 1

Number of Iterations	4-	Size of Neighbourhood		
		3x3	5x5	7x7
0	74.7	74.7	74.7	74.7
1	77.8	78.3	79.0	79.3
3	78.8	79.5	80.1	80.6
11	79.0	79.9	80.5	80.9
19	79.0	79.9	80.5	80.9
Max?	85.9	85.3	84.6	84.2

Point to notice are:

(i)   The performance improves slightly with increasing neighbourhood size
(ii)   Almost 3/4 of the improvement occurs with just the first iteration, and most of the rest has been achieved by the third.
(iii) The final row gives the accuracy of the classification which results when we run the method to convergence using the ground reference data as the first class map. The very large "error" which is evident in this line gives clearly an upper bound to the sort of accuracy we can expect, using the given parameter settings, whatever classmap we use to start the procedure. The total error in the initial classification was 25%. The last line of figures tells us that about three-fifths of this can never be recovered using the present method. We are correctly reclassifying about half of the remaining errant pixels. Kershaw (1986), who has also tested the method with constant β on an aircraft image, has also commented that the method leads to much cleaner looking classmaps, but that the change in overall accuracy may not be great.

6. EXTENSIONS

We can think of several ways in which this simple prior might be modified to increase performance. One would be to keep the general form, but to allow the parameter β to vary

spatially. This is permitted within the framework of MRFs but
there is a formidable problem of estimation involved; maximum
likelihood estimation of the single parameter β is itself
problematic (Besag 1974). Nevertheless, it is a possibility
that should be explored since our experience is that
classification errors tend to cluster in the image, so that
different β's would be appropriate for different parts of the
image. We also intend to investigate the more general form
(3.2) of pairwise Markov prior; this does not greatly increase
the complexity of our implementation.

Besag (1986) has found with his studies on simulated data
that improved accuracy can be obtained by starting with a low
value of β and then increasing this with each iteration; we shall
now suggest an explanation for this.

7. PAIRWISE MARKOV PRIORS AND CLASSIFICATION ERROR

Our use of MRFs is theoretically justified largely by the
fact that the priors we generate from it are legitimate; the
totality of conditional probabilities is consistent with a valid
global probability distribution for the scene. The models we
have used are furthermore simple and appear to work well. One
apparent cause for concern is that when we come to consider the
priors for a pixel we act as though the labels for all the
neighbours are correct, whereas we can be fairly sure that at
least some neighbours of some pixels are wrong. Not only,
therefore, are we using priors derived under an unlikely
assumption about the whole scene, but we are also applying
rules about sets of true labels when in practice some of the
labels concerned will be false.

With these fears in mind let us now consider a pixel Q
which we known to be an interior pixel of a field, so that all
eight nearest pixels are of the same (unknown) class. Suppose
that of these eight, n_i have again been classed as z_i, for
each i; $\Sigma n_i = 8$. The probability of realising $C(Q)$ when the
central pixel belongs to z_i is thus the same as realising $C(Q)$
when all belong to z_i. This we take to be:

$$\Pr(C|z_i) = \prod_j c_{ij}^{n_j} \qquad (7.1)$$

where c_{ij} is the probability of classifying as class j a pixel
which is actually of class i. (The c_{ij}, we may suppose, can
have some spatial variability). We thus have

$$\Pr(z_i|C) = \Pr(C|z_i).\Pr(z_i)/\Pr(C) \propto \Pr(z_i) \Pi c_{ij}^{n_j} \qquad (7.2)$$

so that

$$\Pr(z_i|C) \propto \exp\{\alpha_i + \Sigma\beta_{ij}n_j\} \qquad (7.3)$$

with

$$\begin{array}{l} \alpha_i = \ln \Pr(z_i) \\ \beta_{ij} = \ln C_{ij} \end{array} \qquad (7.4)$$

There is thus a great deal of similarity between this approach and use of the pairwise prior given at (3.2). If $c_{ij} = c_{ji}$ for all i,j then the priors given by (7.3) are equivalent to a set derivable from a MRF. This suggests that we might use error rates, determined over test sites with known ground data, to evaluate appropriate parameters for a Markov prior. The resulting model should be the most appropriate for non-boundary pixels, and may be no worse at boundaries than the somewhat arbitrary choice of a constant β that we have used so far.

Now suppose that we have $c_{ij} = c_{ik}$ for $j,k \neq i$; both are then equal to $(1-c_{ii})/(m-1)$, where m is the number of classes present. We can rewrite the exponent in (7.3) as:

$$\alpha_i + n_i.\ln(c_{ii}) + (8-n_i).\ln\{(1-c_{ii})/(m-1)\} \qquad (7.5)$$

so

$$\Pr(z_i|C) \propto \exp(\alpha_i + \beta_i n_i) \qquad (7.6)$$

where the "effective" β is given by

$$\beta_i = \ln\{c_{ii}(m-1)/(1-c_{ii})\} \qquad (7.7)$$

This is zero if $c_{ii} = 1/m$, and increases with c_{ii}. Thus, the more accurate the class map we are working from, the greater the effective value of β. In an iterative scheme we should certainly hope that the classification does improve with each iteration, and so the optimal value of β, given by (7.7), would increase (for non-boundary pixels). It is possible that it is this effect that was observed by Besag when using a steadily increasing β.

SUMMARY

We have implemented, in an image processor, a contextual classification method using a modified form of Bayesian

discrimination. It is an iterative scheme in which the prior
probabilities are made to depend on the local pattern of
classes from a previous classification. The improvement obtained
in classification accuracy for a real dataset using the pairwise
Markov prior is significant, though less impressive than we
might have hoped for given the fine results the method has
produced in some artificial examples. It was shown how we might
use observed classification errors to evaluate the parameters
in a more general Markov model for the priors.

ACKNOWLEDGEMENTS

The work described was started when the author was involved
with the NERC Special Topic Study: "The Development of Improved
Image Processing and Classification Algorithms for Remotely
Sensed Data". I am grateful to Professor Besag for very
helpful discussions on statistical classifiers.

REFERENCES

Besag, J.E., (1974) "Spatial Interactions and the Statistical
    Analysis of Lattice Systems" (with discussion) *J.R. Statist.
    Soc.*, B, 36, 192-236.

Besag, J.E., (1986) "On the Statistical Analysis of Dirty
    Pictures" (with discussion) *J.R.Statist. Soc.*

Kelly, F.P., (1979) "Reversibility and Stochastic Networks",
    Wiley.

Kershaw, C., (1986) Discussion of paper by J.E. Besag,
    J.R.Statist.Soc., B, 36, 192-236.

Kittler, J. and Foglein, J., (1984) "Contextual Decision Rules
    for Objects in Lattice Configurations"; Proc. 7th Int. Conf.
    Pattern Recognition, Montreal, 270-272.

Strahler, A.H., (1980) "The Use of Prior Probabilities in
    Maximum Likelihood Classification of Remotely Sensed Data",
    Remote Sensing of the Environment, 10, 135-163.

SEARCH PROBLEMS IN LANDSAT-TM IMAGE UNDERSTANDING

G.G. Wilkinson,[*] G. Peacegood[*]
(School of Computing, Kingston Polytechnic)

and

P.F. Fisher[†]
(School of Geography, Kingston Polytechnic)

* Present affiliation: University College London
† Present affiliation: Kent State University, USA.

ABSTRACT

The exploitation of high resolution imagery from the Landsat
Thematic Mapper for cartographic purposes requires complex
problem-solving algorithms. This paper examines both the low-
level and high-level processing requirements for deriving
meaningful line segments from imagery and labelling them for
mapping. Several mathematical formalisms for the high-level
problem are reviewed including: constraint propagation, search
and logical reasoning. It is concluded that a practical
implementation of reasoning as an expert system is an
appropriate approach to develop for this problem.

1. INTRODUCTION

1.1 Current Sensor Capabilities

The current generation of satellite-borne imaging sensors
now provide high quality images of the Earth at resolutions down
to 10m (eg. the SPOT-1 HRV instrument - 10m in panchromatic
mode-, the LANDSAT-4 and -5 TM instrument - 30m in several
spectral bands). Looking forwards into the next 5-10 years, it
is readily apparent that there will be a number of other
satellites collecting Earth imagery at similarly high
resolutions. No instruments are planned to go below 10m
resolution in the foreseeable future (except military systems)
but many will operate in the 10-30m range in a variety of
spectral wavelengths (including visible, infra-red, and
microwave). [see Table 1].

There are many interesting and important applications for
high resolution imagery from these satellites; however, some
of them involve considerable computational difficulties. One
problem of particular interest is the application of high

resolution imagery to national mapping. There appears to be
considerable benefit from using satellites to map hitherto
remote and under-mapped areas of the world besides updating
existing maps of developed areas. This paper considers the
problems involved in mapping from LANDSAT-TM imagery and
possible routes to their solution.

1.2 The Mapping Problem

Many cartographically important features on the Earth's
surface have a linear dimension which is of the order of only
a few metres or so (this particularly applies to man-made
features connected with transportation such as roads, railway-
lines etc).

Platform	Launch Date	Sensor	Resolution
SEASAT-1	1978	SAR	25m
LANDSAT 1	1972	MSS	80m
LANDSAT 2	1975	MSS	80m
LANDSAT 3	1978	MSS	80m
LANDSAT 4	1982	TM	30m
LANDSAT 5	1985	TM	30m
LANDSAT 6/7	1988?	MRS?	15m?
SPOT-1	1986	HRV	10m
ERS-1	1989?	SAR	30m
RADARSAT	1989?	SAR	25m

Ground Resolutions of Recent and Forthcoming Civilian Remote
Sensing Satellites with High Resolution Sensors

Table 1

Although such features are predominantly below the pixel size
for sensors such as the LANDSAT-5 TM they are nevertheless
large enough to make an impact on pixel radiance and to
register their existence. However, the traditional approach
of multispectral classification can not be used reliably to
identify the nature of these linear features in the imagery
because the pixel radiances are corrupted by a superposition
of responses both from the feature of interest and extraneous

background classes. Detection and classification of linear
features for mapping applications therefore relies on a
spatial context-based analysis.

Even constraining the general mapping problem to the more
specific problem of detecting and identifying only the major
linear features of a LANDSAT-TM image, two levels of analysis
are required: "low-level" analysis (for line/edge detection and
enhancement) and "high-level" analysis (for achieving a
consistent labelling scheme for detected features). The low-
level problem of edge detection has received a lot of attention
in the image processing literature - though surprisingly few
algorithms yield useful results with scenes as complex as those
obtained from the Thematic Mapper. This will be discussed
further in section 2. The high-level problem is typical of many
which are being tackled in artificial intelligence research,
particularly for machine vision. However, this problem (or
others like it) has received scant attention in remote sensing
to date. This high-level problem becomes one of some complexity
involving the globally consistent labelling of a large set of
line fragments. This will be discussed further in section 3.

2. THE LOW-LEVEL PROBLEM

Many conventional edge detection approaches are unsatisfactory
when applied to LANDSAT-TM images because there is too much
fine detail and coarse texture. One approach which does seem
to provide useful edge maps is that suggested by Nevatia and
Babu (1980). This method is aimed not only at extracting line
segments from images but also at generating descriptions for
subsequent processing.

The basis of the Nevatia-Babu method which we have adapted
for the TM imagery is the following four steps:

(i)   computation of edge magnitudes $s(x)$ and gradient directions
      $\phi(x)$ by convolution with 5 x 5 masks in 6 orientations.
      An edge exists at image coordinate x if:

$$s(x) > s_\perp (x_+), \ s_\perp (x_-);$$

$$[\phi(x_+) - \phi(x)] \bmod 2\pi \le \frac{\pi}{6};$$

$$[\phi(x_-) - \phi(x)] \bmod 2\pi < \frac{\pi}{6};$$

$$\text{AND } s(X) \ge \varepsilon \text{ (threshold)}.$$

[Note: x_+, x_- represent neighbours of x;

$s_\perp$ represents edge magnitude in normal direction]

(ii)   thinning of edge map derived from above
(iii) linking of edge elements by proximity and orientation
       analysis to bridge small gaps between identified segments
(iv)   approximation of linked line elements by piecewise segments
       (polylines).

Note that this procedure simply derives and encodes line
segments for subsequent higher-level processing. In practice
it is best to set the threshold ε to detect as many meaningful
segments as possible - the higher-level processes must make
sense of them. It would not be appropriate to set high
thresholds at the outset which may result in the rejection of
useful line information. [Ultimately one might conceive of a
feedback system which could sensibly adapt the threshold as a
result of the higher level analysis, though this has not been
attempted yet.] The use of 5 x 5 masks seems to be acceptable
for the LANDSAT scenes - certainly smaller ones are susceptible
to noise and larger ones miss important fine detail.

The output from the adapted Nevatia-Babu process is thus a
collection of polylines extracted from the original satellite
imagery. In general it is very difficult to select thresholds
which generate a "clear" edge map. Even though the thinning
and linking process may remove noise and short spurious
fragments besides closing-up obvious gaps, the output still
usually contains a large number of line fragments which must be
interpreted in a globally consistent manner - this is the crux
of the matter.

3.   THE HIGH-LEVEL PROBLEM

The purpose of this paper is not explicitly to tackle the
low-level problem. There continue to be difficulties
associated with reliable edge detection in satellite scenes
but useful though not perfect results have been achieved with
the Nevatia-Babu method for our purposes. The main aim of
this discussion is to focus attention on the higher-level
problem: that of labelling the line-set in a consistent
manner for mapping applications.

The central problem is the allocation of a label from a
fairly small but well-defined set (which includes most
significant linear surface features) to each member of a much
larger set of polyline-encoded line segments. Since the
allocation of a specific segment has implications for the
possible labellings of other segments in the vicinity, this
problem involves an unavoidable combinatorial explosion. It
must be solved by some form of algorithm to search for the
optimal consistent labelling scheme.

3.1 Appropriate Mathematical Approaches

Simon (1983) identified the three separate but analogous
mathematical models used by the artificial intelligence
community to solve complex combinatorially-explosive problems.
These are:

(i)   constraint satisfaction
(ii)   search
(iii) reasoning.

Each of these methods has been applied to many kinds of
practical problems - though using slightly different formalisms.
Here we shall consider their possible role in the line labelling
problem.

3.1.1 Constraint Satisfaction

The technique of constraint satisfaction works on the general
principle that we begin with a space of objects and eliminate
label classes by applying successive constraints. A good
practical example of this technique is the Waltz Filtering
algorithm applied to the problem of labelling the lines which
form the skeleton of a blocks world vision scene (Waltz 1985).
The constraints are based on the topologically permissible
combinations of line types at object vertices. For such scenes
where the images are of high quality with minimal noise and a
simple structure this approach works well. However, for more
natural scenes this approach is not appropriate.

A more general approach is that of "relaxation labelling"
suggested by Zucker, Hummel and Rosenfeld (1977). This
approach generates a probability vector to indicate the
relative probability that each member of the label set should
be assigned to each object in the image:-

$$\text{set of objects:} \quad a = \{a_1, a_2, a_3, \ldots a_n\}$$

$$\text{set of labels to interpret objects:} \quad \Lambda = \{\lambda_1, \lambda_2, \ldots \lambda_m\}$$

$$\text{probability vector:} \quad \underline{p} = (p_1(\lambda_1), p_1(\lambda_2), \ldots p_1(\lambda_m) \ldots p_n(\lambda_m))$$

The problem-solving algorithm is iterative and is aimed at
updating the probabilities of specific labels being associated
with specific objects. The probability of label λ being
associated with object i on the (k+1)th iteration is given by:

$$p_i^{(k+1)}(\lambda) = f\ (p_i^{(k)}(\lambda),\ p_j^{(k)}(\lambda'))$$

where $p_j^{(k)}(\lambda')$ denotes the probability of label λ' being associated with a neighbouring object; on the previous iteration. The correction applied to $p_i^{(k)}(\lambda)$ in the $(k+1)$th iteration is $q_i^{(k)}$ giving:

$$p_i^{(k+1)}(\lambda) = \frac{p_i^{(k)}(\lambda)\ [1+q_i^{(k)}]}{NF}$$

[NF = normalisation factor] .

The correction $q_i^{(k)}(\lambda)$ is derived from factors r_{ij} which indicate the compatibility between label λ being associated with object a_i and λ' being associated with a_j. It also depends on coefficients d_{ij} which weight the total influence that a_j can have on a_i :-

$$q_i^{(k)}(\lambda) = \sum_j d_{ij}\ [\sum_\lambda r_{ij}\ (\lambda,\lambda')\ p_j^{(k)}(\lambda')]$$

where $r_{ij} = \Lambda_i\ \text{x}\ \Lambda_j \rightarrow [-1,+1]$.

Such relaxation approaches have sometimes been criticised because (i) there is no guarantee of convergence and (ii) the end result may not be the correct solution even if convergence takes place [for example relaxation could generate a solution that $p_1(\lambda_1) > p_1(\lambda_2)$ but λ_2 is the correct labelling not λ_1].

Lloyd (1983) identified an improved "optimisation approach" to relaxation labelling in which a cost function F(p) measuring incompatibilities in labelling is minimised (i.e. a path is followed such that $\frac{dF}{dp} \leq 0$). Nevertheless, the same disadvantages apply.

Essentially we find that the constraint propagation approach, although successful in some situations, appears not to be ideal for the application of interest here. In LANDSAT-TM mapping it would have to be applied as a relaxation process with all its concomitant difficulties.

3.1.2 Search Algorithms

An alternative mathematical description for the line segment

labelling is as a search problem in which image "states" are
defined (consisting of pairings of labels with objects) and
state transitions are computed in a systematic or heuristically-
guided manner:

$$\text{state S1: } \{<a_1, \lambda_1>, <a_2, \lambda_2>, \ldots\}$$

|

Transition

↓

$$\text{state S2: } \{<a_1, \lambda_1'>, <a_2, \lambda_2'>, \ldots\}$$

In practice search can be systematic and exhaustive (though
since the problem is combinatorially explosive this is
unlikely to lead to the desired solution efficiently). It is
preferable however to guide the search in a sensible manner by
selecting transitions which seem to improve the state and take
it closer to a desired goal (possibly with back-tracking).
Such a formalism clearly requires definitions of:

(i)   goal state descriptions
(ii)   successor state generators (operators)
(iii) goal test functions and
(iv)   distance to goal evaluators.

   Even though these may be definable, the selection of a
suitable operator to effect a transition to a more consistent
set of labellings in the global sense may create local
inconsistencies. Decisions would therefore need to be taken on
the acceptability of local problems in the state descriptions.
Conceptually it would not be possible to formulate a complete
and incontrovertible set of acceptable/unacceptable line
segment labellings in a small locality with TM scenes which
necessarily provide a coarse representation of the underlying
cartographically important features.

3.1.3 Reasoning Algorithms

   The third general approach to problem-solving involves the
accumulation of information by <u>inference</u>. A logical framework
must be postulated which permits new statements to be deduced
from axioms and previously deduced statements. This is
usually based on the standard predicate calculus:

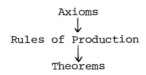

Axioms

↓

Rules of Production

↓

Theorems

This approach is typical of a large class of common systems
known currently of course as "expert systems". These are
normally constructed with antecedent - consequent production
rules of the form:

IF a,b,c THEN x,y

[WITH CERTAINTY p]

These may be used to dynamically up-date a database of deduced
and known facts under control of an "inference engine" (see for
example Gondran 1986). Ideally an expert system for mapping
should contain as much prior knowledge as possible to help
solve the problem presented to it. The advantage of such a
system is that an extensive preliminary knowledge base of facts
and heuristics could be used to substantially narrow the search
space in this formalism.

4.   CONCLUDING REMARKS

It is clear that the problem discussed above - that of
assigning labels to an extensive set of line segments in TM
images - is difficult to solve. However, exploitation of the
new generation of satellite sensors to their best effect requires
that such problems should be solved. We have shown that the
problem can be characterised either as one of global
optimisation preserving local constraints or as one involving
a large state space search in which either heuristics guide the
search strategy or in which logic is used to update a developing
database of consistent facts - a scene description. So far
little work has been done on algorithms of these types in
satellite remote sensing and it is therefore difficult to point
to experimentation which suggests that one approach is better
than another. Needless to say each approach discussed is
complex to implement, yet the value of the information which
they could potentially yield from high resolution imagery is
such that they should be attempted. At Kingston Polytechnic
we have been working on the overall problem of mapping from
space and are currently developing an expert system approach
for deducing facts about suitable line labellings derived from
a Nevatia-Babu low-level edge detection methodology. To
facilitate this approach our line segments (and facts about
their spatial properties) are being built into a frame system
using 'PEARL' (Package for Efficient Access to Representations
in Lisp - Faletti and Wilensky 1982). We hope to be able to
report an implementation of an expert system based around this
problem in due course.

REFERENCES

1. Faletti, J. and Wilensky, R., (1982) "The implementation of
    PEARL - A Package for Efficient Access to Representations in

Lisp", Technical Report, Computer Science Division, Department of EECS, University of California, Berkeley.

2. Gondran, M., (1986) "An Introduction to Expert Systems", McGraw-Hill , London.

3. Lloyd, S.A., (1983) "An Optimisation Approach to Relaxation Labelling Algorithms", Image and Vision Computing, 1, 2, 85-91.

4. Nevatia, R. and Babu, K.R., (1980) "Linear Feature Extraction and Description", Comp. Graphics and Image Proc, 13, 257-269.

5. Simon, H.A., (1983) "Search and Reasoning in Problem Solving", Artificial Intelligence, 21, 7-29.

6. Waltz, D., (1975) In "The Psychology of Computer Vision" (P Winston, ed), pp. 19-92, McGraw-Hill, New York.

7. Zucker, S.W., Hummel, R. and Rosenfeld, A., (1977) "An Application of Relaxation Labelling to Line and Curve Enhancement", IEEE Trans. Computers, C-26, 4, 394-403.